T0201006

Advances in Bioceramics and Porous Ceramics II

Advances in Bioceramics and Porous Ceramics II

*A Collection of Papers Presented at the
33rd International Conference on
Advanced Ceramics and Composites
January 18–23, 2009
Daytona Beach, Florida*

Edited by
Roger Narayan
Paolo Colombo

Volume Editors
Dileep Singh
Jonathan Salem

A John Wiley & Sons, Inc., Publication

Published by John Wiley & Sons, Inc., Hoboken, New Jersey.
Published simultaneously in Canada.

For general information on our other products and services or for technical support, please contact our
Customer Care Department within the United States at (800) 762-2974, outside the United States at
(317) 572-3993 or fax (317) 572-4002.

Wiley also publishes its books in a variety of electronic formats. Some content that appears in print may
not be available in electronic format. For information about Wiley products, visit our web site at
www.wiley.com.

Library of Congress Cataloging-in-Publication Data is available.

ISBN 978-0-470-45756-6

Printed in the United States of America.

10 9 8 7 6 5 4 3 2 1

Contents

POROUS CERAMICS

Preface

This issue contains the proceedings of the "Porous Ceramics: Novel Developments and Applications" and "Next Generation Bioceramics" symposia, which were held on January 27-February 1, 2008 at the Hilton Daytona Beach Hotel in Daytona Beach, FL, USA.

The interaction between ceramic materials and living organisms is a leading area of ceramics research. Novel bioceramic materials are being developed that will provide improvements in the diagnosis and treatment of medical conditions. The Next Generation Bioceramics symposium addressed several leading areas in the use of bioceramics, including rapid prototyping of bioceramics; biomimetic ceramics and biomineralization; in vitro and in vivo characterization of bioceramics; nanostructured bioceramics (joint with the Nanostructed Materials and Nanocomposites symposium)

The link between porous ceramics and bioceramics is very strong, as several biological applications of ceramics require the presence of specific amounts of porosity, which are achieved by carefully controlled processing. Therefore, a joint session was held gathering the participants to both symposia (Bioceramics and Porous Ceramics), in order to stimulate discussion and fruitful interactions between the two communities. Some of the papers in the present volume reflect the interplay between pore morphology and biological behaviour.

The Porous Ceramics symposium aimed to bring together engineers and scientists in the area of ceramic materials containing high volume fractions of porosity, with the porosity ranging from nano- to millimeters. Such solids commonly exhibit cellular architectures and they include foams, honeycombs, fiber networks, connected rods, connected hollow bodies, syntactic foams, bio-inspired structures, meso-porous materials and aerogels. Porous ceramics components are an essential part of numerous devices in various enabling engineering applications, including hydrogen and energy-related technologies, sensors, porous matrix fiber composites, and hot gas filters (e.g., diesel particulate filters).

We would like to thank Greg Geiger, Mark Mecklenborg, Marilyn Stoltz, Marcia Stout, and the staff at The American Ceramic Society for making this proceedings volume possible. We also give thanks to the authors, participants, and reviewers of the proceedings issue. We hope that this issue becomes a significant resource in porous ceramics and bioceramics research that not only contributes to the overall advancement of these fields but also signifies the growing role of The American Ceramic Society in these evolving areas of ceramics research.

PAOLO COLOMBO
Università di Padova (Italy) and The Pennsylvania State University

ROGER JAGDISH NARAYAN
University of North Carolina and North Carolina State University

Introduction

The theme of international participation continued at the 33rd International Conference on Advanced Ceramics and Composites (ICACC), with over 1000 attendees from 39 countries. China has become a more significant participant in the program with 15 contributed papers and the presentation of the 2009 Engineering Ceramic Division's Bridge Building Award lecture. The 2009 meeting was organized in conjunction with the Electronics Division and the Nuclear and Environmental Technology Division.

Energy related themes were a mainstay, with symposia on nuclear energy, solid oxide fuel cells, materials for thermal-to-electric energy conversion, and thermal barrier coatings participating along with the traditional themes of armor, mechanical properties, and porous ceramics. Newer themes included nano-structured materials, advanced manufacturing, and bioceramics. Once again the conference included topics ranging from ceramic nanomaterials to structural reliability of ceramic components, demonstrating the linkage between materials science developments at the atomic level and macro-level structural applications. Symposium on Nanostructured Materials and Nanocomposites was held in honor of Prof. Koichi Niihara and recognized the significant contributions made by him. The conference was organized into the following symposia and focused sessions:

Symposium 1	Mechanical Behavior and Performance of Ceramics and Composites
Symposium 2	Advanced Ceramic Coatings for Structural, Environmental, and Functional Applications
Symposium 3	6th International Symposium on Solid Oxide Fuel Cells (SOFC): Materials, Science, and Technology
Symposium 4	Armor Ceramics
Symposium 5	Next Generation Bioceramics
Symposium 6	Key Materials and Technologies for Efficient Direct Thermal-to-Electrical Conversion
Symposium 7	3rd International Symposium on Nanostructured Materials and Nanocomposites: In Honor of Professor Koichi Niihara
Symposium 8	3rd International symposium on Advanced Processing & Manufacturing Technologies (APMT) for Structural & Multifunctional Materials and Systems

Symposium 9	Porous Ceramics: Novel Developments and Applications
Symposium 10	International Symposium on Silicon Carbide and Carbon-Based Materials for Fusion and Advanced Nuclear Energy Applications
Symposium 11	Symposium on Advanced Dielectrics, Piezoelectric, Ferroelectric, and Multiferroic Materials
Focused Session 1	Geopolymers and other Inorganic Polymers
Focused Session 2	Materials for Solid State Lighting
Focused Session 3	Advanced Sensor Technology for High-Temperature Applications
Focused Session 4	Processing and Properties of Nuclear Fuels and Wastes

The conference proceedings compiles peer reviewed papers from the above symposia and focused sessions into 9 issues of the 2009 Ceramic Engineering & Science Proceedings (CESP); Volume 30, Issues 2-10, 2009 as outlined below:

- Mechanical Properties and Performance of Engineering Ceramics and Composites IV, CESP Volume 30, Issue 2 (includes papers from Symp. 1 and FS 1)
- Advanced Ceramic Coatings and Interfaces IV Volume 30, Issue 3 (includes papers from Symp. 2)
- Advances in Solid Oxide Fuel Cells V, CESP Volume 30, Issue 4 (includes papers from Symp. 3)
- Advances in Ceramic Armor V, CESP Volume 30, Issue 5 (includes papers from Symp. 4)
- Advances in Bioceramics and Porous Ceramics II, CESP Volume 30, Issue 6 (includes papers from Symp. 5 and Symp. 9)
- Nanostructured Materials and Nanotechnology III, CESP Volume 30, Issue 7 (includes papers from Symp. 7)
- Advanced Processing and Manufacturing Technologies for Structural and Multifunctional Materials III, CESP Volume 30, Issue 8 (includes papers from Symp. 8)
- Advances in Electronic Ceramics II, CESP Volume 30, Issue 9 (includes papers from Symp. 11, Symp. 6, FS 2 and FS 3)
- Ceramics in Nuclear Applications, CESP Volume 30, Issue 10 (includes papers from Symp. 10 and FS 4)

The organization of the Daytona Beach meeting and the publication of these proceedings were possible thanks to the professional staff of The American Ceramic Society (ACerS) and the tireless dedication of the many members of the ACerS Engineering Ceramics, Nuclear & Environmental Technology and Electronics Divisions. We would especially like to express our sincere thanks to the symposia organizers, session chairs, presenters and conference attendees, for their efforts and enthusiastic participation in the vibrant and cutting-edge conference.

DILEEP SINGH and JONATHAN SALEM
Volume Editors

Bioceramics

ONE-STEP PREPARATION OF ORGANOSILOXANE-DERIVED SILICA PARTICLES

Song Chen,[1] Akiyoshi Osaka,*[1] Satoshi Hayakawa,[1] Yuki Shirosaki,[1] Akihiro Matsumoto,[1] Eiji Fujii,[2] Koji Kawabata,[2] Kanji Tsuru[1§] .
[1]Graduate School of Natural Science and Technology, Okayama University
Okayama-shi, 700-8530 Japan
[2]Industrial Technology Center of Okayama Prefecture, Okayama-shi, 701-1296 Japan
[§]Now at Faculty of Dental Science, Kyushu University, Fukuoka, 812-8582, Japan
*E-mail: a-osaka@cc.okayama-u.ac.jp

ABSTRACT
    Silica particles and their derivatives with meso-structure attracted much attention, but they were synthesized through complicated multi-step procedure. Considering biomedical application, no surfactants, used in almost all cases above, should be employable due to fear of their toxicity. The present study explored one-step sol-gel preparation of silica particles with biomedical functionalities, starting from Stöber-type systems, and characterized by Transmission Electron Micrograph or ^{29}Si MAS NMR spectroscopy. The Ca-containing particles, derived from the precursor system tetraethoxysilane (TEOS)-H_2O-C_2H_5OH (EtOH)-$CaCl_2$-NH_4OH, consisted of primary particles of ~ 10 nm, and were spherical in shape with the diameter of ~ 1000 nm, where Ca bridged Si-O- on the opposite particle surface. In contrast, the Ca-free particles were smaller with 400 ~ 500 nm in size due to the absence of such bridging effects. In addition, the Ca-containing ones deposited petal-like apatite within one week in Kokubo's simulated body fluid (SBF), which was interpreted in terms of the Ca release from the particles. Amino-modified silica particles were derived from the sol-gel precursor system aminopropyltriethoxysilane (APTES)-TEOS-H_2O-EtOH where APTES behaved not only as the catalyst but also a reactant; i.e., this was a self-catalyzed sol-gel system. Hydrogen bonding among the amino group of APTES on one particle surface and with Si-O- on the other was suggested to work in agglomeration of the primary particles. Bovine serum albumin was covalently fixed on the APTES-silica surface, suggesting their applicability of proteins or other growth factor delivery.

INTRODUCTION
    Sol-gel derived SiO_2 is similar to melt-quenched silica glass in random siloxane bridging network (>Si-O-Si<) but different in involving much free silanol groups (Si-OH) in the amorphous lattice. Moreover, their preparation routes mostly involved some templates or surfactants which might be toxic and inadequate for biomedical applications.[1] A few groups then prepared sol-gel silica particles from simple system like tetraalkoxysilane-ethanol (EtOH)-H_2O, where tetraalkoxysilane included tetramethoxysilane or tetraethoxysilane (TEOS). For example, Kortesuo prepared silica xerogels (dry-gels without calcination) from the system TEOS-polyethyleneglycol-H_2O-CH_3COOH and explored their ability of delivering toremifene citrate.[2] Kneuer et al.[3] proposed organo-modified hybrid silica particles, or amino-functionalized silica particles, as DNA carriers, where N-(2-aminoethyl)- or N-(6-aminohexyl)-3-aminopropyltrimethoxysilane was employed for introducing amino-functionality. Such application of sol-gel silica is recognized because of good biocompatibility of Si-OH groups.[4,5] No significant adverse effects are observed when those silica materials are applied in the human body. Rather, silicate components eluted in the surrounding medium were found to stimulate several genes,[6] or those remaining in the gel body favored cell proliferation.[7] As particle size effects are taken into account, the silica gel particles, small enough (~10 μm), are more convenient and attractive, because they can be directly introduced by injection into human tissues as Shirosaki et al. demonstrated with ceramic particles dispersed in hydrogels.[8] Especially, particles less than 500 nm in size are highly likely endocytosed by the living cells, and the therapeutic drugs held by them are

directly released into the cells. Moreover, such injection route deserves the lowest level of tissue invasion that contributes to the patients' comfort.

By virtue of the above advantages, various silica particles with solid, porous, and hollow structures have been prepared and applied in, e.g., drug delivery system[9] and immunosassy.[10] Their size, morphology, and homogeneity could be well controlled by means of the conventional sol-gel route. However, lower chemical reactivity of the Si-OH groups with functional groups like $-NH_2$ or $-COOH$ groups hardly led to direct covalent bonds with biologically active proteins, enzymes or anti-body, if the silica particles were simply mixed or in contact with their solutions. Consequently, the loading efficiency of those biological factors was very low if only the physical absorption was predominant. Therefore, actually, most applications are related to the functionalized silica particles, not the naked or original silica particles.

Among possible agents for surface modification of silica, aminosilanes seem most attractive and important, because their $-NH_2$ groups could be reacted with the $-COOH$ groups of enzymes or other peptides to form amide bonds, or form RGD peptides (R: arginin, G: glycine, and D: aspartic acid peptides) via cross-linking with, for example, carbodiimide (EDC).[11] What is more, no additional catalytic additives should be needed when aminosilanes like aminopropyltriethoxysilane (APTES) are involved in the precursor systems. That is, with such silanes pH of the systems becomes alkaline enough to initiate hydrolysis and condensation reactions of the relevant components, which may lead to silica nanoparticles similar to those in the Stöber-type systems. In the conventional route, amino-modified silica particles will be prepared in two-steps: the preparation of silica particles, and the aminosilane modification of the resultant silica particles. After Li et al.,[12] successful incorporation of APTES depended on high amount of Si-OH groups. The silica nanoparticles were commonly treated with "Piranha solution" (H_2SO_4/H_2O_2) or HNO_3 to introduce high amount of Si-OH groups and then refluxed in the APTES/toluene solution for a few hours. Such route is complex and dangerous since "Piranha solution" is very corrosive and toluene is also toxic. The exploration of novel and concise route for such modification is necessary.

An advantage of the Si-OH groups or hydrated silica layers to serve nucleation sites for biologically active apatite has been commonly described in the literature. For example, Li et al.[13] pointed out that pure silica gels via a sophisticated sol-gel route developed by Nakanishi et al.[14] deposited apatite layer in the Kokubo's simulated body fluid (SBF)[15] that had the same inorganic ion components as the human blood in similar concentration. Such apatite layers exhibit good affinity with bone tissue, and hence accelerate recovery of bone defects. Thus, the Si-OH containing silica particles might deposit the bioactive apatite when soaked in SBF and can be used as bioactive fillers. Unfortunately, the rate of apatite deposition was very low if only Si-OH groups existed.[16] Tsuru et al.[17] found that Ca(II) released from the silicate materials could significantly increase the super-saturation degree of SBF and promoted bioactive apatite deposition, pushing the equilibrium (eq. (1)) toward apatite formation.

$$5Ca^{2+} + 3PO_4^{3-} + OH^- \rightarrow Ca_5(PO_4)_3OH \tag{1}$$

Thus, apatite could reasonably be deposited on Ca(II)-involved silica particles. Moreover, as in silicate glass or crystals, calcium ions are expected to bridge adjacent $^-O-Si<$ units, $>Si-O^-\cdot\cdot Ca^{2+}\cdot\cdot^-O-Si<$, which may control the particle size of the resultant silica.

On the above bases, the one-step sol-gel method was applied in the modified Stöber sol-gel precursor system to yield two functionalized silica particles: Ca-free and Ca-containing silica particles and amino-functionalized silica particles. The Ca-containing ones were derived from the precursor solution in the system $TEOS-H_2O-EtOH-CaCl_2-NH_4OH$, while the amino-functionalized ones were derived from the system $TEOS-APTES-EtOH-H_2O$. Apatite formation on the particles in SBF was examined with X-ray diffraction or with scanning and transmission electron micrographs. Bovine serum albumin (BSA) was also fixed on the amino-functionalized silica particles. Mechanisms of secondary particle formation and particle size change as well as effects of amino groups were discussed.

EXPERIMENTAL

Preparation of the Ca-containing and amino-functionalized silica particles
    The Ca-containing silica particles were prepared from the modified Stöber precursor system[18] TEOS-EtOH-H$_2$O-NH$_4$OH-CaCl$_2$. CaCl$_2$ and TEOS solutions were prepared beforehand: CaCl$_2$ aqueous solution and TEOS/ethanol solution. Appropriate amounts of CaCl$_2$ (0 ~ 0.15 mmol) were dissolved in 389 mmol of water to obtain CaCl$_2$ aqueous solutions with various concentrations. The TEOS/ethanol solution was obtained by adding 12 mmol of TEOS into 120 mmol ethanol. Those two solutions were mixed in varied ratios to prepare precursor solutions, held in tightly capped one-necked flasks (50mL), which were then transferred in an ultrasonic bath. As soon as 3 mL of 28mass% ammonium hydroxide solution was added to initiate the hydrolysis and condensation of TEOS, the reaction mixture was irradiated with an ultrasound for 30 min at room temperature to produce opaque suspension. The resultant particles were separated by centrifugation at 3,500 rpm for 5 min, and then washed with water for 3 times before dried at 105 °C overnight. Table I shows typical two compositions of starting materials and pH in the precursor solutions.

Table I.   The starting systems similar to Stöber et al.[18] and pH in the precursor solutions.

Samples	CaCl$_2$ (mmol)	TEOS (mmol)	H$_2$O (mmol)	EtOH (mmol)	pH
Silica	0	12	389	120	12.5
Ca-Silica	0.05	12	389	120	12.5

    The amino-functionalized silica nanoparticles were prepared from the precursor system TEOS-EtOH-H$_2$O-APTES using a novel one-step sol-gel route, as originally proposed by Chen et al.[19] Unlike ammonia in the above Stöber precursor system, here, APTES not only provided the functional amino groups but also served as the base self-catalyst. Both APTES (0, 0.45, and 4.5 mmol) and TEOS (4.5 mmol) were mixed in the ethanol (440 mmol)/water (280 mmol) solution, held in 50-mL one-necked flasks at room temperature. The mixture was kept stirring for 1 h. Like in the above Stöber sol-gel system,[18] the hydrolysis and condensation of the silanes gave opaque suspension. The resultant particles were collected by centrifugation at 2,500 rpm for 5 min, washed with water 3 times, and then dried at 105 °C overnight. Table II shows the compositions of the starting materials and pH in the starting solution. Both systems above involved water in great excess over the stoichiometric amount to fully hydrolyze all ethoxy groups into silanol ones: >Si-OEt → >Si-OH.

Table II.   The starting systems for the amino-modified silica, and pH in their precursor solutions.

Samples	APTES (mmol)	TEOS (mmol)	H$_2$O (mmol)	EtOH (mmol)	pH
AMSi0	0	4.5	280	440	6.9
AMSi045	0.45	4.5	280	440	10.9
AMSi45	4.5	4.5	280	440	11.2

Characterization
    Size and morphology of the samples were observed under a field-emission type scanning electron microscope (FE-SEM, JSM-7500, JEOL, Japan). Infrared spectra were taken on a Fourier transform infrared spectrometer (FT-IR, Model 300, JASCO, Japan) using the KBr pellet method. ^{29}Si magic

angle spinning (MAS) nuclear magnetic resonance (NMR) spectra and cross-polarization (CP)-MAS NMR spectra were recorded with a Fourier transform (FT)-NMR spectrometer (UNITYINOVA300, Varian, Palo Alto, CA, USA) equipped with a CP-MAS probe.

The Ca-release and Si-release characteristics were measured for the Ca-containing silica particles where 10 mg particles were soaked in 10 mL of saline, whose pH was adjusted to 7.4 at 36.5 °C with tris-(hydroxymethyl)aminomethane (Tris, 50 mM) and 1 mol/L HCl. Every other day, the particles were collected by centrifugation at 3,500 rpm for 5 min. The Ca(II) and Si(IV) concentration of the supernatant was measured by inductively coupled plasma emission spectroscopy (ICP, SPS-7700, Seiko, Japan). Both Ca-free and Ca-containing silica particles were soaked in 30 mL SBF at 36.5 °C up to 7 d, and then collected every other day by centrifugation at 3,500 rpm for 5 min. After rinsing with water, the particles were dried at 105 °C and their structures were further examined by thin film X-ray diffractometry (TF-XRD, model RAD IIA, Rigaku, Tokyo; Cu kα, 30kV-20mA) as well as scanning electron micrograph (FE-SEM; S-4700, Hitachi, Tokyo) and transmission electron micrograph (TEM; JEM-2010, JEOL, Tokyo).

Bovine serum albumin (BSA) was fixed on AMSi045 and AMSi45. Prepared was aqueous solution (20 mL) involving 1-ethyl-3-3-dimethylaminopropylcarbodiimide hydrochloride (EDC, 50 mg) and N-hydroxysuccinimide (NHS, 50 mg), to which BSA and the particles, 20 mg each, were added. After the mixtures were stirred for 24 h at room temperature, the silica particles were separated from the solution by centrifugation at 2,500 rpm for 5 min, washed with water 3 times, and finally dried at 60 °C overnight.

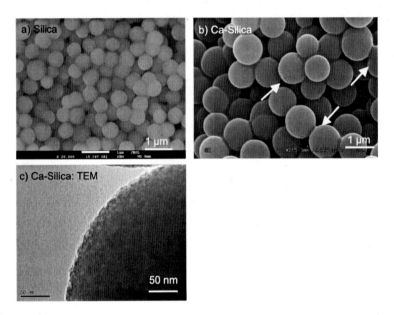

Fig. 1.   FE-SEM images of (a) silica particles and (b) Ca-silica particles; (c) a TEM image of a Ca-silica particle (Table I). The arrows in (b) show surface irregularities (necking and defects).

RESULTS

Ca-containing silica particles: microstructure and apatite deposition

The resultant silica particles were spherical in shape. The FE-SEM images in Fig. 1 show that the Ca-free silica particles (Fig. 1a) were 400 ~ 500 nm in diameter, while the Ca-containing ones (Fig. 1b) were 800 ~ 1000 nm. Fig. 1c shows a typical TEM image of the Ca-silica sample in Table I; as the Ca-silica image indicates, the particles found in the SEM images consisted of much smaller primary particles (~10 nm). Similar TEM images were taken for other samples from the systems with larger $CaCl_2$ contents in the present study. Size distribution profiles for the secondary particles, not shown here, taken with a Particle size analyzer (Honeywell, UPA-150, Microtrac Inc., USA) were centered at 500 nm with ~200 nm width (at half height maximum) for the Ca-free silica, and centered at 900 nm with 300 nm width for the Ca-silica particles. Those data well agreed with the SEM images in Fig. 1. It is thus indicated that the addition of Ca(II) in the starting solution favors the growth of the

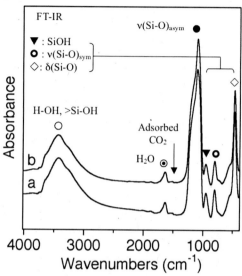

Fig. 2. The Ca-free silica particles (a) and Ca-silica particles (b) gave basically same FT-IR profiles.

secondary particles. The image in Fig. 1b indicates that some particles were slightly fused together, showing a little crater-like surface defect (arrows).

Fig. 2 shows the FT-IR spectra of (a) silica particles and (b) Ca-containing silica particles. No significant differences were found between spectra (a) and (b), and this confirmed that both particles consisted of similar molecular groups. The broader peak around 3430 cm^{-1} was assigned to the stretching vibration of O-H in the >Si-OH groups and the adsorbed water molecules, which also give a small but sharp one at 1640 cm^{-1}. The strongest one at 1111 cm^{-1} and a smaller one at 806 cm^{-1} was $v(Si-O)_{asym}$ and $v(Si-O)_{sym}$ due to the Si-O-Si bonds, while the bands at 960 cm^{-1} were assigned to Si-OH groups. The strongest peak at 1111 cm^{-1} in both spectra indicated the present particles, regardless of the presence of Ca(II), had similar silicate networks and most Si-OH groups have condensed into Si-O-Si groups. Indeed, both yielded very similar ^{29}Si MAS NMR spectra (not presented here), that is, they had similar distribution in Q^n groups, $Q^4/Q^3/Q^2 = 70/28/2$ in mol % for the Ca-free silica, and 69/30/1 for the Ca-silica: here n stands for the number of bridging oxygen atoms in a SiO_4 tetrahedron.

Fig. 3(a) plots the release profile of Ca(II) for the Ca-silica particles and (b) represents that of Si (IV) for both Ca-free silica and Ca-silica when the corresponding silica samples were soaked in saline solution. In (a), least square fitting curves were indicated, assuming that the Ca(II) release proceeded in diffusion-controlled and reaction controlled mechanisms: Diffusion model: $[Ca] \approx 0.013 \, d^{1/2}$ ($\chi^2 = 4.5 \times 10^{-5}$); Reaction model: $[Ca] \approx 0.0055 \, d$ ($\chi^2 = 1.5 \times 10^{-5}$). Empirically, the concentration of Ca(II) increased with the soaking time to confirm the involvement of Ca(II) in the Ca-silica particles. Yet

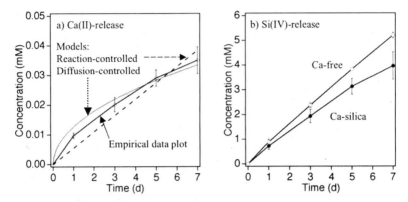

Fig. 3. (a): Ca(II) release profile for the Ca-silica particles. The release profiles due to reaction- and diffusion-controlled mechanism are also plotted. (b): Si(IV) profile for silica and Ca-silica.

either model could not well reproduce the empirical data plot. Fig. 3(b) shows a slight difference in Si(IV) release between the Ca-free silica and Ca-silica samples, responsible to the Ca(II) involvement. The observation is relevant to the secondary particle microstructure and degradation of the secondary particles, which are to be discussed later.

Although increase in Ca(II) stimulates apatite precipitation[17] (eq. (1)), the present Ca-silica is inferior in apatite-forming activity as relatively small amounts of Ca(II) got into saline (Fig. 3(a)). Fig. 4(left) shows the XRD patterns of the Ca-free silica particles (a) and the Ca-containing silica particles (b) after both were soaked in SBF for one week. The Ca-free silica particles showed no sign of apatite deposition, but only exhibited an amorphous XRD profile, except a faint peak at 31° (arrow). In contrast, the Ca-containing silica particles gave two weak but distinct peaks at 26° and 32° in profile

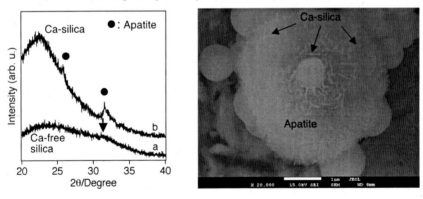

Fig. 4. (Left) XRD pattern of the Ca-free silica particles (a) and Ca-silica particles (b) after soaked in SBF for 7 d. See text for the arrow.
Fig. 5. (Right): a FE-SEM image of the Ca-silica particles after soaked in SBF for 7 d, with petal-like apatite crystallites. Bar: 1 μm.

(b). The 26° peak was assigned to the (002) diffraction and the 32° one was due to the envelope of the (211), (112) and (300) diffractions of apatite.[13,14,16,17] Although the peak intensity was lower than that for common materials with the apatite depositing ability,[13,14,16,17] the SEM photograph of the Ca-silica particles soaked in SBF for 7 (Fig. 5) shows petal-like crystalline deposits, with the same morphology as those apatite particles on those materials above. Those crystallites looked embracing the silica particles. From those results, the Ca-containing silica particles are surely active to be fixed with living bone when embedded in the bone tissues.

Amino-functionalized silica particles: microstructure and protein immobilization

The FE-SEM images of the particles AMSi045 and AMSi45 in Figs. 6 (a) and (b), respectively, illustrates some differences among them in size and morphology. The particles of both samples were spherical but they were largely fused together, which gives an impression that those are much more agglomerated than the previous Ca-free and Ca-silica. Yet, the component particles of AMSi045 remain more particle-like than those of AMSi45 which show highly fused peanut-shell morphology. That is, increase in APTES brought greater degree of agglomeration. The individual particles for AMSi045 seemed larger in diameter (300 ~ 400 nm) than those for AMSi45 (200 ~ 300 nm), though high irregularity in shape prevents definite measurement of the size. Moreover, the surface of AMSi45 particles appeared rougher than that for AMSi045. The TEM observation indicated that both samples consisted of ~ 10 nm primary particles. It was noted in the course of preparation that, at the lower APTES amounts (e.g., 0.45 mmol), the reaction in the precursor solution was quite mild and the opaque suspension was obtained after 30 min. In contrast, at the higher APTES amounts (> 0.45 mmol), the reaction in the precursor solutions became very vigorous and white sediments precipitated on the wall and at the bottom of the flask within few min. Such reaction conditions lead to the difference in agglomeration detected above.

Though they showed a little difference in morphology, ^{29}Si MAS NMR spectra indicated that both samples had very similar structural constitution in terms of the fraction of T^n and Q^n units. Fig. 7 shows the ^{29}Si MAS NMR spectrum of AMSi045, and a similar one was obtained for the other sample. The NMR spectrum was extended in two regions, -80 ~ -120 ppm for Q^n units and -40 ~ -80 ppm for T^n ones. The Gaussian fitting after Carroll et al.[20] was employed to deconvolute the profile into five clear peaks, i.e., T^2 units $NH_2(CH_2)_3Si(-O-Si)_2(OH)$ at -58 ppm, T^3 units $NH_2(CH_2)_3Si(-O-Si)_3$ at -66 ppm, Q^2 units $Si(-O-Si)_2(OH)_2$ at -90 ppm, Q^3 units $Si(-O-Si)_3(OH)$ at -100 ppm, and Q^4 units $Si(-O-Si)_4$ at -109 ppm. The presence of both T and Q units confirmed that the hydrolysis and

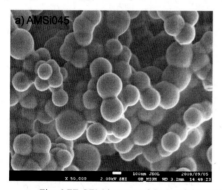 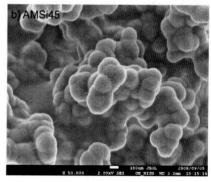

Fig. 6 FE-SEM images of the particles AMSi045 (a) and AMSi45 (b). (Bar: 1 μm)

condensation of APTES and TEOS took place in the precursor solutions. The strongest intensity of both Q^3 and Q^4 units and the medium one of T^3 indicated that most Si-OH in the particles have condensed into the siloxane network bonds (Si-O-Si).

BSA fixation is confirmed by detecting vibrational bands characteristic of proteins and the cross-linking agent EDC. Fig. 8 shows the FT-IR spectra for AMSi045 and its derivatives. Spectrum (a) for AMSi045 had the bands for the silica network at 1070 and 800 cm^{-1}, assigned to v(Si-O)$_{asym}$ and v(Si-O)$_{sym}$. A shoulder peak at 960 cm^{-1} was due to the Si-OH groups, which was detected in Fig. 2 as an independent peak for Ca-free and Ca-silica. The broader band extending from 2800 to 3500cm^{-1} was due to –OH of >Si-OH and adsorbed H_2O, which gave a sharper structure in Fig. 2. The distinct –NH$_2$ peak at 1543 cm^{-1} shows the presence of APTES in AMSi045. It follows the result from the above ^{29}Si NMR spectrum. In addition, both peaks at 2941 and 2873 cm^{-1} were attributed to –CH$_2$ groups or methylene skeleton introduced by APTES. Spectra (b) and (c) for the samples after contact with the BSA solution indicate a few IR bands related to BSA whose IR spectrum was represented as (d). Of the spectra for two AMSi045 samples, spectrum (b) showed no additional bands other than those found for the silica structure. On the other hand, spectrum (c) for the sample treated with EDC together with APTES gave a doublet band with medium intensity characteristic of amide bonds, which was commonly denoted as amide I (~1680 cm^{-1}) and amide II (~1590 cm^{-1}), and both were basically due to a C=O stretching vibration, v(C=O). A trace peak at 3305 cm^{-1} was assigned to –N-H bonds of BSA that was absent in spectrum (a) or (b) in Fig. 8. Those results indicate that BSA was immobilized on the particle

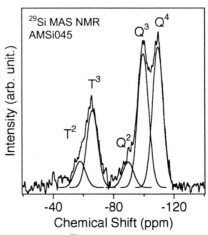

Fig. 7. A ^{29}Si MAS NMR spectrum for AMSi045. Each component peak was derived from Gaussian deconvolution.[20]

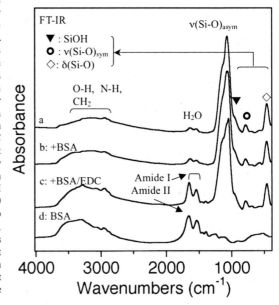

Fig. 8. FT-IR spectra of AMSi045 (a), AMSi045 after soaked in the BSA solution without EDC (b) and with EDC (c) as well as BSA (d).

surface as EDC was employed as an effective cross-linker. It agrees with the results from Szurdoki[21] and Mao[22] who used EDC as the cross-linker for proteins and natural polymers.

DISCUSSION

Ca-containing silica particles: formation mechanism and apatite deposition

Bogush and Zukoski,[23,24] considered that the Stöber silica particles consisted of ~ 10 nm primary particles, and Green et al.[25] proposed the microstructure as they measured small-angle X-ray scattering. From Table 1, the reaction solution was basic with pH 12.5. The silanol groups have an isoelectric point of about $2 \sim 3$,[26] and hence, in such alkaline solution, they should be in the forms of hydrated >SiO⁻ (aq). As they were un-polymerized, almost all of those units remained on the primary particle surface, yielding a negatively charged hydrated layer. The layer then causes repulsive interactions among the adjacent primary particles, but the >SiO⁻ (aq) forms strong hydrogen bonds with the water molecules. This overwhelms the repulsion to agglomerate the primary particles to SEM observable larger secondary particles.

The presence of $CaCl_2$ in the starting solutions stimulated the secondary particle growth; ~500 nm for the Ca-free particles and ~ 1000 nm for the Ca-silica ones. Zerrouk et al.[27] studied the interfacial behavior of mono-disperse (~15 nm) silica sols in terms of their surface charge as a function of pH and salt concentration. They proposed the concept of Ca^{2+} coagulation critical concentration in the range pH 7.5~9, assuming that Ca(II) promoted the agglomeration. In the present study, pH of the starting solution well exceeded their range but was 12.5. Since the higher was pH, the more >SiO⁻ (aq) yielded, one would expect even smallest amount of Ca(II) should form >Si-O⁻ •• Ca^{2+} •• ⁻O-Si< links. Thus, it is reasonable to suggest that the Ca^{2+} ions also contributed to the agglomeration of the ~ 10 nm primary particles. If

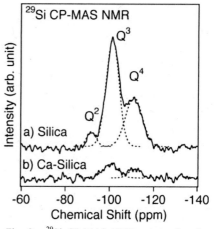

Fig. 9. ²⁹Si CP-MAS NMR spectra for a) the Ca-free silica and b) Ca-silica particles. The calcium involvement drastically decreased the CP-MAS NMR intensity (Chen et al.[28])

like this, the Ca(II) must be involved in the silica particles. Indeed, the release curve of Ca(II) in Fig. 3(a) confirmed it. Fig. 9 compares the ²⁹Si CP-MAS NMR spectra for the Ca-free silica and Ca-silica samples.[28] Under the cross-polarization (CP) mode, the nuclear magnetic moment energy of H is transferred to another nuclei in the neighborhood. Thus, the drastic decrease in NMR intensity due to the Ca(II) incorporation means that almost all Si atoms in the Ca-silica were relatively far apart from H than in the Ca-free silica. This means very little amount of >Si-OH or H_2O molecules was present among the primary particles or in the primary particles themselves. In other words, above results conclude that the calcium ions expelled H from the >Si-OH units (or Si-O-(aq) involved in the hydrogen bonds) to form >Si-O⁻ •• Ca^{2+} •• ⁻O-Si< links.

The Ca(II) release then takes place in two mechanisms: one is to be associated with the degradation of the secondary particles due to the hydrolysis of the >Si-O⁻ •• Ca^{2+} •• ⁻O-Si< links, and the other is due primarily to Ca(II) diffusion, associated with the hydrolysis, without degradation of the silica secondary particles. The Ca(II) release profile in Fig. 3(a) has moderately been approximated with either model in Fig. 3, yet the diffusion-controlled model seems better than the reaction-controlled model in terms of kai-square (χ^2). Diffusion model: [Ca] 0.013 d$^{1/2}$ ($\chi^2 = 4.5\text{x}10^{-5}$); Reaction model: [Ca]$\approx$ 0.0055 d ($\chi^2 = 1.5\text{x}10^{-5}$). Moreover, the Si(IV) release was definitely reaction controlled because of the profiles indicated linear dependence on time, though a small deviation was detected for the Ca-silica sample in prolonged contact with saline. After the Pourbaix diagram,[29] dissolution of silica as $HSiO_3^-$ practically takes place at pH~10 or higher: with log ($HSiO_3^-$) = -15.21 + pH, and log ($HSiO_3^-/H_2SiO_3$) = -10.00 + pH, 0.1mM $HSiO_3^-$ will be attained pH > ~10. In the saline with 7.4 in pH, detectable dissolution unlikely occurs, but, if the presence of Si(IV) is evidenced in Fig. 2, degradation of the secondary particles to liberate primary particles, and this is responsible for the Si(IV) species. In association with the degradation, therefore, the surface layer that involves Ca(II) as bridging the primary particles is to be hydrolyzed, and the calcium ions are liberated from the >Si-O$\cdot\cdot$Ca bonds and diffuse to the top surface. This makes the diffusion-controlled model dominant in the release profile analysis, with keeping some validity of the reaction-controlled model. Yet, detailed study is needed before elucidating a definite conclusion on the mechanism.

Amino-functionalized silica particles: mechanism of formation

The amino group of APTES is susceptible to hydrolysis or protonation, and hence the system becomes basic due to the liberated hydroxyl group as eq. (3):

$$(EtO)_3Si(CH_2)_3\text{-}NH_2 + H_2O \rightarrow (EtO)_3Si(CH_2)_3\text{-}NH_3^+ + OH^- \qquad (3)$$

Indeed, Table 2 showed pH was 10.89 for the precursor system APTES-TEOS-EtOH-H_2O. Under such basic conditions, the OH$^-$ ions directly attacked Si atoms having the highest positive charge,[30] i.e., the Si atom in the $(EtO)_3Si$- group of both TEOS and APTES molecules was subjected to coordination expansion, taking 5-hold coordination in the reaction intermediate. Then, the Si atom of APTES has three active sites, like that in alkali disilicate. Since the latter Si(IV) forms an amorphous silicate network or alkali disilicate glass, APTES should form silicate gel particles. TEOS with

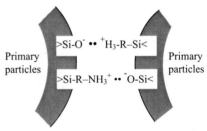

Fig. 10. Two amino-modified silica primary particles are agglomerated via the hydrogen bonding between >Si-O$^-$ and ^+H_3N-R-Si< on the facing surfaces.

four-functional Si(IV) is known to yield silica gel bodies. Contrary to the expectation, the present study confirmed that APTES would not form gels. After van Blaaderen and Vrij,[31] the hydrolyzed APTES tended to form six- or five-membered intra-molecular rings. Such ring formation should, due to their bulkiness, sterically suppress condensation of APTES itself from yielding three-dimensionally grown gels. It is then considered that the formation of amino-modified silica particles requires >Si-OH groups derived from TEOS. In the present alkaline precursor solution, two hydrated species were present: positively charged $-NH_3^+$(aq) and negatively charged >Si-O$^-$(aq). After Chen et al.,[19] the particle surface was rich in amino groups. The presence of those hydrated ions with opposite charges favors agglomeration of the primary particles due to hydrogen bonding. Fig. 10 schematically represents such interaction leading to the agglomeration. The difference in morphology between AMSi045 and AMSi45 might be attributed to more basic level for the latter system, which would lead to highly vigorous reaction and higher degree of chances in which the >Si-O$^-$ and $^+H_3$-R-Si< should be present toward each other on the facing surfaces. Yet, detailed understanding of the agglomeration mechanism should need further study.

CONCLUSION

From one-step sol-gel procedure, two series of silica particles were prepared, i.e., Ca-containing silica and amino-modified silica particles, as well as pure silica particles, from the starting solution systems TEOS-H_2O-EtOH-$CaCl_2$-NH_4OH for the former silica particles, and from TEOS-H_2O-EtOH-APTES for the latter. The Ca-silica particles were spherical in shape and consisted of ~ 10 nm primary particles. Their size increased with the amount of $CaCl_2$ in the precursor solutions from 400 ~ 500 nm for the Ca-free silica particles to 800 ~ 1000 nm for the Ca-silica ones. The ^{29}Si MAS NMR analysis indicated that the NMR intensity of the Si atoms was considerably reduced due to the cross polarization, and hence calcium ions formed >Si-O⁻ •• Ca^{2+} •• ⁻O-Si<. That bond was considered to bridge the primary particles together, and responsible for the particle growth. Analysis of Si(IV) release into saline indicated that the secondary particles were degraded to liberate the primary particles. The Ca(II) was released dominantly due to the diffusion controlled mechanism since a parabolic approximation fitted better the empirical data. The Ca(II) released from the Ca-silica particles into SBF promoted the deposition of petal-like apatite crystallites within one week. Thus, the Ca-silica particles are applicable to the bioactive fillers for bone regeneration.

In the course of evolving the amino-modified silica particles, APTES worked not only as a reactant but also the catalyst which resulted in pH ~ 11. The FE-SEM image revealed that the particles were spherical in shape less than 500 nm in diameter, consisting of ~ 10 nm primary particles. From the analyses of ^{29}Si MAS NMR spectrum and FT-IR spectra, >Si-O⁻ (aq) groups and $^+H_3$-R-Si< (aq) groups were presented on the particle surface, and strong hydrogen bonds formed between them was to promote the agglomeration of the ~ 10 nm primary particles into the secondary particles. Moreover, in the presence of carbodiimide, bovine serum albumin was covalently fixed on the particle surface. Thus, the amino-functionalized silica particles have the high potential applications as carriers for enzymes or DNA.

REFERENCES

[1]Q.-M. Zhang, K. Ariga, A. Okabe, and T. Aida, A Condensable Amphiphile with a Cleavable Tail as a "Lizard" Template for the Sol-Gel Synthesis of Functionalized Mesoporous Silica, *J. Am. Chem. Soc.*, **126**, 988-989 (2004).
[2]P. Kortesuo, M. Ahola, S. Karlsson, I. Kangasniemi, A.Yli-Urpo, and J. Kiesvaara, Silica Xerogel as an Implantable Carrier for Controlled Drug Delivery—Evaluation of Drug Distribution and Tissue Effects After Implantation, *Biomaterials*, **21**, 193-198 (2000).
[3]C. Kneuer, M. Sameti, E. G. Haltner, T. Schiestel, H. Schirra, H. Schmidt, and C.-M. Lehr, Silica Nanoparticles Functionalized with Aminosilanes as Carriers for Plasmid DNA, *Int. J. Pharm.*, **196**, 257-261 (2000).
[4]L. Ren, K. Tsuru, S. Hayakawa, and A. Osaka, Novel Approach to Fabricate Porous Gelatin–Siloxane Hybrids for Bone Tissue Engineering, *Biomaterials*, **23**, 4765-4773 (2002).
[5]Y. Shirosaki, K. Tsuru, S. Hayakawa, A. Osaka, M. A. Lopes, J. D. Santos, and M. H. Fernandes, In vitro Cytocompatibility of MG63 Cells on Chitosan-Organosiloxane Hybrid Membranes, *Biomaterials*, **26**, 485-493 (2005).
[6]L. L. Hench, Stimulation of Bone Repair by Gene Activating Glasses, Key Engineering Materials, **254-256**, 3-6 (2004).
[7]K. Deguchi, K. Tsuru, T. Hayashi, M. Takaishi, M. Nagahara, S. Nagotani, Y. Sehara, G. Jin, H. Zhang, S. Hayakawa, M. Shoji, M. Miyazaki, A. Osaka, N.-H. Huh, and K. Abe, Implantation of a New Porous Gelatin-Siloxane Hybrid into a Brain Lesion as a Potential Scaffold for Tissue Regeneration, *J. Cerebral Blood Flow and Metabolism*, **26**, 1263-1273 (2006).
[8]Y. Shirosaki, C. M. Botelho, M. A. Lopes, and J. D. Santos, Synthesis and Characterization of

Chitosan-Silicate Hydrogel as Resorbable Vehicle for Bonelike® Bone Graft, *J. Nanosci. Nanotechn*, **8**, 1–6 (2008).

[9]Z. Z. Li, L. X. Wen, L. Shao, and J. F. Chen, Fabrication of Porous Hollow Silica Nanoparticles and their Applications in Drug Release Control, *J. Control. Release*, **98**, 245-254 (2004).

[10]Z. Q. Ye, M. Q. Tan, G. L. Wang, and J. L. Yuan, Development of Functionalized Terbium Fluorescent Nanoparticles for Antibody Labeling and Time-resolved Fluoroimmunoassay Application, *Talanta*, **65**, 206-210 (2005).

[11]M. Arroyo-Hernández, R. J. Martín-Palma, J. Pérez-Rigueiro, J. P. García-Ruiz, J. L. García-Fierro, and J. M. Martínez-Duart, Biofunctionalization of Surfaces of Nanostructured Porous Silicon, *Materials Science and Engineering: C*, **23**, 697-701 (2003).

[12]X. Li, Y. He, and M. T. Swihart, Surface Functionalization of Silicon Nanoparticles Produced by Laser-Driven Pyrolysis of Silane Followed by HF-HNO$_3$ Etching, *Langmuir*, **20**, 4720-4727 (2004).

[13]P.-J. Li, C. Ohtsuki, T. Kokubo, K. Nakanishi, N. Soga, T. Nakamura, T. Yamamuro, Apatite Formation Induced by Silica Gel in a Simulated Body Fluid, *J. Am. Ceram. Soc.*, **75**, 2094-97 (1992).

[14]K. Nakanishi, Pore Structure Control of Silica Gels Based on Phase Separation, *J. Porous Mat.*, 4, 67-112 (1997).

[15]T. Kokubo and H. Takadama, How useful is SBF in predicting in vivo Bone Bioactivity? *Biomaterials*, **27**, 2907–2915 (2006).

[16]L. Ren, K. Tsuru, S. Hayakawa, and A. Osaka, Sol–Gel Preparation and in vitro Deposition of Apatite on Porous Gelatin–Siloxane Hybrids, *J. Non-Cryst. Solids*, **285**, 116-122 (2001).

[17]K. Tsuru, Y. Aburatani, T, Yabuta, S. Hayakawa, C. Ohtsuki, and A. Osaka, Synthesis and In Vitro Behavior of Organically Functionalized Silicate Containing Ca Ions, *J. Sol-Gel Sci. Technol.*, **21**, 89-96 (2001).

[18]W. Stöber, A. Fink, and E. Bohn, Controlled Growth of Monodisperse Silica Spheres in the Micron Size Range, *J. Colloid Interf. Sci.*, **26**, 62-69 (1968).

[19]S. Chen, A. Osaka, S. Hayakawa, K. Tsuru, E. Fujii, and K. Kawabata, Novel One-Pot Sol-Gel Preparation of Amino-Functionalized Silica Nanoparticles, *Chem. Lett.*, **37**, 1170-1171 (2008).

[20]S. A. Carroll, R. S. Maxwell, W. Bourcier, S. Martin, and S. Hulsey, Evaluation of Silica-Water Surface Chemistry Using NMR Spectroscopy, *Geochim. Cosmochim. Acta*, **66**, 913-26 (2002).

[21]F. Szurdoki, E. Trousdale, B. Ward, S. J. Gee, B. D. Hammock, and D. G. Gilchrist, Synthesis of Protein Conjugates and Development of Immunoassays for AAL Toxins, *J. Agric. Food Chem.*, **44**, 1796−1803 (1996).

[22]J. S. Mao, H. F. Liu, Y. J. Yin, and K. D. Yao, The Properties of Chitosan–Gelatin Membranes and Scaffolds Functionalized with Hyaluronic Acid by Different Methods, *Biomaterials*, **24**, 1621-1629 (2003).

[23]G. H. Bogush and C. F. Zukoski, Studies of the Kinetics of Precipitation of Uniform Silica Particles Through the Hydrolysis and Condensation of Silicon Alkoxides, *J. Colloid Interf. Sci.*, **142**, 1-18 (1991).

[24]G. H. Bogush and C. F. Zukoski, Uniform Silica Particle Precipitation: an Aggregative Growth Model, *J. Colloid Interf. Sci.*, **142**, 19-34 (1991).

[25]D. L. Green, J. S. Lin, Y. F. Lam, M. Z. Hu, D. W. Schaefer, and M. T. Harris, Size, Volume, Fraction, and Nucleation of Stöber Silica Nanoparticles, *J. Colloid Interf. Sci.*, **266**, 346-358 (2003).

[26]H. Choi and I.-W. Chen, Surface-modified Silica Colloid for Diagnostic Imaging, *J. Colloid Interf. Sci.*, **258**, 435-437 (2003).

[27]R. Zerrouk, A. Foissy, R. Mercier, and Y. Chevallier, J. C. Morawski, Study of Ca^{2+} Induced Silica Coagulation by Small-Angle Scattering, *J. Colloid Interface Sci.*, **139**, 20-29 (1990).

[28]S. Chen, A. Osaka, S. Hayakawa, K. Tsuru, E. Fujii, and K. Kawabata, Microstructure Evolution in Stöber-Type Silica Nanoparticles and their in vitro Apatite Deposition, *J. Sol-Gel Sci. Technol.*, **48**,

322–335 (2008).

[29] J. Van Muylder, J. Besson, W. Kunz and M. Porubaix, in *Atlas of Electrochemical Equilibria in Aqueous Solutions*, ed. M. Pourbaix, Pergamon Press, Oxford, Chapter IV, §17.2

[30] L. L. Hench and J. K. West, The Sol-Gel Process, *Chem. Rev.*, **90**, 33-72 (1990).

[31] A. van Blaaderen and A. Vrij, Synthesis and Characterization of Monodisperse Colloidal Organo-Silica Spheres, *J. Colloid Interf. Sci.*, **156**, 1-18 (1993).

FABRICATION OF HYBRID THIN FILMS CONSISTING OF CERAMIC AND POLYMER USING A BIOMIMETIC PRINCIPLE

Langli Luo and Junghyun Cho
Dept. of Mechanical Engineering & Materials Science and Engineering Program
T. J. Watson School of Engineering
State University of New York (SUNY) at Binghamton
Binghamton, New York, USA

ABSTRACT

The concept of biomimetic processing has drawn strong interests in thin film processing but few studies focused on this aspect on titania films that have numerous promising applications. Natural materials, often incorporating organic additives into the inorganic structure, that yield to a specific structure with extraordinary properties are difficult to reproduce in a way that they form in nature due to their slow reaction kinetics. In this study, we deposit the ceramic-polymer hybrid thin films by combining titania (TiO_2) and poly(acrylic acid) (PAA) in an aqueous solution at temperatures about 40-70°C. The resultant thin films were uniform with the thickness ranging from 100 nm to 200 nm. Film microstructures were adjustable by maintaining different molar ratios between TiO_2 and PAA. Due to the addition of PAA, corresponding mechanical property changed, in which with more polymer concentration, it exhibited higher toughness to resist crack propagation.

INTRODUCTION

There have been immense interests in biomimetic or bioinspired processing because of the requirements for low cost, ambient condition and environmental friendly processing in latest engineering techniques. Particularly, the capability to process at near room temperature and in aqueous solution is the most notable merit where tremendous potential can be seen. In the past decade, great efforts have been made to mimic the natural formation of functional structures in both bulk materials and thin films. Mann suggested a clever way of the biomimetic pathway for assembling inorganic thin films[1] and summarized the principles of biomineralization.[2] It can be basically divided into four steps: (i) supramolecular preorganization via construction of an organized reaction environment; (ii) interfacial molecular recognition, where the preformed organic supramolecular systems provide a framework for the assembly of the inorganic phase; (iii) assembly of the mineral phase through crystal growth and termination enabling the formation of textures and shapes; (iv) formation of higher order architectures involving larger scale cellular activity. Recent review by Gao and Koumoto explained the mechanism related to bioinspired ceramic thin films processing.[3]

Among such concepts, using an organic template such as SAM (self-assembled monolayer) has been considered a possible candidate for constructing the inorganic functional structures. For example, several studies demonstrated successful fabrication of oxide thin films such as TiO_2, ZrO_2, and SiO_2, with the presence of SAM at temperatures below 100°C. Furthermore, SAM has shown the great potential in patterning and selective deposition of ceramic oxide thin films.[4] However, regarding microstructure and related properties, ceramic thin films have not gained much attention from the SAM. Recent studies also showed that the ceramic films can form from aqueous precursor solution even without SAM[5] as this process is rather based on *in-situ* precipitation in solution. In natural biomimenrals, various functional inorganic structures can be fabricated with the aid of versatile organic additives such as amino acids and proteins. This structure evolution was thought to be responsible for the formation of biominerals such as nacre and sea urchin spine.[6,7]

Ceramic thin films have prominent potential in the field of thin film transistors and solar cells in addition to traditional electronic semiconductors and optical coatings.[8,9]. Also, protective coatings for MEMS or electronic devices in harsh environment can be seen as another application. More importantly, the relaxation of inorganic and organic composite thin film can benefit the stringent requirement for electronic packaging due to the incorporation of polymers.[10] Poly(acrylic acid) (PAA) is a common polyelectrolyte which is fully soluble in water and also environmentally benign. It has shown to help form hierarchical structures in the biomimetic inorganic salts,[5] specific inorganic structure such as capsules[11] and layered thin films.[12]

For these applications, controlling the formation of bulk precipitates and their interactions in aqueous solution is essential for ceramic film to be more reliable and tailorable towards desirable microstructures. Therefore, the purpose of this study is to deposit the nanocomposite films consisting of *in-situ* precipitated titania and polymers by mimicking the inorganic structures found in the aforementioned biomineralization. We investigate the effect of this polymer on the formation of thin films and on microstructure developments and property changes.

EXPERIMENTAL PROCEDURE

For deposition of TiO_2 - PAA nanocomposite thin films, precursor solution was prepared with PAA additives in chilled water. To maintain a mild hydrolysis process in precursor solution, we tried to inhibit extreme reactivity of $TiCl_4$. First, 0.1 M hydrochloric acid was added into chilled distilled water followed by the addition of PAA (25% water solution, Average M.W. 240,000) (Alfa Aesar, Wald Hill, MA). The PAA water solution was diluted 10 times to avoid inaccurate micropipetting due to its high viscosity. Then, titanium chloride ($TiCl_4$, 99.99%, Alfa Asear, Wald Hill, MA) was syringed drop by drop to the chilled acid solution and stored in a refrigerator. The pH value of precursor solution was 1.29.

Both slide and cover glasses were employed as a substrate for deposition of thin films. The pretreatment of substrate included ultrasonic cleaning with acetone and distilled water in sequence for 20 minutes each. Then, cleaned substrates with inert holders were kept in the oxygen plasma for 20 minutes for further cleaning. The deposition process resembles the chemical bath deposition which typically involves the interaction between aggregation of molecules or small particles in solution and on the substrate through the nucleation and growth either homogeneously or heterogeneously. We tried to mimic the natural mineralization process, but they are often too slow to proceed. Thus, we have to increase the solution temperature to 40~70℃ to expedite the deposition process.

We also varied the $TiCl_4$/PAA molar ratio to tackle the function of polyelectrolytes in solution and polymers in solid thin films. Substrates were placed vertically in the slots on the teflon holder. The round holder were put into a beaker of the same size, and then kept in a magnetic stirred water bath at the set temperatures. The deposition time was typically set to 6 hrs for 4 steps each with 1.5 hr. The multistep deposition was used to ensure that the nanoparticles of uniformly distributed sizes offered by a fresh precursor solution can be deposited on the substrate. Finally, the as-deposited thin films were rinsed with distilled water and dried by blowing nitrogen gas.

Microstructures of the nanocomposite thin films were characterized by field-emission scanning electron microscope (FE-SEM; Zeiss Supra 55) at a low accelerating voltage (2kV) with an in-lens detector. Both as-deposited plan view samples and fractured cross-sectional samples were prepared, both of which were carbon coated to prevent electron charging during SEM observation. Microhardness tester (Leco M-400) was employed to characterize the mechanical property of the nanocomposite thin films. The post indentation marks have been analyzed through a Zeiss optical microscope to measure the indented area and crack lengths.

RESULTS AND DISCUSSION

Thin Film Formation

Solution-based deposition may involve two nucleation mechanisms: 1) heterogeneous nucleation at the interface between the substrate and solution, and 2) homogeneous nucleation by forming stable nuclei in a supersaturated solution.[13] In fact, heterogeneous nucleation is not favorable in a bare titania deposition process because the precursor solution is highly supersaturated and the interface between the substrate and solution may not provide ideal nucleation sites to lower the free energy to ensure the heterogeneous nucleation. When polyelectrolyte PAA is, however, added to the precursor solution, the interaction between PAA molecules and Ti(IV) aqueous species[5] can trigger the heterogeneous nucleation process to ensure the deposition of uniform nanoparticles.

PAA plays a crucial role in this processing because of its promotion for the hydrogen bonding[14] with the substrate, as well as the adsorption of PAA anions onto the TiO_2 particles. In fact, at a low polymer concentration this adsorption may not be sufficient for precipitation. For example, in the case of ZnO, Foissy[15] claimed that zinc polyacrylate can precipitate at the maximum adsorption isotherm. The same phenomenon was also observed for PAA on calcium carbonate for the precipitation and adsorption. Only when the polymer was sufficiently in excess of calcium ions did complete adsorption take place.[16] In our case, we varied the molar ratio between $TiCl_4$ and PAA from 2:1 to 80:1 in order to investigate the role of PAA in the deposition process. The schematic illustration of a possible mechanism to describe formation of nanocomposite thin film is shown in Fig.1.

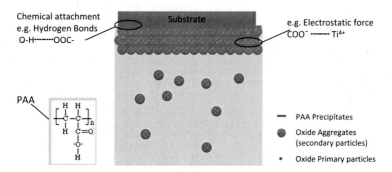

Figure 1. Schematic illustration of the mechanism of TiO_2-PAA thin film formation and PAA molecule structure. Here, oxide particles can be either primary particles nucleated and grown, or their aggregates.

Basically, the organic additive such as PAA is expected to have two roles: 1) anchored to the surface of the substrate to form the thin film, and 2) control or suppress the crystal growth of small particles and their aggregates (i.e., secondary particles), as in the nacreous layer.[6] The surplus -OH groups on the surface of substrate has a connection with the –COO group of PAA molecules, which has a strong interaction with particles in the solution that results in enhancing the attachment. The electrostatic force between PAA ions and Ti(IV) ions at high polymer concentration, or the adsorption of polyelectrolyte onto TiO_2 particles at low polymer concentration can be responsible for the connection between PAA and TiO_2 to form this ceramic-polymer nanocomposite structure.

Thin Film Microstructure

Microstructures of nanocomposite thin films are determined by assembling nanobuilding blocks including TiO_2 particles and PAA molecules. The role of multifunctional organic species both as a template and an organizer for specific structures was not clearly identified for the TiO_2 case. There seemed to have two interactions concerned in the thin film process. First, there is interaction between Ti(IV) and polymer species in the precursor solution that controls the formation of aggregates, and secondly, there is interaction between the aggregates and the surface of substrate that controls microstructure of the nanocomposite films.

When the PAA concentration is comparable (i.e., half of that of TiO_2), the adsorption of PAA on titania precipitates can lead to the precipitates of Ti-PAA complex. With large polymer concentration at 40°C, film formation of TiO_2 and PAA was observed on the substrate that exhibited a non-continuous, interconnected film structure as shown in Fig. 2a. The film structure did not seem to be fully developed, showing the 'gel' like structure. As the polymer concentration decreased, the non-uniform behavior was reduced. In Fig. 2b, with a lower concentration of polymer (1:10), a continuous and denser film structure appeared, thereby suggesting the benefit of right amount of polymer in the film formation. Titania primary particles have an average diameter of 20 nm in the film structure that is consistent with our previous study.[5] TiO_2 particles were well connected and organized by the PAA at this concentration and there were no significant aggregation.

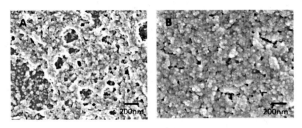

Figure 2. SEM images of microstructures in TiO_2-PAA thin films with different polymer concentration: a) [Ti/PAA] =2:1; b) [Ti/PAA] =10:1.

In order to accelerate the precipitation of TiO_2 particles, solution temperature was increased to 50°C or higher. At 50°C, film deposition rate was much enhanced due to higher supersaturation that also leads to more particulate and aggregated structure in the film. Because of higher solution temperature, the optimum amount of PAA was lower than the above 10:1 ratio. A sequence of Ti/PAA molar ratio of 20:1, 40:1 and 80:1 has been applied to a 6 hrs deposition process. As shown in Fig. 3, more aggregates (secondary particles) consisting of primary particles of about 10 nm were clearly shown in the film structure.

It is evident that the size of aggregated particles decrease with the decrease of PAA concentration. The size of aggregated particles decreased from 200 nm for higher PAA concentration (20:1) to less than 100 nm for the smallest PAA concentration (80:1).

Cross-sectional view (Fig. 3, right column), obtained from manual fracture, shows a similar structure discussed in the plan-view images. For the same amount of polymer concentration, the thickness of the nanocomposite film was larger due to high deposition rate. It is noted that much denser structure was observed near the substrate regardless of the amount of the polymer.

Polymer in the film tends to attract more particles into a unit whose size seems to be controllable by its amount. The secondary particles become larger to reduce the surface energy by attracting each other. However, there seemed to have no preferred orientation in the TiO_2 aggregates unlike those in the nacreous layer in nature. Microstructure with large aggregated particles causes large roughness and surface area that could be desirable for the application of thin film solar cells. On the other hand, dense films with lower PAA concentration could be potentially used as a protective and/or a buffer layer for the electronic packaging or MEMS devices.

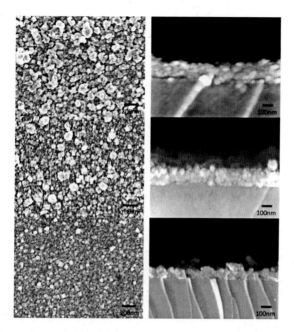

Figure 3. Plan-view and cross-sectional SEM images of nanocomposite thin films at different polymer concentrations: Top: [Ti/PAA] =20:1; Middle: [Ti/PAA] =40:1; Bottom: [Ti/PAA] =80:1

In the formation of TiO_2-PAA nanocomposite thin films, PAA anchored on the surface of the substrate may enhance the heterogeneous nucleation and also attract more nuclei in the precursor solution to settle down and grow at very initial stage of deposition. And during the deposition, it plays another role of suppressing the growth of TiO_2 particles and organizing them into a fine and dense film. It resembles the way of an organic matrix that performs as a template and an organizer in natural biomineralization process but in an expedited way.

MECHANICAL PROPERTY OF THIN FILM

Microhardness test of TiO_2-PAA nanocomposite thin films was performed at 10g and 25g to assess the effect of polymer addition on mechanical properties of ceramic films. Although the indentation depth was larger than thickness of the film, the measured indentation data can offer useful information in a relative sense. When there was no polymer, the apparent hardness measured with the indentation load was very similar to that of glass substrate. However, as the polymer concentration increases in the film, the apparent hardness becomes reduced due to the effect from the polymer, thereby transforming the film softer and possibly more compliant. Table 1 shows the results of apparent Vickers hardness (Hv) of nanocomposite thin films with different PAA concentration. The thickness of these thin films was all similar.

During indentation, crack lengths were also generated when higher indentation loads were applied. The result shows a marked difference between bare TiO_2 films and TiO_2 films with the polymer. Adding the polymer to the TiO_2 film increases resistance to crack propagation. Table 1 shows crack length (c) comparison, and Fig. 4 shows an example of the representative crack morphology observed from the bare TiO_2 film and the TiO_2-PAA nanocomposite film. The fracture toughness of this film is inversely proportional to $c^{3/2}$. [17]

    This observation proved the potential for a high-toughness nanocomposite thin film which can be beneficial for ceramic film applications that require crack-free and strain-tolerance. Even though hardness will decrease, the nanocomposite films can have significant merits due to improvement in toughness. More work is currently in progress to optimize the microstructure of the nanocomposite film for further enhancement of toughness. In order to more accurately measure the mechanical properties of the film, nanoindentation will be utilized for future analysis, in which case the indentation depth will be confined within the film.

Table 1. Apparent Vickers hardness and crack propagation of nanocomposite thin films at different PAA concentrations

[TiO_2/PAA]	TiO_2 only	80:1	40:1	20:1
Hv (GPa)	5.4±0.6	4.3±0.5	3.7±0.4	3.1±0.4
Crack length, c (μm)	100.9±3.2	92.7±3.8	88.3±3.9	84.8±3.1

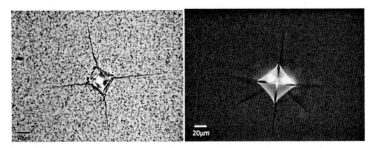

Figure 4. Microhardness test sample under a 1kg load: a) bare TiO_2 thin film grown on silicon substrate; b) TiO_2-PAA thin film grown on glass substrate.

CONCLUSION

    TiO_2-PAA nanocomposite thin films were prepared through *in-situ* precipitation of the corresponding ceramic precursor and polymer electrolyte in an aqueous solution at a near room temperature (40-70°C). The size of nanocrystallites, which are nucleated and grown, is usually 5-10 nm but they are quickly aggregated to 50-200 nm depending on the PAA concentration. With the increase of PAA concentration, the aggregates grew larger and the surface roughness increased. With the temperature increase in solution, more particulate structure formed due to contribution from the bulk precipitation and its aggregation. The characterization of mechanical property of these thin films that contain various polymer concentrations showed the potential for tougher ceramic film through the hierarchical organization of the nanoparticles formed in low temperature aqueous solution, which is frequently found in biomineralization.

ACKNOWLEDGMENTS

This research was sponsored by the New York State Office of Science, Technology and Academic Research (NYSTAR), Center for Electronic Imaging Systems (CEIS) at University of Rochester, and Center for Advanced Microelectronics Manufacturing (CAMM) at State University of New York at Binghamton.

REFERENCES

[1] I. A. Aksay, M. Trau, S. Manne, Biomimetic Pathways for Assembling Inorganic Thin Films, *Science*, New Series, **273**, 892-898, (1996).

[2] S. Mann, Biomineralization: Principles and Concepts in Bioinorganic Materials Chemistry, Oxford University Press, Inc., New York, (2001).

[3] Yanfeng Gao and Kunihito Koumoto, Bioinspired Ceramic Thin Film Processing: Present Status and Future Perspectives, *Crystal Growth and Design*, **5**, 1983-2017, (2005).

[4] Masuda, Y., Sugiyama,T., Koumoto, K., Micropatterning of anatase TiO_2 thin films from an aqueous solution by a site-selective immersion method, *J. Mater. Chem.*, **12**, 2643- 2647(2002).

[5] G. Zhang, B. K. Roy, L. Allard, and J. Cho, Titanium Oxide Nanoparticles Precipitated from Low-Temperature Aqueous Solutions: I. Nucleation, Growth, and Aggregation, *J. Am. Ceram. Soc.*, **91**, 3875-3882, (2008)

[6] Y. Oaki and H. Imai, The Hierarchical Architecture of Nacre and Its Mimetic Material, *Angew. Chem. Intl. Ed*, **44**, 6571-6575, (2005).

[7] Y. Politi, T.Arad, E. Klein, S. Weiner, L.Addadi, Sea Urchin Spine Calcite Forms via a Transient Amorphous Calcium Carbonate Phase, *Science*, **306**, 1161-1164, (2004).

[8] Nomura K, Ohta H, Ueda K, Thin-film transistor fabricated in single-crystalline transparent oxide semiconductor, *science*, **300**, 1269-1272, (2003).

[9] Park NG, van de Lagemaat J, Frank AJ, Comparison of dye-sensitized rutile- and anatase-based TiO2 solar cells, *J. Phys. Chem. B*, **104**, 8989-8994, (2000).

[10] N. Maluf, An Introduction to Microelectromechanical Systems, Artech House, Inc., Norwood, MA, (2000).

[11] Y. Hu, J. Ge, Y. Sun, T. Zhang, and Y. Yin, A Self-Templated Approach to TiO_2 Microcapsules, *Nano Lett.*, **7**, 1832-1836, (2007).

[12] T. Kato, A. Sugawara, N. Hosoda, Calcium Carbonate- Organic Hybrid Materials, *Adv. Mater.*, **14**, 869-877, (2002).

[13] G. Zhang, J. Y. Howe, D. W. Coffey, D.A. Blom, L. F. Allard, J. Cho, A Biomimetic Approach to the Deposition of ZrO_2 Films on Self-assembled Nanoscale Templates, *Mater. Sci.&Eng. C*, **26**, 1344-1350, (2006).

[14] Y. Oaki, H. Imai, Morphological Evolution of Inorganic Crystal into Zigzag and Helical Architectures with an Exquisite Association of Polymer: A Novel Approach for Morphological Complexity, *Langmuir*, **21**, 863 – 869, (2005).

[15] C. Dange, T.N.T. Phan, V. André, J. Rieger, J. Persello, A. Foissy, Adsorption mechanism and dispersion efficiency of three anionic additives [poly(acrylic acid), poly(styrene sulfonate) and HEDP] on zinc oxide, *J. of Colli. & Inter. Sci.*, **315**, 107-115, (2007).

[16] C. Geffroy, A. Foissy, J. Persello, B. Cabane, Surface Complexation of Calcite by Carboxylates in Water, *J. of Colli. & Inter. Sci.*, **211**, 45- (1999).

[17] G. R. Anstis, P. Chantikul, B. R. Lawn, and D. B. Marshall, A Critical Evaluation of Indentation Techniques for Measuring Fracture Toughness: I, Direct Crack Measurements, *J. Am. Ceram. Soc.*, **64**, 533-538 (1981).

STRUCTURAL INVESTIGATION OF NANO HYDROXYAPATITES DOPED WITH Mg^{2+} AND F^{-} IONS

Z. P. Sun[1], Z. Evis[1,2]
[1]*Middle East Technical University, Department of Micro and Nanotechnology, Ankara, 06531, Turkey*
[2]*Middle East Technical University, Department of Engineering Sciences, Ankara, 06531, Turkey*

ABSTRACT
Pure and Mg^{2+} & F^{-} doped nano hydroxyapatites (HA) were synthesized by a precipitation method and sintered at 1100°C to investigate their densification and structure. Different amounts of Mg^{2+} and F^{-} ions were doped in to HAs. X-ray diffraction was used to identify the presence of phases and lattice parameters of HA and tri-calcium phosphates (TCP) present in the samples. Densification of HA was improved by the addition of these ions. In most of the doped samples, β-tri-calcium phosphate (TCP) formation was observed as a second phase with the addition of Mg. Structure of the pure and doped samples were hexagonal and rhombohedral for HA and TCP, respectively.

1. INTRODUCTION
Hydroxyapatite (HA, Ca$_{10}$(PO$_4$)$_6$(OH)$_2$) has been widely used as an implant material for bones owing to its excellent biocompatibility to human tissues because it has similar structure with the mineral part of bone [1-3]. Nevertheless, its poor mechanical properties such as strength, hardness and toughness limit the applications of HA in load-bearing areas [4]. Therefore, it has been used in non-load bearing fields such as ossicles in middle ear [3, 4].

HA has Ca and P elements in its hexagonal structure. These elements are the same elements present in the inorganic part of the bone. There have been many attempts to improve the various properties of synthetic HA [5-10]. The special attention was on nano crystalline HA to achieve improvements both biologically and mechanically. Bone minerals are in nano scale; therefore, the studies and researches on nano HA have been accelerated to improve its properties in all aspects.

Biological HA crystals contain many ionic impurities such as; K^{+}, Mg^{2+}, Na^{+}, CO$_3^{2-}$ and F^{-} [11-14]. Substitution of these ions enhance the biological, mechanical and chemical properties of nano HA [13]. Among the dopants of 5 wt % Mg^{2+}, Zn^{2+}, La^{3+}, Y^{3+}, In^{3+} and Bi^{3+}, following ions (Zn^{2+}, In^{3+}, Bi^{3+}) were reported as best effective dopants enhancing the osteoblast attachment to HA [13].

Magnesium is one of the most important bivalent ions, which affects the bone fragility. It can substitute for Ca sites in HA. In bone tissue, Mg stimulates the transformation of immature (amorphic) bone into a more crystalline form. The translocation of Mg into mineral tissues prevents fractures by increasing the bone elasticity [15,16]. Its deficiency effects all stages of skeletal metabolism, causing cessation of bone growth, decrease of osteoblastic and osteoclastic activities, increase in bone fragility. Therefore, Mg incorporation into the HA structure is of great interest for developing of artificial bone substitutes [17].

β-TCP, also known as β-whitlockite, is a slow degrading, bioresorbable calcium phosphate ceramic. It is reported that β-TCP is stable below 1125°C and above this temperature it goes under phase transition to α-TCP leading to an expansion in the crystal lattice. Mg ions stabilize the β-TCP upon heating above 800°C along with the formation of HA and improve the thermal stability of TCP [16,18].

Fluoride doping is also frequently used in order to improve the thermal stability and biological properties of HA. It can substitute for the OH^{-} sites in HA. Fluoride substituted HA has influenced physical and biological characteristics. Fluoride is necessary for dental and skeletal development; it is known to be very important in suppressing dental caries [19, 20]. It has been reported that doping low amounts (0.5-0.05 wt%) of either Mg^{2+} or F^- increased in vivo stability of HA [21].

In this study, HA doped with Mg^{2+} and/or F^- were synthesized by a novel precipitation method and sintered at 1100°C for 1 hour. Mg^{2+} and F^- ions were doped in different amounts by mole % and in HA for the first time. The densities of the materials were determined by Archimedes method. Presence of phases in sintered samples was investigated by x-ray diffraction (XRD). Lattice parameters and particle size of the samples were calculated from XRD results.

2. EXPERIMENTAL PROCEDURE

Pure-HA was synthesized by a precipitation method [22-24]. Calcium nitrate tetra hydrate $(Ca(NO_3)_2 .4H_2O)$ and di-ammonium hydrogen phosphate $((NH_4)_2HPO_4)$ were mixed into each other to obtain pure HA. 0.5 M $Ca(NO_3)_2.4H_2O$ and 0.3 M $(NH_4)_2HPO_4$ was separately prepared in distilled water. The Ca/P ratio was kept 1.67 to obtain the correct stoichiometry of HA. Ammonia solution was added into both solutions to bring the pH to 11-12 level. Calcium nitrate solution was dropwisely added into continuously stirring di-ammonium hydrogen phosphate solution. During stirring, the mixture was initially heated until boiling in order to increase the reaction rate and it was left to stirring overnight at room temperature (RT). All the samples were aged for ~ 2 days. After the aging, the mixture was filtered by using a fine filter paper to obtain a wet cake. It was dried overnight in an oven at 200°C. After the drying, calcination was performed at 600°C for 10 minutes. Next, the sintering was performed at 1100°C for 1 hour.

For doped HA, the main precursors were same. Additionally, magnesium nitrate $(Mg(NO_3).6H_2O)$ and ammonium fluoride (NH_4F) were used. Magnesium nitrate was added into calcium nitrate solution in 1, 2.5 and 7.5 moles % of calcium nitrate. Ammonium fluoride was added into di-ammonium phosphate solution in the compositions of 2.5 and 7.5 moles % of di-ammonium phosphate. The procedure used in preparation of pure HA was performed to prepare doped HA. The proportions of elements additions were adjusted according to mole percentage of precursors. After the solutions were mixed, the products were stirred, filtered, calcined and sintered similar to pure HA.

The list of the produced samples and their designations are given in Table 1. Each sample was designated referring to its sample IDs. The Ca/P ratios in Table 1 refer to the mixing ratios (by mole) of the Ca and P precursors. Samples were classified in 2 different groups according to the amounts of Mg^{2+} and F^- dopants.

The bulk densities (ρ_{bulk}) of the materials were measured by Archimedes method. Following formula was used to calculate the density of the samples [25]:

$$Density\,(g\,/\,cm^3) = \frac{Wt_{(air)}}{Wt_{(air)} - Wt_{(water)}} \times \rho_{(water)} \qquad (1)$$

where; $\rho_{(water)}$ is the density of water. The theoretical densities of HA and β-TCP are 3.156 g/cm^3 and 3.07 g/cm^3, respectively. The theoretical densities (ρ_t) of the HA/ β-TCP composites were calculated according to their weight percentages (wt% of HA and wt% of β-TCP) in the samples by the following formula [26];

$$\rho_t = \frac{1}{\dfrac{W_{\beta\text{-TCP}}}{3.07} + \dfrac{W_{HA}}{3.156}} \qquad (2)$$

The fractional densities (ρ) are calculated by dividing the measured bulk densities (ρ_{bulk}) to theoretical densities (ρ_t).

XRD was performed on the samples using a Rigaku DMAX 2200 machine with a Cu-Kα radiation at 40 kV/ 40 mA and samples were scanned from 20° to 50° in 2θ with a scan speed of 2.0° /min. Joint Committee on Powder Diffraction Standards (JCPDS) files were used to determine the phases present in the powders. The amount of phases present in pure and doped HA was calculated by using relative intensity measurements of highest diffracted peaks by the following formulas [24];

$$\frac{W_{HA}}{W_{\beta-TCP}} = \frac{R}{R_0} = \frac{I_H / I_{\beta-TCP}}{1.755} \tag{3}$$

$$W_{HA} + W_{\beta-TCP} = 1 \tag{4}$$

where; W_{HA} and $W_{\beta-TCP}$ are the amounts by weight % of the phases. The ratio R_0 of the peak heights of HA to that of β-TCP was taken as 1.755 [24]. R is the ratio of the intensities of the HA (I_H) to that of β-TCP ($I_{\beta-TCP}$) measured from the XRD peaks.

The hexagonal lattice parameters of hexagonal pure and doped HA/β-TCP were calculated by successive approximations [27]. The lattice parameters of the rhombohedral β-TCP were expressed in the hexagonal setting. The unit cell volumes of HA and β-TCP were calculated by the following formulas, respectively:

$$V = 2.589 \cdot \left(a^2\right) \cdot c \tag{5}$$

$$V = 0.866 \cdot \left(a^2\right) \cdot c \tag{6}$$

where; V (Å^3) is the volume of unit cell, a and c are the hexagonal lattice parameters in Angstroms.

The particle sizes of the sintered pure and doped HAs were determined from XRD results. The peaks from the planes (002) and (300) were used to calculate particle sizes for HA and the (0210) reflection is used for the particle size measurement for β-TCP. Scherrer equation was used to determine the particle size (D) values:

$$D = \frac{K\lambda}{\beta_{1/2}\cos\theta} \tag{7}$$

In this formula, "K" expresses a constant changing with crystal properties and was chosen as 0.9. "λ (=1.514Å)" and "θ" are x-ray wave-length and the diffraction angle, respectively. The value of "$\beta_{1/2}$" is measured as the width at half maximum intensity of XRD peaks for the planes (002), (300) of HA and (0210) of β-TCP.

3. RESULTS AND DISCUSSION

The fractional densities of pure and doped HAs sintered at 1100°C are given in Table 2. High densities were achieved after the sintering at 1100 °C except for the sample VII with a fractional density of 80.91%. Sample VI exhibited the highest density. Addition of Mg^{2+} alone did not show a significant improvement to the densification behavior of the HAs. By the addition of 1 mole % Mg^{2+} (sample II), the fractional density was decreased. However, as the Mg^{2+} amount increased to 2.5 mole % (sample III), density was increased. On the other hand, by the addition of 2.5 mole % F$^-$ ion together with Mg^{2+} in group 2 (sample VI), the fractional density was increased significantly as the amount of Mg^{2+} was increased to 1 and 2.5 mole%. However for 7.5 mole% Mg^{2+} doping, the presence of F$^-$ ion reduced the density of the sample VII significantly when compared to sample IV with no F$^-$ addition.

In group 2, the densities of the samples decreased sharply compared to group 1 when the mole % of Mg^{2+} was increased from 2.5 to 7.5. The theoretical densities of HA and β-TCP are 3.156 and 3.07 g/cm^3, respectively. Therefore, it is logical to assume that the densities should decrease as the weight % of β-TCP phase increases, in the samples VI and III (~15 weight% β-TCP) the densities were slightly increased in both F^- doped and undoped samples with constant Mg amount, in comparison to the samples V and II which are pure HA. An enhancement in densification by the incorporation of F^- ions into sample VI was observed when compared with sample III with no F^- addition.

The XRD results of the sintered samples are presented in Figures 1 and 2. In the samples I, II, V, no TCP formation was observed after the sintering at1100°C. In ambient atmosphere, HA leans to decompose by Reaction 8 [28]. Water is given off at the end of this reaction, pores are generated in the material. As a result of Mg doping, β-TCP phase is formed along with HA even at 1100°C (Figures 1, 2).

$$Ca_{10}(PO_4)_6(OH)_2 \rightarrow 3Ca_3(PO_4)_2 + CaO + H_2O \tag{8}$$

Table 3 shows the presence of the phase formations and their amounts. Although Mg^{2+} is known to trigger β-TCP formation besides stabilizing it, no β-TCP phase was observed in the sample II (Figure 1) and in sample V (Figures 2).

Hexagonal lattice parameters and unit cell volumes of the samples are given in Table 4. As seen from this table, hexagonal lattice parameters 'a' and 'c' decreased for all sintered materials in comparison with the lattice parameters of the standard HA (JCPDS file 09-0432). Because ion sizes of both Mg^{2+} and F^- are smaller than those of Ca^{2+} and OH^- ions, it is expectable that the addition of these ions resulted in the shrinkage of lattice volumes (Table 4). The effective ionic radii of Mg^{2+} and Ca^{2+} ions are 0.72 Å and 1.0 Å, respectively [29]. The F^- and OH^- ions have an effective ionic radii of 1.33 Å and 1.37 Å, respectively [29].

The amount of Mg ions associated on the surface of HA is usually much more than the Mg ions which are incorporated into its crystalline structure [30]. However, as a result of the change in the lattice parameters (Table 5) of the Mg and/or F doped HA samples, it can be concluded that Mg ions are also incorporated into the lattice structure and not just adsorbed on the HA crystalline surface.

After the sintering, very small particle sizes (< 50 nm) were obtained for all samples (Table 5). For β-TCP, the smallest particle size is obtained for sample-VI in (0210) plane (Table 5). As the Mg content increased, the particle sizes for (002) and (300) planes were decreased in groups 1 and 2. The reduction in the particle sizes were sharper as the amount of Mg increased from 1 mole% to 2.5 mole% in both F^- doped and undoped samples. In group 2, particle sizes along the (300) plane were increased with the addition of F^- ion.

XRD patterns presented in Figures 1 and 2 present the selective effect of increase in Mg^{2+} amount with no F^- addition and with constant amount of (2.5 mole%) doped F^-, respectively. No β-TCP formation is observed in 1 mole% Mg doped sample II and V. In the samples VI and VII in Figure 2 and in the samples III and IV in Figure 1, as the Mg content increased, a slight shift to higher 2θ angles was observed in β-TCP peaks, indicating that the lattice parameters of β-TCP decreased, which can also be verified from Table 4. A similar observation was reported by H.-S. Ryu et al. [31] for MgO addition into HA/TCP mixtures. The decrease in the lattice parameters of β-TCP, occurs as a result of the lattice contraction due to the replacement of Ca atoms by Mg atoms with smaller radii. Except for sample IV, peak shifts with the increase in Mg content was not observed in HA peaks, which indicates that Mg incorporated preferentially into TCP phase of HA/TCP mixture [31]. However, as the Mg amount was increased, the diffraction peaks of HA became broader in both samples VII and IV with 7.5 mole% Mg , which indicates poor crystallinity.

CONCLUSION

Density and phase transformations of HAs were studied in terms of influences of Mg^{2+} and F^- ions. High densities were achieved for each group. Although ~ 15 wt% β-TCP is formed besides HA for sample VI, this sample exhibited the highest density. Addition of Mg ions resulted in formation of β-TCP and led to decomposition of HA. As a result of the Mg addition, β-TCP phase is formed together with HA. Lattice parameters and unit cell volumes decreased with the additions of Mg^{2+} and F^- ions due to these ions being smaller than Ca^{2+} and OH^- ions.

REFERENCES

[1] Legeros RZ, Legeros JP (1993) In: Hench LL, Wilson J (eds.) Bioceramics, vol 1. World Scientific, Singapore, New Jersey, London, Hong Kong, p.139
[2] Legeros RZ (1981) Progr Crystal Growth and Charact 4:1
[3] Hench LL (1991) J Am Ceram Soc 74:1487
[4] DeGroot K, DePutter C, Smitt P, Driessen A (1981) Sci Ceram 11:433
[5] With GD, Dijk HJAV, Hattu N, Prijs K (1981) J Mater Sci 16:1592
[6] Fang Y, Agrawal DK, Roy DM, Roy R (1995) Mater Lett 23:147
[7] Ioku K, Yoshimura M, Somiya S (1990) Biomaterials 11:57
[8] Li J, Fartash B, Hermansson L (1998) Biomaterials 16:417
[9] Niihara K (1985) J Ceram Soc Jpn 20:12
[10] Uematsu K, Takagi M, Honda T, Uchida N, Saito K (1989) J Am Ceram Soc 72:1476
[11] Guyton AC (1991) Textbook of medical physiology, 8th ed. Philadelphia: W.B. Saunders Company.
[12] Kaplan FS, Hayes WL, Keaveny TM, Boskey A, Einhorn TA, Iannotti JP (1994) In: Simon SP (editor) Orthopaedic basic research, Columbus, OH: American Academy of Orthopaedic Surgeons, p.127
[13] Webster TJ, Schlueter EAM, Smith JL, Slamovich EB (2004) Biomaterials 25:2111
[14] Narasaraju TSB, Phebe DE (1996) J Mater Sci 31:1
[15] Machoy MA (1995) J Inter Soc Fluoride Res 28:175
[16] Kannan S, Ferreira JMF (2006) Chem Mater 2006 18:198
[17] Percival M (1999) Appl Nutr Sci Rep 5:1
[18] Kannan S, Ventura JM, Ferreira JMF (2007) Ceram Inter 33:637
[19] Kim HW, Kong YM, Koh YH, Kim HE, Kim HM, Ko JS (2003) J Am Ceram Soc 86:2019
[20] Nakade O, Koyama H, Arai J, Ariji H, Takad J, Kaku T (1999) Arch Oral Biol 44:89
[21] Gineste L, Gineste M, Ranz X, Ellefterion A, Guilhem A, Rouquet N, Frayssinet P (1999) J Biomed Mater Res 48:224
[22] Jarcho M, Bolen CH, Thomas MB, Babock J, Kay JF, Doremus RH (1976) J Mater Sci 11:2027
[23] Evis Z (2006) J Ceram Soc Jpn 114:1001
[24] Evis Z (2007) Ceram Inter 33:987
[25] Kuwahara H, Mazaki N, Takahashi M, Watanabe T, Yang X, Aizawa T (2001) Mater Sci Eng A 319-321:687
[26] Evis Z, Ergun C, Doremus RH (2005) J Mater Sci 40:1127
[27] Cullity BD (1978) Elements of X-ray diffraction, 2nd edn., Addison-Wesley Publishing company, Massachusetts
[28] Evis Z, Doremus RH (2007) Mater Chem Phys 105:76
[29] Shannon RD (1976) Acta Crystallog A 32:751
[30] Bertoni E, Bigi A, Cojazzi G, Gandolfi M, Panzavolta S, Roveri N (1998) J Inorg Biochem 72:29
[31] Ryu H-S, Hong KS, Lee JK, Kim DJ, Lee JH, Chang B-S, Lee C-K, Chung S-S (2004) Biomaterials 25:393

Structural Investigation of Nano Hydroxyapatites Doped with Mg^{2+} and F^- Ions

Table 1: List of the samples investigated in this study.

Group No	Sample ID	Ca/P ratio	Mg %	F %
1	I	1.67	—	—
	II	1.67	1	—
	III	1.67	2.5	—
	IV	1.67	7.5	—
2	I	1.67	—	—
	V	1.67	1	2.5
	VI	1.67	2.5	2.5
	VII	1.67	7.5	2.5

Table 2: Densities of the samples after the sintering.

Group No	Sample ID	Bulk density (g/cm³)	Theoretical density	% Density
1	I	2.964	3.1560	93.92
	II	2.907	3.1560	92.12
	III	2.923	3.1428	93.01
	IV	2.825	3.1151	90.70
2	I	2.964	3.1560	93.92
	V	3.022	3.1560	95.76
	VI	3.031	3.1429	96.44
	VII	2.520	3.1146	80.91

Table 3: Presence of HA and β-TCP phases in the samples.

Group No	Sample ID	Ca/P ratio	Mg %	F %	Phase Formations	$W_{\beta-TCP}\%$	$W_{HA}\%$
1	I	1,67	—	—	No TCP formation	—	100,00
	II	1,67	1	—	No TCP formation	—	100,00
	III	1,67	2,5	—	TCP formation	14,98	85,02
	IV	1,67	7,5	—	TCP formation	46,93	53,08
2	I	1,67	—	—	No TCP formation	—	100,00
	V	1,67	1	2,5	No TCP formation	—	100,00
	VI	1,67	2,5	2,5	TCP formation	14,83	85,17
	VII	1,67	7,5	2,5	TCP formation	47,50	52,50

Table 4: HA and β-TCP lattice parameters of the samples.

Sample IDs	HA (09-0432)		a= 9.418 Å c= 6.884 Å V= 1580.85Å³		β-TCP (09-0169)		a= 10.429 Å c= 37.38 Å V= 3520.81Å³	
	a (Å)	c (Å)	V (Å³)	ΔV (Å³)	a (Å)	c (Å)	V (Å³)	ΔV (Å³)
I	9.3996	6.8639	1570.1	10.8	—	—	—	—
II	9.4037	6.8706	1573.0	7.9	—	—	—	—
III	9.4037	6.8674	1572.3	8.6	10.3628	37.7537	3511.0	9.8
IV	9.4100	6.8686	1574.6	6.2	10.3364	37.1675	3438.9	81.9
V	9.3990	6.8704	1571.4	9.5	—	—	—	—
VI	9.3970	6.8678	1570.1	10.7	10.3682	37.3723	3479.2	41.6
VII	9.3990	6.8881	1575.4	5.4	10.3625	37.2995	3468.6	52.2

Table 5: HA and β-TCP particle sizes.

Sample ID	$D_{(002)}$ (nm)	$D_{(300)}$ (nm)	$D_{(0210)}$ (nm)
I	52.21	48.01	—
II	47.91	45.26	—
III	42.89	39.72	36.35
IV	40.50	39.77	37.76
V	56.90	47.08	—
VI	42.42	43.43	28.21
VII	39.23	39.81	41.67

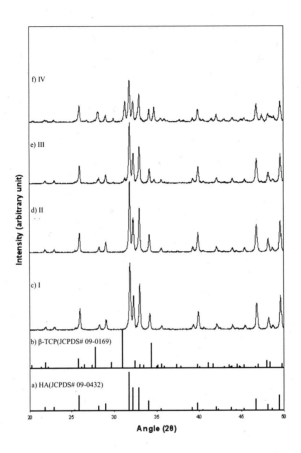

Figure 1: XRD results of the samples sintered at 1100°C: a) standard HA (JCPDS# 09-0432); b) standard β-TCP (JCPDS# 09-0169); c) Sample I; d) Sample II; e) Sample III; f) Sample IV.

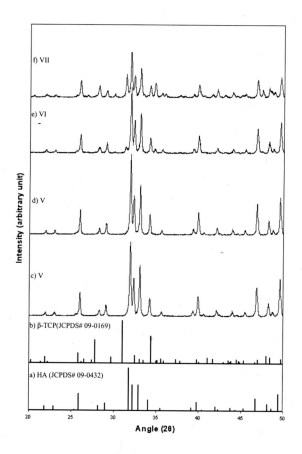

Figure 2: XRD results of the samples sintered at 1100°C: a) standard HA (JCPDS# 09-0432); b) standard β-TCP (JCPDS# 09-0169); c) Sample I ; d) Sample V; e) Sample VI; f) Sample VII.

NOVEL BIOCERAMICS FOR BONE IMPLANTS

P.I. Gouma[*1](Ph.D), K. Ramachandran[1](MS), M. Firat[2](Ph.D), M. Connolly[2](MS), R.Zuckermann[2](Ph.D), Cs. Balaszi[3](Ph.D), P. L. Perrotta[4](M.D. Ph.D) and R. Xue[1](BE)

[1]Department of Materials Science and Engineering, State University of New York at Stony Brook, Stony Brook, New York 11794-2275, USA.
[2]Biological Nanostructures Facility at The Molecular Foundry, Lawrence Berkeley National Laboratory, Berkeley, CA 94720, USA.
[3]Hungarian Academy of Sciences, Research Institute for Technical Physics and Materials Science, Ceramics and Nanocomposites Department, Konkoly-Thege M. út 29-33, 1121 Budapest, Hungary
[4]Department of Pathology, West Virginia University, Morgantown, WV 26506-9203

ABSTRACT

Electrospun cellulose acetate (eCA) and its hybrid with nanoaggregates of hydroxyapatite (eCA-nHA) were developed as potential and novel materials for bone tissue engineering purposes. Cultured human osteoblasts were seeded on eCA and eCA-nHA scaffolds, after which cell proliferative capacity and viability were studied using complementary assays. The interactions between the cells and the scaffolds were further characterized by scanning electron microscopy (SEM). These studies demonstrated that the scaffolds supported cell growth, as evidenced by continued cell proliferation for up to 3 days of exposure. Moreover, the osteoblasts remained viable and metabolically active while exposed to the scaffolds. SEM images detailed the tight interactions between the cell membranes and electrospun fibers. The presence of the hydroxyapatite nanoaggregates appeared to enhance osteoblast attachment, while allowing the cells to spread out along the fibers. These studies show that both eCA fiber and eCA-nHA fibrous nanocomposite scaffolds hold promise for bone tissue engineering applications.

INTRODUCTION

Bone tissue engineering is of significant importance because many serious bone injuries heal slowly and result in permanent changes in bone structure and function [1]. There are about 1 million patients with skeletal defects and injuries a year that require bone-graft reconstruction [2]. Cell-based approaches for bone grafting that utilize scaffolds fabricated from synthetic biomaterials such as metals, ceramics, polymers and composites are promising alternatives to autogenic and allogeneic bone-grafting materials [3]. Scaffolds used for bone tissue engineering share several ideal characteristics. First, materials must be compatible with cells and tissue they contact. Osteoconductive materials can serve as scaffolds on which bone cells (osteoblasts) can attach, grow, and migrate. The ability of materials to stimulate immature bone cells to grow and mature is known as osteoinductive stimulation. Finally, materials should possess high porosity and interconnected pores in order to support cell and bone growth in three-dimensions [4, 5].

Electrospinning (ES) was invented by Formhals [6] in 1934 to fabricate fibrous structures of polymers by applying high voltage to viscous polymeric based solutions. It has recently emerged as a leading technique to construct fibrous structures for tissue engineering [7, 8] because the process has the unique capability to fabricate three-dimensional scaffolds that mimic biological extracellular matrix (ECM). Moreover, scaffolds with tunable porosity and specific morphologies can be generated via electrospinning [9]. The importance of producing structures that can replace the ECM is based on the critical role the ECM plays in providing structural support for cells [7]. Natural polymers have been proposed as alternatives to synthetic biomaterials currently being used to build scaffolds [5]. One such natural polymer, cellulose acetate (CA), is a plentiful organic compound found in the many plant cell walls [10] that has been used to support cardiac and microvascular cell growth [9, 11]. One advantage of cellulose-based materials is that their low water solubility allows better control over scaffold design [11].

Hydroxyapatite (HA), a bioactive material known to promote differentiation of osteoblastic cells *in vitro* [12], has been used in bone tissue engineering because of its favorable biocompatibility and osteoconductive properties. Nanoaggregates of HA are crystalline structures similar to bone apatites that are recognized for their excellent mechanical properties and bioactivity [13]. Human osteoblasts are often used to evaluate the potential of a material as a bone scaffold because these cells are critical to bone formation [14]. Preferred materials would not affect the ability of osteoblasts to proliferate and to attach to other cells. Thus, the major goal of this study was to assess the potential of electrospun CA (eCA) and eCA-nHA scaffolds for supporting osteoblast attachment and proliferation.

MATERIALS AND METHODS

Preparing HA powders

HA powder was prepared from raw eggshells calcined at 900°C. Short duration (30 min) heat treatment of eggshells caused their color change from white to black, thereafter; they became white after 3 h heating. The color change suggests that most of the organic materials were removed. Powders were then crushed in an agate mortar and reacted with phosphoric acid powder (50:50 wt % H_3PO_4) in an exothermic reaction. Products were milled in ethanol solvent for 10 h by ball mill, after which polyethylene glycol (PEG) was added to portions of the powder. Batches were sieved with 100 μm mesh and green compacts were formed through dry pressing at 220 MPa. HA samples were then sintered in air at 900°C for 2 h while limiting the heating rate to 2°C/min. The structural and morphological evolution of HA particles has been recently detailed [15]. The sintered HA was crushed to powder form in order to introduce HA into the CA solution.

Preparing CA and CA-HA composites for electrospinning

CA (29,000 g/mol) powder with an acetyl content of 40% (Fluka Chemie CH-9471, Buchs, Switzerland) was mixed with acetone at room temperature for 1 h to form a 15 weight % solution. This viscous polymeric spinning solution was used to produce CA scaffolds. Mixtures containing 15% w/v CA and 3.75% w/v HA nanoparticles were combined in a solvent containing 50:50 (%v) of acetone and acetic acid, respectively, for use in fabricating CA-HA scaffolds. The mixture was sonicated for 1 hour to prevent precipitation of CA.

Electrospinning conditions

The electrospinning apparatus consists of a high-voltage power supply that provides up to 40 kV, two electrodes, a metering pump, a glass syringe attached to a 22 gauge conducting needle and an aluminum foil collector. One of the two electrodes is attached to the syringe needle and the other is attached to the collector. For electrospinning CA scaffolds, a flow rate of 160 μL/min and voltage of 19 kV were used. A flow rate of 41.6 μL/min and voltage of 12 kV were used for electrospinning the CA-nHA composite scaffolds to ensure the deposition of single layer fibrous mats on the collector.

Cell culture and cell seeding
SaOS2 cells were cultured in Dulbecco's modified Eagle's medium (DMEM; Sigma-Aldrich, Milwaukee, WI) supplemented by 10% fetal bovine serum (FBS; Sigma-Aldrich), 1% L-glutamine (Sigma-Aldrich) and a 1% antibiotic/antimycotic formulation (Invitrogen, Carlsbad, CA). Cells were incubated at 37°C in a humidified atmosphere containing 5% CO_2. The scaffolds were cut into circular discs about 15 mm in diameter. Three scaffolds of each type were placed in 24-well tissue-culture polystyrene plates (Becton Dickinson, Franklin Lakes, NJ); these were used for the viability assays. Polytetrafluoroethylene rings (PTFE rings, McMaster-Carr, Atlanta, GA, part #013TEF) were placed on the scaffolds to immobilize them in the wells. The disc-shaped scaffolds and rings were then sterilized in 70% ethanol for 30 min, washed with nuclease-free water, and immersed in DMEM overnight at 37°C in a humidified atmosphere containing 5% CO_2. The cell suspensions were seeded in triplicate on the eCA and eCA-nHA scaffolds, as well as empty control wells, in order to study cell attachment and proliferation. The seeding density of cells was approximately 68,000 cells/well for the PicoGreen assay and 50,000 cells/well for the MTS assay. The cells were incubated with the scaffolds for 1 and 3 days before analysis.

PicoGreen Assay
The proliferative ability of cells was studied using a fluorescent nucleic acid dye that quantifies double stranded DNA (dsDNA) (Quant-iT™ PicoGreen® Kit, Invitrogen). A standard curve was prepared using 1 μg/mL to 50 ng/mL DNA. The PicoGreen® reagent (100 μL) was mixed with 100 μL of a series of DNA concentrations into a 96-well plate (Corning, Corning, NY). The samples were incubated at room temperature for 5 min in the dark. Fluorescence was then measured with an excitation wavelength at 488 nm and an emission wavelength at 525 nm using a SpectraMax® fluorescence microplate reader (Molecular Devices, Sunnyvale, CA).
For the attachment studies, culture media was removed from each plate containing scaffolds and the control plates after 1 and 3 days of incubation as described above. The scaffolds with the stability rings were moved to empty wells, whereas the control plates were examined directly to determine the degree of non-specific binding to the plates. After adding 1 mL of a Tris buffer (10 mM Tris-Cl, 1 mM EDTA) to each well, the plates were sealed and the cells were lysed through two freeze/thaw cycles. Then, 100 μL of the PicoGreen® reagent was added to 100 μL of the cell lysate in a 96-well plate before measuring fluorescence in the fluorescence microplate reader [16].

MTS Assay
The metabolic activity of cells was monitored using a tetrazolium compound (MTS) that is chemically reduced by metabolically active cells into formazan (CellTiter 96® Aqueous One Solution Cell Proliferation Assay, Promega, Madison, WI). The amount of formazan produced is an indicator of the cell viability. The measurement of formazan absorbance was performed in 96 well plates after 1 and 3 days incubation [17]. Standard curves were prepared by diluting a series of cell suspensions from 15.7 cells/mL to 157,000 cells/mL. An aliquot (1.0 mL) of each dilution was transferred to wells of a 24-well tissue culture plate in triplicate. Subsequently, 150 μL of the MTS solution was added to each suspension. The plate was incubated at 37°C in a humidified atmosphere containing 5% CO_2 in the dark for 4 h, after which 1.0 mL of the kit's solubilization/stop solution was added to each well. Plates was sealed and incubated overnight. Absorbance was read at 570 nm wavelength and also at 650 nm as a reference wavelength using the plate fluorimeter.

Morphological Analysis of Cultured Cells
The culture medium was removed from each well at specified times and the cell-exposed eCA and eCA-nHA scaffolds were rinsed twice with phosphate-buffered saline (Sigma-Aldrich). The cell-scaffold constructs were fixed in 3% gluteraldehyde solution (Electron Microscopy Science, Hatfield,

PA), dehydrated through a graded series of ethanol solutions, 100% hexamethyldisilazane (HMDS, Sigma) followed by air drying [18]. The cell-scaffold constructs were then attached to aluminum stubs, sputter-coated with gold, and then examined under a LEO-Gemini Schottky FEG scanning electron microscope. The working distance used for this study was varied between 8-10 mm and the operating voltage was varied from 5-20 kV. The fiber diameter, pore diameter, and cell diameter data were determined using ImageJ software [19].

RESULTS

eCA and eCA-nHA scaffolds support cell growth
The proliferative capacity of the osteoblasts was measured using the PicoGreen® fluorescence assay, which quantifies the amount of double stranded DNA (dsDNA) in solution. These measurements were taken after exposing the osteoblasts to the two different scaffolds for 1 day and 3 days (Figure 1). Fluorescent intensity increased an average of 25%, 48%, and 30% for the control samples, eCA scaffolds, and eCA-nHA scaffold, respectively ($p<0.05$). The degree of increase from day 1 to day 3 was not significantly different between the scaffold-exposed cells and the controls.

Cells remain viable on electrospun scaffolds
The viability of the osteoblasts exposed to the electrospun scaffolds was determined using the MTS assay. The amount of formazan produced is proportional to number of living cells in culture since the chemical reduction of formazin to a colored product is dependent on viable cells. For our purposes, the number of cells that have attached to the eCA and eCA-nHA scaffolds is directly proportional to the absorbance reading at 570 nm. As for the PicoGreen®, cellular metabolic activity was monitored after 1 day and 3 days of incubation (Figure 2). Overall, increases in cell viability measured 113%, 48%, and 60% for the control samples, eCA, and eCA-nHA scaffolds, respectively. Control osteoblasts exhibited higher viability at day 3 than either of the cells exposed to scaffolds.

Cultures osteoblasts interact with electrospun scaffolds
SEM micrographs of electrospun scaffolds revealed that the eCA fibers were ribbon-shaped. The average width of each fiber was 3.1 ± 0.5 μm, and the thinner fibers measured 0.8 ± 0.3 μm. The eCA-nHA hybrid mats consisted of fibers with a wire-like appearance having an average diameter of 440 ± 170 nm. The average size of hydroxyapatite nanoclusters in the matrix was 970 ± 650 nm. The pore diameter in the eCA scaffold varied between 8-12 μm with a porosity of around 30%. The pore diameter in the eCA-nHA scaffolds varied between 3-7 μm, with porosities relatively less than those of the eCA scaffolds.

The osteoblasts attached themselves along the thinner fibers of the eCA scaffolds (Figure 3a) and typically displayed a spherical shape. The average diameter of the cells was 8.5 ± 1.4 μm. The cells were typically found to be attached to a single synthetic fiber (Figure 3b). On higher magnification (Figure 3c), thin processes are seen encircling the eCA fiber. These appear to be portions of the cell membrane that form an anchor around the fiber. It cannot be determined if the cell incorporates within the fiber itself.

In contrast, the osteoblasts seeded on the eCA-nHA composite scaffolds attached themselves to several fibers that formed an interconnecting network. Only a portion of the cells display a spherical shape. The average diameter of the cells was 6.88 ± 1.53 μm. Visualization at higher voltages highlight the HA aggregates as rough conglomerates of material (Figure 4a). The aggregates are located at the edge of the cells, and do not seem to penetrate the cell. Lower voltage images (Figure 4b) show the cell interacting with several fibers that form the meshwork. It is not clear if this is due to the smaller fiber sizes comprising the eCA-nHA scaffold, the different shape of these fibers, or the different consistency of the mats. What is clear is that the majority of osteoblasts visualized on the eCA-nHA scaffold (80% based on SEM imaging information) displayed a flatter (i.e. less spherical) morphology. The flatter cells are seen to spread out along the mesh-like fiber network, and several fibers are visualized interlocking individual fibers on the cell surface (Figures 5). Again, the nHA aggregates are well dispersed throughout the eCA fibers, but the cells do not appear to incorporate these aggregates (Figure 5).

DISCUSSION

This study explores the feasibility of using eCA and eCA-nHA composite hybrid mats as scaffolds for bone tissue engineering applications. The MTS viability assay demonstrated that the cells exposed to these scaffolds maintain the ability to proliferate for up to 3 days. These findings were supported by quantifying the amount of dsDNA (e.g. PicoGreen assay), which also showed measurable increases in the number of viable cells exposed to either scaffold. These findings show that these scaffolds are able to support and maintain cell proliferation.

Structural characterization studies carried out using scanning electron microscopy (SEM) suggest that the osteoblasts preferentially attach to thinner fibers. This was most clearly seen in the case of the eCA scaffolds, in which single cells associated along the thinner surface of ribbon-like fibers. However, it is unclear if there is an optimum fiber diameter to support cell attachment based on our data. Cells exposed to the more complex eCA-nHA scaffolds interacted with multiple fibers. This observation is partially explained by the significantly smaller diameter and rounder shape of eCA-nHA fibers as compared to the ribbon-like eCA fibers; it appears fiber size has a direct influence on the cell attachment behavior. Interestingly, the anchorage points for the cells to the fibers could be visualized by SEM. Furthermore, the hydroxyapatite nanoclusters were consistently located at the edge of cells, which may provide additional anchoring sites for cell attachment to the eCA-nHA hybrids. Cell density also appeared to be higher for cells exposed to the eCA-nHA scaffolds.

Osteoblasts are anchorage-dependent cells whose attachment to scaffolds depends on the relatively high surface area and the porosity of the scaffold [20]. The electrospinning process produces fibers with small diameters and correspondingly high surface area-to-volume ratios that are suitable for osteoblast attachment, while allowing for cell migration. Yoshimo et al. [20] have electrospun microporous, non-woven polycaprolactone scaffolds for bone tissue engineering purposes that have complex three-dimensional fibrous meshwork consisting of randomly oriented fibers with diameters ranging from 20 nm to 5 µm. Mesenchymal stem cell-derived osteoblasts seeded on such scaffolds have been reported to migrate into the scaffold and produce an extracellular matrix (ECM) of collagen throughout the scaffold, as confirmed by histology and immunohistochemical studies [20]. Based on these prior studies, we suspect that the areas of the cells observed to anchor to the fibers (Figures 3a, 3c) are composed of common ECM constituents like type I collagen secreted by the cells.

The mechanisms responsible for interactions between cells including osteoblasts and the engineered fibers are partially understood. Relatively abundant proteins such as fibronectin, laminin, vitronectin, and collagen may first adsorb to a material [21]. Once such proteins coat the material's surface, cells are able to interact with the proteins through specific molecules on the surface of the cell. Other non-specific and poorly understood interactions between materials and cells also may take place, including covalent bonding between peptides and nanomaterials [22].

One of the most important characteristics of a useful scaffold is its ability to interact with cells without doing harm. It is recognized that the degree of cell attachment directly influences cell motility, cell proliferation rate, and ultimately, the cell's phenotype [23]. Hunter et al. [23] have shown that the shape of a cell regulates its ability to proliferate while anchored to scaffolds. Specifically, more spherical cells appear to divide at lower rates than more flattened cells that spread along the matrix [23]. This observation suggests that spherically shaped cells have suffered some degree of damage and hence, are less capable of proliferating. This is further supported by studies showing that as cells become spherical in shape, there is a corresponding decrease in DNA synthesis [24]. Thus, scaffolds should support the attachment of anchorage-dependent cells without significantly affecting cell shape.

In our study, cultured osteoblasts attached to CA fibers had a spherical morphology and did not spread along the fibers, suggesting that the cells attached to CA scaffolds may divide at a lower rate. It has been reported that biomaterials that retard cell spreading might favor differentiation of osteoblasts to a fibroblast phenotype [23]. HA has been shown to facilitate cell spreading when incorporated into starch-based polymer scaffolds [12]. This observation concurs with our results demonstrating that cells attached to eCA-nHA hybrid fibers appear flattened and well spread out; these cells are expected to be more capable of DNA synthesis and cell division based on prior studies [23, 24]. Overall, our studies suggest that both the composition and structure of the eCA-nHA composite scaffolds play important roles in facilitating cell spreading and differentiation.

Pores found in tissue-engineered scaffolds are the spaces in which cells reside [22]. Pore properties such as size, shape, and volume are key parameters that determine the usefulness of a scaffold. High porosities have been reported to provide more structural space to accommodate cells. For bone regeneration, pore sizes between 100 and 350 μm and porosities of more than 90% are preferred [23]. A stiffness value of -0.41 N/m has been reported for electrospun CA scaffolds using a polymer concentration of 16.5% [25]. This suggests that electrospun fibrous CA scaffolds are very flexible and the pores can dynamically expand to accommodate cell growth into the scaffold.

CONCLUSIONS

Cellulose acetate and hydroxyapatite incorporated cellulose acetate composites were assessed as novel biomaterials for developing bone tissue engineering scaffolds. Cell culture studies reveal that osteoblast growth can be supported by these scaffolds. Structural characterization studies demonstrate subtle differences in the appearance of osteoblasts attached to the different scaffolds. Cells on eCA scaffold assumed a more spherical morphology with less evidence of spreading, while those on eCA-nHA scaffolds were flatter in appearance and more spread out along the fibers. The latter observation is attributed to the fiber size and the presence of HA nanoclusters, and we believe that these characteristics enhance cell differentiation, and thus, cell functionality. The HA is an important part of hybrid scaffolds used in biomaterials, while the eCA fibers provide a surface area for the attachment of cells, without causing cell damage. Novel scaffolds of this type utilizing natural polymers and nanocomposites with tailored fibrous architectures hold promise as bone tissue engineering materials.

REFERENCES

1. Salgado AJ, Continho OP, Reis RL. Bone tissue engineering: state of the art and future trends. Macromol Biosci 2004;4:743-765.
2. Yaszemski MJ, Oldham JB, Lu L, Currier BL. Clinical needs for bone tissue engineering technology. Davis JE, editor. Bone Engineering. em squared, Toronto, 2000, p 541-547.
3. Meijer GJ, de Bruijn JD, Koole R, van Blitterswijk C. Cell-based bone tissue engineering. PLoS Med 2007;4:260-264.
4. van Lenthe GH, Hagenmuller H, Bohner M, Hollister SJ, Meinel L, Müller R. Nondestructive micro-computed tomography for biological imaging and quantification of scaffold–bone interaction *in vivo.* Biomaterials 2007;28:2479-2490.
5. Salgado AJ, Gomes ME, Chou A, Coutinho OP, Reis RL, Hutmacher DW. Preliminary study on the adhesion and proliferation of human osteoblasts on starch-based scaffolds. Mat Sci Eng C-Bio S 2002;20:27-33.
6. Formhals, A. Process and Apparatus for Preparing Artificial Threads. US patent No. 1,975,504, 1934.
7. Li WJ, Laurencin CT, Caterson EJ, Tuan RS, Ko FK. Electrospun nanofibrous structure: A novel scaffold for tissue engineering. J Biomed Mater Res 2002;60:613-621.
8. Li MY, Mondrinos MJ, Gandhi MR, Ko FK, Weiss AS, Lelkes PI. Electrospun protein fibers as matrices for tissue engineering. Biomaterials 2005;26:5999-6008.
9. Han D, Gouma PI. Electrospun bioscaffolds that mimic the topology of extracellular matrix. Nanomed-Nanotechnol Biol Med 2006;2:37-41.
10. Kennedy JF, Phillips GO, Wedlock DJ, Williams PA. Cellulose and its derivatives: Chemistry, Biochemistry and Applications. New York: Halsted Press; 1985.
11. Entcheva E, Bien H, Yin L, Chung CY, Farell M, Kostov Y. Functional cardiac cell constructs on cellulose-based scaffolding. Biomaterials 2004;25:5753-5762.
12. Marques AP, Reis RL. Hydroxyapatite reinforcement of different starch-based polymers affects osteoblast-like cells adhesion/spreading and proliferation. Mat Sci Eng C-Bio S 2005;25:215-229.
13. Zhang N, Nichols HL, Tylor S, Wen X. Fabrication of nanocrystalline hydroxyapatite doped degradable composite hollow fiber for guided and biomimetic bone tissue engineering. Mat Sci Eng C-Bio S 2006;27:599-606.
14. El-Amin SF, Botchwey E, Tuli R, Kofron MD, Mesfin A, Sethuraman S, Tuan RS, Laurencin CT. Human osteoblast cells: isolation, characterization, and growth on polymers for musculoskeletal tissue engineering. J Biomed Mater Res A 2006;76:439-449.
15. Balázsi Cs, Wéber F, Kövér Zs, Horváth E, Németh Cs. Preparation of calcium-phosphate bioceramics from natural resources. J Eur Ceram Soc 2007;27:1601-1606.
16. Quant-iT PicoGreen® dsDNA Reagent, Technical manual, (Invitrogen); available at http://probes.invitrogen.com/media/pis/mp07581.pdf. 2005.
17. Promega CellTiter 96® Non-Radioactive Assay Technical Bulletin, http://www.promega.com/tbs/tb112/tb112.pdf.
18. Wutticharoenmongkol P, Sanchavanakit N, Pavasant P, Supaphol P. Novel Bone Scaffolds of Electrospun Polycaprolactone Fibers Filled with Nanoparticles. J Nanosci Nanotechno 2006;6:514-522.
19. Rasband WS. ImageJ, U. S. National Institutes of Health, Bethesda, Maryland, USA, http://rsb.info.nih.gov/ij/, 1997-2007.
20. Yoshimoto H, Shin YM, Terai H, Vacanti JP. A biodegradable nanofiber scaffold by electrospinning and its potential for bone tissue engineering. Biomaterials 2003;24:2077-2082.
21. Woo KM, Chen VJ, Ma PX. Nano-fibrous scaffolding architecture selectively enhances protein adsorption contributing to cell attachment. J Biomed Mater Res A 2003;67A:531-537.

22. Ma ZW, Kotaki M, Inai R, Ramakrishna S. Potential of nanofiber matrix as tissue-engineering scaffolds. Tissue Eng 2005;11:101-109.
23. Hunter A, Archer CW, Walker PS, Blunn GW. Attachment and proliferation of osteoblasts and fibroblasts on biomaterials for orthopedic use. Biomaterials 1995;16:287-295.
24. Folkman J, Moscona A. Role of cell shape in growth control. Nature 1978;273:345-349.
25. Rubenstein, D.A. Development of a Novel Bioassay Chamber to Optimize Autologous Endothelial Cell Viability and Density on Topological and Topographical Substrates. [PhD thesis] Department of Biomedical Engineering, Stony Brook University, NY, 2007.

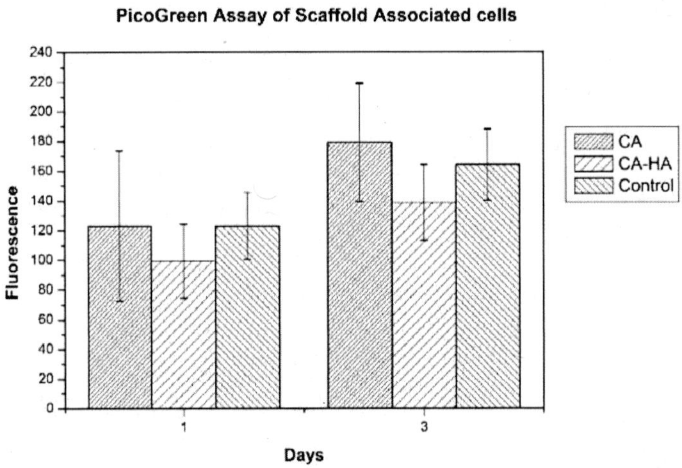

Figure 1: PicoGreen assay fluorescence spectrums for controls and cells exposed to eCA and eCA-nHA scaffolds for 1 and 3 days. Error bars represent means ± SD.

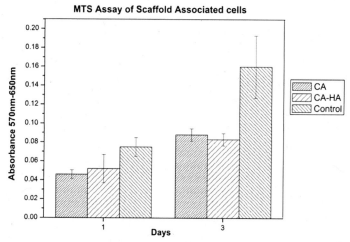

Figure 2: MTS Assay absorbance spectrum for controls and cells exposed to eCA and eCA-nHA scaffolds for 1 and 3 days. Error bars represent means ± SD.

Figure 3: SEM micrographs showing the morphology of a single cell attached to an eCA fiber after 1 day of seeding at different magnifications: (a) 3kx (b) 15kx (c) 30kx. Note the stepwise appearance of the interaction between the cell and the fiber

(a) (b)

Figure 4: SEM micrographs showing the morphology of a spherical cell attached to several eCA-nHA fibers (day 3 of seeding) at 20kx and different voltages: (a) 20 kV; circles highlight the hydroxyapatite aggregates at the areas where the cell contacts the fibers (b) 5 kV; image details how the cell attaches to the fibers.

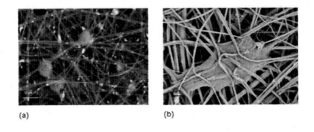

(a) (b)

Figure 5: SEM micrographs showing the morphology of a flat and spread out cell attached to eCA-nHA fibers on day 3. (a) 5kx and 20 kV displaying morphology of elongated cells (b) at 10kx and 5 kV; higher magnification of the flattened cell elaborating a large number of pseudopodia overlying the fine eCA fibers.

20 YEARS OF BIPHASIC CALCIUM PHOSPHATE BIOCERAMICS DEVELOPMENT AND APPLICATIONS

Guy DACULSI[1], Serge BAROTH[1,2], Racquel LeGEROS[3]

[1] INSERM, Université de Nantes, UMR U791, Faculté de Chirurgie Dentaire, Place Alexis Ricordeau, 44042 Nantes Cedex 01, France. guy.daculsi@univ-nantes.fr

[2] Biomatlante SAS, ZAC des 4 Nations, 5 rue Edouard Belin, 44360 Vigneux de Bretagne, France

[3] Department of Biomaterials and Biomimetics, New York University College of Dentistry, 345 East 24th Street, New York, New York 10010, USA

ABSTRACT

We developed 20 years ago, biphasic calcium phosphate ceramics (BCP). The BCP concept is determined by a balance of the more stable HA phase and more soluble TCP. BCP is gradually resorbed, seeding new bone formation.

The main attractive feature of BCP is the ability to form a direct bone bonding resulting in a strong interface. This interface is the result of a sequence of events involving interaction with cells; and dissolution/precipitation processes.

The efficiency of BCP concept was due to: (1) partial dissolution of the CaP ceramic macrocrystals inducing an increase in the calcium and phosphate ions concentration in the local microenvironment; (2) formation of carbonate hydroxyapatite CHA, (3) incorporation of these microcrystals with the collageneous matrix and non collagenic proteins like growth factors and others, in the newly formed bone. The BCP/bone interface represent a dynamic process, including physico-chemical processes, crystal/proteins interactions, cells and tissue colonization, bone remodelling.

BCP have the advantage of bioactivity control by changing the HA and β-TCP ratio. BCP are osteoconductive and osteoinductive through appropriate critical geometry of microporosity. Besides the medical and dental applications, BCP has a potential for other applications such as delivery system for drugs, antibiotics, hormones; carriers for growth factors; scaffolds for tissue engineering.

Specific matrices for combination with bone marrow or stem cells; and the need of material for Minimal Invasive Surgery (MIS) induced the development of injectable BCP granules. This paper summarizes 20 years of development of BCP and derivatives.

Although bone tissue possesses the capacity for regenerative growth, the bone repair process is impaired in many clinical and pathological situations. Large bone loss caused by trauma and tumor resection and/or aging, require reconstructive surgery and/or bone regeneration. At present, bone surgeons have three different possibilities when it comes to replacing bone.

1. **Autogenous bone grafts** are considered as the gold standard for bone replacement, because of their osteoconductive, osteogenic and osteoinductive properties, as well as the absence of immunological rejection. Autografts contain viable osteogenic cells, bone matrix proteins and support bone growth [1]. Cortical and cancellous bone grafts are the materials of choice, but their clinical use involves some difficulties. Septic complications, and unavailability of native bone were known. In addition, the amount harvested from the iliac crest or other sites is limited and complications have been observed at the additional surgery site. In addition, autologous bone grafts have shown considerable resorption and limited viability because of the lack of vascularization.

2. **Allogenic bone grafts** obtained from tissue banks also have limitations because of the possible transmission of non-conventional agents or viruses and the risk of immunological incompatibility [2,3]. Allogenic bone grafts have both osteoinductive properties (they release morphogenic bone proteins that act on bone cells) and osteoconductive properties, but lack osteogenic properties because of the absence of viable cells [4]. The harvesting and conservation of allogenic grafts are further limiting factors [5-7]. In view of the limitations of biologically-derived grafts, synthetic bone substitutes have been developed and used clinically for several years.

3. In this context, there was a critical need to develop implant technologies to promote bone healing. **Bone substitutes** are produced in various compositions and shapes. These biomaterials can be used alone to fill bone cavities, serving as a scaffold for bone regeneration from the peri-implant region. Bone substitutes can also be used to supplement autogenous bone or in combination with bone marrow aspirates. The ideal biomaterial should have a variety of forms and sizes, all with sufficient strength for use in load-bearing sites, and also be biocompatible, biodegradable and substituted by newly-formed bone [1]. Natural coral granules, bovine porous demineralized bone, bioactive glass ceramic and calcium phosphate ceramics such as hydroxyapatite (HA), β-tricalcium phosphate (β-TCP) or biphasic calcium phosphate (BCP), and mixtures of the first two have been developed [8]. These materials differ in composition and physical properties from each other and from bone [9-12]; and must be taken in consideration for more efficient bone ingrowth at the expense of the biomaterials and to adapt to new development of dedicated biomaterials.

The bioactive concept developed 20 years ago was based on the association of micro and macrostructure and specific chemical formulation of biphasic calcium phosphate ceramics (BCP) [8, 13-15]. The concept is determined by an optimum balance of the more stable phase of HA and more soluble TCP phase. The material is soluble and gradually dissolves in the body, seeding new bone formation as it releases calcium and phosphate ions into the biological medium [16-19]. Biphasic calcium phosphate (BCP) bioceramics consists of a mixture of hydroxyapatite (HA), $Ca_{10}(PO_4)_6(OH)_2$ and beta-tricalcium phosphate (β-TCP), $Ca_3(PO_4)_2$ of varying HA/β-TCP ratio. Legeros in US and Daculsi in France initiated basic studies on preparation of BCP and their *in vitro* properties early in 1986. At the present time, BCP is commercially available in Europe, Brazil, Japan, as a bone-graft or bone substitute materials for orthopaedic and dental applications under various trade marks and several manufacturers.

Chemical nature of BCP

BCP is obtained when a synthetic or biological calcium deficient apatite (CDA) is sintered at temperatures above 700°C. The extent of calcium deficiency (Ca/P molar ratio < 1.67) depends on the method of preparation (by precipitation, hydrolysis or mechanical mixture), the reaction pH and temperature in the preparation of the unsintered apatite. The calcium deficiency determines the HA/β-TCP ratio in the BCP. The HA/β-TCP ratio in the BCP determines its reactivity [20-23] the lower the ratio, the higher the reactivity (expressed *in vitro* as the extent of dissolution in an acid buffer). The BCP is obtained from the CDA according to the following reaction:

$$Ca_{10-x}M_x(PO_4)_{6-y}(HPO_4)_y(OH)_2 \rightarrow Ca_{10}(PO_4)_6(OH)_2 + Ca_3(PO_4)_2$$
$$\text{(CDA)} \qquad\qquad \text{(HA)} \qquad\qquad \text{(β-TCP)}$$

The apatite is considered calcium deficient when the Ca/P ratio is lower than the stoechiometric value of 1.67 for pure calcium hydroxyapatite.

BCP ceramics may be also prepared by mechanically mixing two types of synthetic apatite or commercial calcium phosphate reagents: one preparation that when sintered above 900°C results in the formation of only HA, and the other preparation, only β-TCP [24,25]. However, this process (mechanical mixing of two types of synthetic apatites) is not recommended for production of BCP with reproducible HA/β-TCP ratio and homogeneous crystal distribution. FT-IR spectra and X-ray diffraction profiles and of BCP with two different HA/β-TCP ratios (20/80 and 60/40) are shown in Fig. 1.

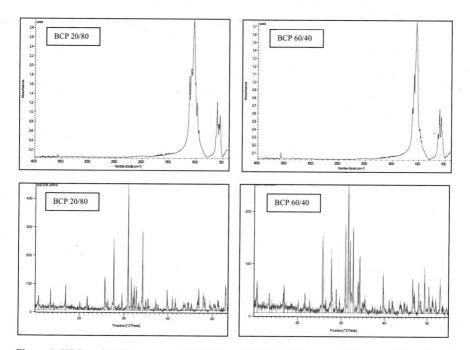

Figure 1: XRD and FTIR spectra of MBCP® with HA/ β-TCP ratio of 20/80 and 60/40 60

Micro and Macrostructure of BCP

Two physical properties of bioceramics are believed to be very important for optimum biological performance, including bioceramic-cell interactions, bioceramic resorption, bioceramic-tissue interface and new bone formation. These fundamental physical properties are interconnecting macroporosity and appropriate microporosity [26,27] .

Macroporosity (figure 2) in the BCP ceramics is introduced by incorporating volatile materials (e.g., naphthalene, hydrogen peroxide or other porogens). The mixture was heated at temperature below 200°C and followed by sintering at higher temperatures [24-30]. Macroporosity is formed resulting from the release of the volatile materials. Microporosity (Figure 3) is a consequence of the temperature and duration of sintering [30]: the higher the temperature, the lower the microporosity content and the lower the specific surface area.

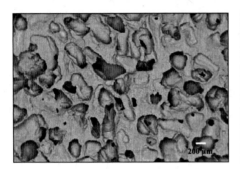 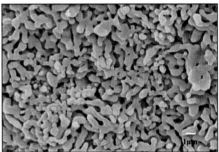

Figure 2: SEM of MBCP® block showing interconnected macroporous structure

Figure 3: Higher magnification of MBCP® block showing crystal size and microporous structure

Presently, commercial BCP products of different or similar HA/β-TCP ratios are manufactured in many parts of the world and their successful use in medicine and dentistry were published [31-41]. The total porosity (macroporosity plus microporosity) of these products is reported to be about 70% of the bioceramic volume. The current BCP commercial products with HA/β-TCP ratios ranging from 60/40 to 75/25, present similar % macroporosity (50 to 60%), but the % microporosities are very different (varying from 3% to 25%). Low % microporosity and surface area can result in lower bioactivity and dissolution property. A microporosity of at least 20%, with specific surface area higher than 2 m^2/g is recommended for optimal BCP efficacy.

The ideal pore size for a bioceramic material should approximate that of bone. It has been demonstrated that microporosity (diameter < 10 μm) allows body fluid circulation whereas macroporosity (diameter > 100 μm) provides a scaffold for bone-cell colonization. It was reported that BCP ceramic with an average pore size diameter of 565 μm (compared to those with average pore size diameter of 300 μm) and 40% macroporosity (compared to 50% macroporosity) had greater bone ingrowth [42].

Biological properties:

The interest of BCP concept is the equilibrium of the resorption and the bone ingrowth at the expense of the bioceramics. The bioactive bone graft materials such as BCP ceramics, have the ability to form a strong direct bond with the host bone resulting in a strong interface compared to bio inert or bio tolerant materials, which form a fibrous interface [42-46].

The formation of this dynamic interface is believed to result from a sequence of events involving interactions with cells and formation of carbonate hydroxyapatite CHA (similar to bone mineral) by dissolution/precipitation processes. The BCP materials elicit responses from bone cells and related cells *in vitro* and *in vivo,* similar to those elicited by bone. These materials promote cell attachment, proliferation and expression. The first biological events after BCP ceramics implantation are biological fluid diffusion, followed by cell colonization. These cells are macrophages, in early steps, followed by mesenchymal stem cells, osteoblasts, osteoclasts, that act the macropores of the implants or at the surface of BCP particles (figure 4). The resorbing cells formed both at the surface of the newly formed bone and the bioceramic surface look like osteoclasts and are mostly TRAP positive. Simultaneously to the resorption of the MBCP, lamellar bone ingrowth was observed at the expense of the bioceramics. Haversian system and vascularisation were observed both into macropores (figure 5)

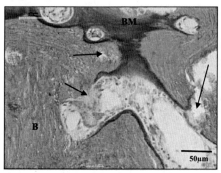

Figure 4: Osteoclasts like cells at the surface of MBCP particles (arrow), bone (B), BM (bone matrix in formation)

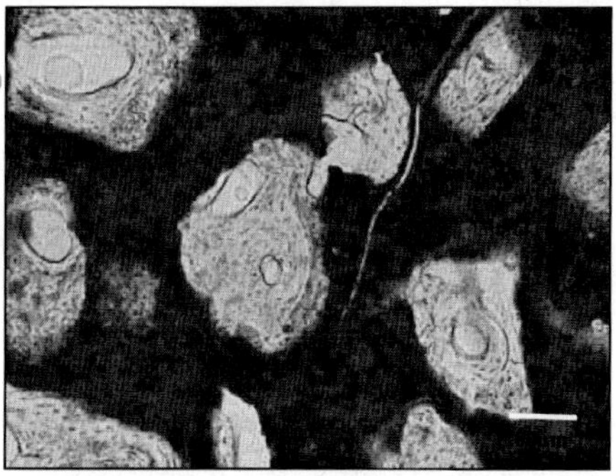

Figure 5: Polarized microscopy of the newly formed bone into MBCP® macropore after 3 months for spine arthrodesis in dog.

or between intergranular spaces. After a couple of months bone remodelling was observed, and the first bone formed into the macropores was resorbed forming new haversian system (figure 6). Architectured newly formed bone is obtained with adaptation to the site of implantation (cortical bone or spongious bone with interconnected bone trabeculae).

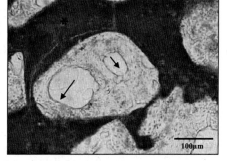

Figure 6: Bone remodelling by haversian system (arrow) into MBCP® (*) macropore after 3 months for spine arthrodesis in dog.

Osteoinductivity or osteogenicity are the property of the material to induce bone formation *de novo* or ectopically (in non-bony sites). Bioceramics (calcium phosphates, bioactive glass) do not usually have osteoinductive properties [26]. However, several reports indicated osteoinductive properties of some calcium phosphate bioceramics [47-49]. Reddi [50] explains this apparent osteoinductive property as the ability of particular ceramics to concentrate bone growth factors that are circulating in the biologic fluids, and these growth factors induce bone formation.

Surface microstructure appears to be a common property of materials inducing ectopic bone formation. Micropores have a critical role on the ceramic-induced osteogenesis. For example, it was reported that bone formation occurred in muscles of dogs inside micro-macroporous calcium phosphate ceramics but bone was not observed inside macroporous ceramics without micropore [51]. It was also reported that metal implants coated with microporous layer of octacalcium phosphate could induce ectopic bone in muscles of goats while a smooth layer of carbonated apatite on these porous metal implants was not able to induce bone formation [52]. In all of the previous experiments, ectopic bone formation occurred inside macroporous ceramic blocks.

The osteogenic/osteoinductive properties of BCP ceramics was due to the formation of microcrystals with Ca/P ratios similar to those of bone apatite crystals observed after implantation of the MBCP. The abundance of these crystals was directly related to the initial HA/ß-TCP ratio in the BCP: the lower the ratio the greater the abundance of the microcrystals associated with the BCP crystals [53,54]. Using high resolution TEM, we demonstrated that the formation of these bone apatite-like microcrystals after implantation of calcium phosphates (HA, BCP) were non-specific, i.e., not related to implantation site (osseous or non-osseous sites), subjects of implantation, and types of CaP ceramics [16-18]. The crystals growth was due to secondary nucleation and hetero epitaxy of biological HA on synthetic BCP crystals.

The coalescing interfacial zone of biological apatite and residual BCP ceramic crystals (mostly HA), provide a scaffold for further bone-cell adhesion and stem cells differentiation in osteogenic lines, and further bone ingrowth. The bone repair or bone regeneration process involves dissolution of calcium phosphate crystals and then a precipitation of carbonate hydroxyapatite (CHA) needle-like crystallites in micropores close to the dissolving crystals. The coalescing zone constitutes the new biomaterial/bone interface, which includes the participation of proteins and CHA crystals originating from the BCP ceramic crystals. This has been described as a coalescing zone and dynamic interface [55].

These observations suggest that biphasic calcium phosphate with macropores present suitable chemical environment associated to efficient architecture able to catch mesenchymal stem cells and to induce their phenotype to osteogenic cell lines. These observations have been also described by other groups in Netherlands [56]. This property can be used for artificial bone in irradiated implantation site. Irradiation produces irreversible effects on normal tissues, involving damages on their reparation properties. Nevertheless quality of life of patients who undergo radiotherapy could be improved by bone reconstructions. A preclinical study performed in irradiated dogs demonstrated bone ingrowth at the expense of structured implants of micro macroporous biphasic calcium phosphate filled by autologous bone marrow after implantation in irradiated soft and bone tissue [57].

The resorbing process involves dissolution of calcium phosphate crystals and then a precipitation of CHA needle-like crystallites in micropores close to the dissolving crystals. The coalescing zone constitutes the new biomaterial/bone interface, which includes the participation of proteins and CHA crystals originating from the CaP materials, but does not include the biomaterial surface. The following events of bone ingrowth and the newly formed bone progressively replace the initially formed CHA from the CaP biomaterials.

The process of cell colonization, adhesion, phagocytosis and osteoclastic resorption, ECM elaboration and mineralization, bone ingrowth and bone remodeling associated with the biological apatite precipitation during CaP ceramics dissolution, are continuously in progress. Consequently the interface is not static but dynamic, in constant evolution, taking into account bone physiopathology, biomechanical factors and bone maturation. The processes involve a well-organized and mineralised bone ingrowth at the expense of the artificial bone (figure 4).

This concept of bioactivity could also be applied to Injectable Bone Substitute [58]. Calcium phosphate materials are also used as components or fillers in polymeric composites [59, 60] and in cements [61]. The hydraulic cements are not macroporous and numerous studies have demonstrated the necessity of macropores for bone osseous-conduction.

The need for a material for minimal invasive surgery (MIS) motivated the development of BCP granules combined with polymer or calcium phosphate cement for injectable/mouldable Bone Substitutes. Three types of injectable/mouldable bone substitutes have been developed by INSERM Nantes University.

The bioactive concept of BCP has been applied to a composite associating hydrosoluble polymer and BCP granules [6]. We have elaborated such injectable bone substitute ready to use and able to be largely invaded by osseous-conduction due to osteogenic cells (figure 7).

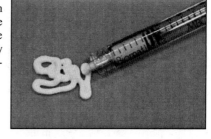

Figure 7: Syringe of MBCP gel ®

These materials are perfectly biocompatible and potentially resorbable and, thanks to their initial plasticity, they can fit bone defects very easily, without necessity to elaborate shaping of implantation site [64-65].

MBCP gel® do not have mechanical properties like the hydraulic bone cements. However bone cells are able to invade the spaces created by the disappearance of the polymer carrier. Bone ingrowth takes place all around the granules at the expense of the resorption of the BCP granules. In time, the mechanical property is increased due to the presence of the newly formed bone [65-67]. The spaces released by the polymer was invaded by macrophages, mesenchymal stem cells, microvascularisation. The BCP granules act as a scaffold for bone matrix formation at least at 1 week after implantation (figure 8) and bone trabeculae appeared after 2 weeks, becoming larger and well interconnected on time. At 12 weeks bone architectecture was equivalent to normal spongious bone (figure 9).

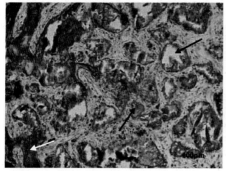 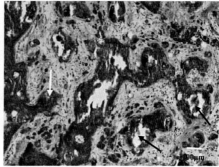

Figure 8: Light microscopy of MBCP gel, after 1 week of implantation in rabbits, black arrow BCP granules, white arrow bone trabeculae

Figure 9: Light microscopy of MBCP gel in rabbits after 12 weeks of implantation black arrow BCP granules, white arrow bone trabeculae

Tricos T® is a fibrin/BCP composite was obtained by mixing fibrinogen, thrombin components fibrin sealant (Tisseel®Baxter BioSciences BioSurgery, Baxter) and BCP granules (TricOs® , Baxter). TricOs® is a BCP with a 60HA/40β-TCP ratio. Granules of 1 to 2 mm in diameter presenting both macroporosity (50-55%) and microporosity (30-35%) were used. To enhance the working time, a low thrombin concentration (4U) was used. The ratio Tisseel/TricOs volume was 1/2. Numerous preclinical studies have been performed in rabbits, and goats, both for biocompatibility and biofunctionality in sinus lift augmentation and bone filling in long bone [68, 69].

Several calcium phosphate bone cements are commercially available and more are being investigated since the concept was first introduced by LeGeros et al in 1982 [70] and the first patent obtained by Brown and Chow [71] in 1986. All the current CPCs are reported to have good mechanical properties and reasonable setting times. However, after setting, these materials remain dense and do not provide rapid bone substitution due to the lack of macroporosity. Numerous studies have reported on the applications of currently available commercial calcium phosphate cements, CPC [72,73]. A BCP-based calcium phosphate cement, **MCPC®,** was recently developed [74-76]. The powder component of the MCPC comprises a settable and resorbable matrix (which includes α-TCP, stabilized amorphous calcium phosphate s-ACP and monocalcium phosphate monohydrate, MCPM). A sieved fraction of macroporous biphasic calcium phosphate (BCP) granules ranging between 80 and 200 μm were incorporated into the matrix. The cement liquid is an aqueous solution of Na_2HPO_4. After setting of MCPC in distilled water at 20°C, the mechanical strength obtained were 10MPa± 2 at 24hours and 15MPa ± 2 after 48 hours. The cohesion time for injectability was reached after 20 mn. *In vivo,* the components of the cement resorb at different rates allowing the formation of interconnecting macroporosity when the cement sets, facilitating bone ingrowth and substitution of the cement with the newly forming bone [36] (figure 10).

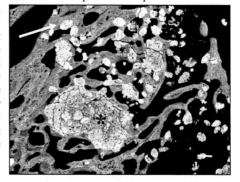

Figure 10 : Light microscopy of the newly formed bone trabeculae at the expense of calcium phosphate cement MCPC® and scaffold effect of the residual BCP (arrow) of implantation in femoral epiphysis of Beagle dog . * Not yet resorb ACP matrix

In vivo, resorption of the ACP component of the MCPC produced macroporosity and facilitated bone ingrowth between and at the surface of BCP granules extending to the core of the implanted site. Light microscopy revealed bone ingrowth at the expense of MCPC matrix extending from the surface of the implant to the core of the defect. MBCP granules act as scaffold for bone osteoconduction . The *in vivo* biological events of the MCPC was different of classical CPC cement which was shown to have centripetal dissolution and bone ingrowth from its surface, while in MCPC it was evidenced large resorption of the ACCP matrix and trabeculae bone ingrowth in the core of the cements by osteoconduction process on the BCP granules [76].The efficiency of the new MCPC in promoting new bone formation and increasing the mechanical strength with time are attributed to the formation of macrostructure due to the resorption of the matrix and the presence of osteoconductive scaffolds (BCP granules).

Granules for Drug Delivery:

Calcium phosphate bioceramics have been very often proposed for adsorption of bioactive factors and for drug delivery system. However recently Smucker et al [77,78] reported for the first time a study demonstrating enhanced posterolateral spinal fusion rates in rabbit using different concentrations of a synthetic peptide (B2A2-K-NS) coated on micro porous BCP granules 60HA/40βTCP ratio. The peptide is a synthetic receptor-targeted that appears to amplify the biological response to *rh*BMP-2). This study indicates more evidence of mature/immature bone ingrowth across the inter-transverse process spaces than did the controls. Microporous macroporous BCP bioceramic granules for peptide adsorption and local delivery will combine the osteoconductive/osteogenic properties of the BCP bioceramics and the osteoinductive properties of the peptide and growth factors.

CONCLUSION

The bioactive concept based on the dissolution/transformation processes of HA and TCP can be applied to both Bulk, and Injectable Biomaterials. The Biphasic Calcium Phosphate concept based on the mixture of HA and β-TCP in the different forms have the same evolution and adaptation to the tissues : (1) partial dissolution of the CaP ceramic macrocrystals cause an increase in the calcium and phosphate concentrations in the local microenvironment; (2) formation of CHA (either by direct precipitation or by transformation from one CaP phase on an other or by seeded growth) incorporating ions (principally carbonate) and non collagenic proteins from the biological fluid during its formation; (3) association of the carbonate-apatite crystals with an organic matrix; and (4) incorporation of these microcrystals with the collagenous matrix in the newly formed bone (in osseous sites). The events at the CaP biomaterial/bone interface represent a dynamic process, including physico-chemical processes, crystal/protein interactions, cells and tissue colonization, bone remodelling, finally contributing to the unique strength of such interfaces. These type of artificial bone revealed from a long time in preclinical and in clinical trials the efficiency for bone filling, performance for bone reconstruction and efficacy for bone ingrowth at the expense of the micro macroporous biphasic calcium phosphate bioceramics.

ACKNOWLEDGMENT

The individual and collaborative studies were supported by research grants from the INSERM U225, CJF 93-05 and E 99-03 and CNRS EP 59 [Dr. G. Daculsi, Director] and from the National Institute for Dental Research of the National Institutes of Health Nos. DE04123 and DE07223 and special industrial Calcium Phosphate Research Funds and Grants (Dr. R.Z. LeGeros, Dr G. Daculsi).

BIBLIOGRAPHY

1. Bauer T, Muschler G. Bone grafts materials. An overview of the basic science. *Clin Orthop Rel Res* **371**:10-27. (2000).

2. Farrington M, Mathews I, Foreman J, Cafrey E. Bone graft contamination from a water de-ionizer during processing in a bone bank. *J Hosp Infect* **32**:61-64. (1996).

3. Marthy S, Richter M. Human immunodeficiency virus activity in rib allografts. *J Oral Maxillofac Surg* **56**:474-476. (1998).

4. Stevenson S, Horowitz M. The response to bone allografts. Current concepts review. *J Bone Joint Surg* **74A**:939-942 (1992).

5. Tomford W, Starkweather R, Goldman M. A study of the clinical incidence of infection in the use of banked allograft bone. *J Bone Joint Surg (Am)* **63** :244-248 (1981).

6. Hofmann C, Von Garrel T, Gotzen L. Bone bank management using a thermal disinfection system (Lobator SD-1). A critical analysis. *Unfallchirurg* **99**:498-508 (1996).

7. Farrington M, Mathews I, Foreman J, Richardson K, Cafrey E. Microbiological monitoring of bone grafts: two years experience at a tissue bank. *J Hosp Infect* **38**:261-271 (1998).

8. Daculsi G, Legeros R, Nery E, Lynch K, Kerebel B. Transformation of biphasic calcium phosphate in vivo: ultrastructural and physico-chemical characterization. *J Biomed Mater Res* **23**:883 -894. (1989).

9. De Groot K. Ceramics of calcium phosphates: preparation and properties. In bioceramics of calcium phosphate. CRC Press, Boca Raton,100-114. (1983)

10. Hench L.L. Bioceramics : From concept to clinic, *J Am Ceram Soc* **74** : 1487-1510. (1994)

11. Jarcho M. Calcium phosphate ceramics as hard tissue prosthetics. *Clin Orthop* **157**: 259-278 (1981).

12. Daculsi G., Bouler J.M.., Legeros R.Z. Adaptive crystal formation: in normal and pathological calcification, in synthetic calcium phosphate and related biomaterials. *Int. Rev.Cytology*, **172**:129-191 (1996)

13. Daculsi G, Laboux O, Malard O, Weiss P. Current state of the art of biphasic calcium phosphate bioceramics. *J Mater Sci Mater Med.* Mar;**14**(3):195-200. (2003)

14. LeGeros R.Z., Daculsi G. The in vivo behaviour of biphasic calcium phosphate. Histological, ultrastructural and physico-chemical characterization. In: Handbook of Bioactive Ceramics, Calcium Phosphate and Hydroxylapatite Ceramics, Yamamuro T., Hench L.L., Wilson-Hench J.W. eds. Amsterdam, CRC Press (1990)

15. Daculsi Guy Biphasic calcium phosphate concept applied to artificial bone, implant coating and injectable bone substitute. *Biomaterials* **19** : 1473-1478 (1998)

16. Heughebaert M., LeGeros R.Z., Gineste M., Guilhem A. Hydroxyapatite (HA) ceramics implanted in non-bone forming site. Physico-chemical characterization. *J Biomed Mat Res* **22**: 257- 268 (1988)

17. Daculsi G., LeGeros R.Z., Nery E., Lynch K. and Kerebel B. Transformation of biphasic calcium phosphate in vivo: Ultrastructural and physico-chemical characterization. *J Biomed Mat Res* **23**: 883-894(1989)

18. Daculsi G., LeGeros R.Z., Heugheaert M., Barbieux. I. Formation of carbonate apatite crystals after implantation of calcium phosphate ceramics. *Calcif Tissue Int* **46**: 20-27 (1990)

19. LeGeros R.Z.) Calcium Phosphates in Oral Biology and Medicine. Monographs in Oral Sciences, Vol. 15, H. Myers, ed., S. Karger, Basel. (1991

20. Metsger SD, Driskell TD, Paulsrud JR Tricalcium phosphate ceramic: a resorbable bone implant: Review and current uses. *J Am. Dent. Assoc* **105**:1035-1048 (1982).

21. LeGeros RZ Calcium phosphate materials in restorative dentistry. *Adv. Dent. Res* **2**: 164-183 (1988).

22. Ellinger RF, Nery EB, Lynch KLHistological assessment of periodontal osseous defects following implantation of hydroxyapatite and biphasic calcium phosphate ceramics: a case report. J Periodont Restor Dent 3:223 (1986).

23. LeGeros RZ, Nery E, Daculsi G, Lynch K, Kerebel B. In vivo transformation of biphasic calcium phosphate of varying b-TCP/HA ratios: ultrastructural characterization. Third World Biomaterials Congress (abstract no. 35) (1988).

24. LeGeros RZ, Lin S, Rohanizadeh R, Mijares D, LeGeros JP Biphasic calcium phosphates: Preparation and properties. J. Mater. Sci.: Mat in Med 14:201-210 (2003).

25. Bouler JM, Trécant M, Delécrin J, Royer J, Passuti N, Daculsi G Macroporous Biphasic Calcium Phosphate Ceramics : Influence of Five Synthesis Parameters on Compressive Strength J. Biomed. Mater. Res. 32 : 603-609 (1996)

26. LeGeros RZ Properties of osteoconductive biomaterials: calcium phosphates. Clin Orthopaed Rel Res 395:81-90 (2002).

27. Daculsi G., High performance of new interconnected MicroMacroporous Biphasic Calcium Phosphate matrices MBCP2000 for bone tissue engineering Proceedings 20th European Conference on Biomaterials, Nantes France (2006).

28. Hubbard, W.G. Physiological calcium phosphate as orthopedic implant material. PhD Thesis, Milwaukee, Marquette University (1974).

29. Schmitt, M PhD Thesis, Nantes, University of Nantes (2000).

30. Daculsi G, Legeros R, Grimandi G., Soueidan A, Aguado E, Goyenvalle E., Legeros J. Effect of Sintering Process on Microporosity, and bone growth on Biphasic Calcium Phosphate Ceramics. Key Engineering Materials vols 361-363,pp 1139-1142, Trans Tech Publication Switzerland (2008).

31. Daculsi, G. Passuti, N. Martin, S., Deudon, C. LeGeros, R.Z. Macroporous calcium phosphate ceramic for long bone surgery in human and dogs. J Biomed. Mater. Res. 24: 379 396 (1990).

32. Daculsi, G., Bagot D'arc, M., Corlieu , P., Gersdorff, M. Ma croporous Biphasic Calcium Phosphate efficiency in mastoid cavity obliteration. Ann Orol. Rhinol Laryngol 101:669-674 (1992).

33. Gouin, F., Delecrin, J., Passuti, N., Touchais, S., Poirier, P, Bainvel, J.V. Comblement osseux par céramique phosphocalcique biphasée macroporeuse : A propos de 23 cas. Rev Chir Orthop 81:59-65 (1995).

34. Ransford A.O., Morley T., Edgar M.A., Webb P., Passuti N., Chopin D., Morin C., Michel F., Garin C., Pries D. Synthetic porous ceramic compared with autograft in scoliosis surgery. A prospective, randomized study of 341 patients." J Bone Joint Surg Br 80(1): 13-18 (1998).

35. Cavagna, R., Daculsi, G., Bouler, J-M., Macroporous biphasic calcium phosphate: a prospective study of 106 cases in lumbar spine fusion. Long term Effects Med Impl 9 : 403-412 (1999).

36. Wykrota LL, Garrido CA, Wykrota FHI Clinical evaluation of biphasic calcium phosphate ceramic use in orthopaedic lesions. In LeGeros RZ, LeGeros JP (Eds) Bioceramics 11, Singapore, World Scientific, pp 641-644. (1998).

37. Clemencia Rodríguez M, Jean A., Kimakhe S., Sylvia Mitja S., Daculsi G., Five Years Clinical Follow up Bone Regeneration with CaP Bioceramics *Key Engineering Materials* vols 361-363,pp 1339-1342, Trans Tech Publication Switzerland(2008).

38. Maillac N. Moreau F., Daculsi G., Bone ingrowth for Sinus lift augmentation with Micro Macroporous Biphasic Calcium Human cases evaluation using microCT and histomorphometry. *Key Engineering Materials* vols 361-363,pp 1347-1350, Trans Tech Publication Switzerland (2008),

39. Delecrin J, Takahashi S, Gouin F, Passuti N A synthetic porous ceramic as a bone graft substitute in the surgical management of scoliosis: a prospective, randomized study. *Spine* **25**:563-569 (2000).

40. Cho D.Y., Lee W.Y., Sheu P.C., Chen C.C., Cage containing a biphasic calcium phosphate ceramic (Triosite) for the treatment of cervical spondylosis, *Surgical Neurology* 63:497-503, (2005)

41. Rouvillain J.L., Lavallé F., Pascal-Mousselard H. Catonné Y., Daculsi G.Clinical and histological evaluation of micro macroporous biphasic calcium phosphate bioceramic wedges in open tibial osteotomy for valgisation, Knee , in press (2009)

42. Gauthier O., BoulerJ-M., Aguado,E., Pilet P, Daculsi G Macroporous biphasic calcium phosphate ceramics: influence of macropore diameter and macroporosity percentage on bone ingrowth. *Biomaterials* **19** (1-3):133-9 (1998).

43. De Groot K. Ceramics of calcium phosphates: preparation and properties. In bioceramics of calcium phosphate. CRC Press, Boca Raton,100-114. (1983)

44. Hench L.L., Bioceramics : From concept to clinic, *J Am Ceram Soc* 74 : 1487-1510 (1994)

45. Daculsi G., LeGeros R.Z., Deudon C. Scanning and transmission electron microscopy and electron probe analysis of the interface between implants and host bone. *Scan. Micr* **4**: 309-314. (1990)

46. Hench L.L., Splinter R.J., Allen W.C., and Greelee T.K. Bonding mechanisms at the interface of ceramic prosthetic materials, *J Biomed Mater Res* 2: 117-141 (1971)

47. Ripamonti U. The morphogeneis of bone in replicas of porous hydroxyapatiteobtained by conversion of calcium carbonate exosk eletons of coral. *J Bone Joint Surg* Am **73**:692-703 (1991)

48. Kuboki Y., Takita H., Kobayashi D (1998). BMP-induced osteogenesis on the surface of hydroxyapatite with geometrically feasible and nonfeasible structures: topology of osteogenesis. J Biomed Mater Res 39:190-199.

49. Le Nihouannen D, Daculsi G, Saffarzadeh A, Gauthier O, Delplace S, Pilet P, Layrolle P. Ectopic bone formation by microporous calcium phosphate ceramic particles in sheep muscles. *Bone* **36**(6):1086-93 (2005)

50. Reddi AH Morphogenesis and tissue engineering of bone and cartilage: Inductive signals, stem cells and biomimetic biomaterials. *Tissue Eng* **6**:351-359 (2000).

51. Yuan H., Kurashina K., Joost de Bruijn D., Li Y., de Groot K., Zhang X. A preliminary study ofn osteoinduction of two kinds of calcium phosphate bioceramics. *Biomaterials* **20**:1799-1806 (1999)

52. Barrere F, van der Valk CM, Dalmeijer RA, Meijer G, van Blitterswijk CA, de Groot K, Layrolle P. Osteogenicity of octacalcium phosphate coatings applied on porous titanium. *J. Biomed Mater Res* **66**A:779. (2003).

53. Daculsi G., LeGeros R.Z., Mitre D. Crystal dissolution of biological and ceramic apatites. *Calcif Tissue Int* **45**: 95-103 (1989)

54. Albee F.H. Studies in bone growth. Triple calcium phosphate as a stimulus to osteogenesis. *Ann Surg* **71**: 32-36. (1920)

55. Daculsi G., Dard M. Bone-Calcium-Phosphate ceramic interface *Osteosynthese International* **2** :153-156 (1994)

56. Yuan H., Kurashina K., Joost de Bruijn D., Li Y., de Groot K., Zhang X. A preliminary study on osteoinduction of two kinds of calcium phosphate ceramics. *Biomaterials* **20**:1799-1806 (1999)

57. Malard O., Guicheux J., Bouler J.M., Gauthier O., Beauvillain de Montreuil C., Aguado E., Pilet P., LeGeros R., Daculsi G. Calcium phosphate scaffold and bone marrow for bone reconstruction in irradiated area : a dog study. *Bone* **36**:323-330 (2005).

58. Daculsi G. Biphasic calcium phosphate concept applied to artificial bone, implant coating and injectable bone substitute. *Biomaterials* **19**:1473-1478 (1998)

59. Bonfield W. Hydroxyapatite-reinforced polyethylene as an analogous material for bone replacement. In Bioceramics: Materials Characteristics Versus In Vivo Behavior, P. Ducheyne and J. E. Lemons (eds) *Ann NY Acad Sci* **523**: 173-177 (1988)

60. Ducheyne P., Marcolongco M., and Schepers E. Bioceramic composites. In: An Introduction to Bioceramics, L.L. Hench and J. Wilson (eds). World Scientific Publishers, London, pp. 281-297 (1993)

61. Constanz B.R., Ison I.C., Fulmer M.T., Poser R.D., Smith S.T. VanWagoner M. Ross J. Goldstein S.A. (1995) Skeletal repair by in situ formation of the mineral phase of bone. *Science* 267: 1796-1799.

62. Daculsi G., Weiss P., Delecrin J., Grimandi G., Passuti N., Guerin F. Composition pour biomatériau - procédé de préparation. Patent n° 94-01-414 1994235. (1994)

63. Daculsi G., Weiss P., Bouler J.M., Gauthier O., Aguado E. Bcp/hpmc composite : a new concept for bone and dental substitution biomaterials. *Bone*, **25**:59-61 (1999)

64. Millot F., Grimandi G., Weiss P. Daculsi G. Preliminary In Vitro and In Vivo studies of a New Injectable Bone Substitute. *Cells and Mat.* **9**:21-30 (1999)

65. Gauthier O, Bouler JM, Aguado E, Legeros Rz, Pilet P, Daculsi G. Elaboration conditions influence physicochemical properties and in vivo bioactivity of macroporous biphasic calcium phosphate ceramics. *J of Mater Sc : Mat in* Med ,**10**:199-204 (1999)

66. Gauthier O, Bouler Jm, Weiss P, Bosco J, Daculsi G, Aguado E. (1999). Kinetic study of bone ingrowth and ceramic resorption associated with the implantation of different injectable calcium-phosphate bone substitutes. *J Biomed Mater Res* **47**(1) : 28-35 1999

67. Gauthier O., Bouler J.M., Weiss P., Bosco J., Aguado E., Daculsi G. Short-term effects of mineral particle sizes on cellular degradation activity after implantation of injectable calcium phosphate biomaterials and consequences for bone substitution. *Bone*, **25**:71-74 (1999)

68. LeNihouannen DL, Saffarzadeh A, Aguado E, Goyenvalle E, Gauthier O, Moreau F, Pilet P, Spaethe R, Daculsi G, Layrolle P.J Osteogenic properties of calcium phosphate ceramics and fibrin glue based composites. *J Mater Sci Mater Med.* **18**(2):225-235 (2007).

69. Le Guehennec L, Layrolle P, Daculsi G A review of bioceramics and fibrin sealant. *Eur Cell Mater.* **13**;8:1-11 (2004).

70. LeGeros RZ, Chohayeb A, Shulman A. Apatitic calcium phosphates: possible restorative materials. *J Dent. Res* **61** (spec issue):343 (1982)

71. Brown WE, Chow LC. "A new calcium phosphate water-setting cement" in Brown PW (ed). *Cement Research Progress 1986.* American Ceramic Society, Westerville, OH pp. 352-379 (1987).

72. Niwa S, LeGeros RZ "Injectable calcium phosphate cements for repair of bone defects". In: Lewandrowski K-U, Wise D.L., Trantolo DJ., Gresser J.D. *Tissue Engineering and Biodegradable Equivalents. Scientific and Clinical Applications.* Marcel Dekker, New York, pp. 385-400. . (2002).

73. Constanz B.R., Ison I.C., Fulmer M.T., Poser R.D., Smith S.T. Van Wagoner M. Ross J. Goldstein S.A. *Science* **267**: 1796-1799 (1995)

74. Khairoun I, LeGeros RZ, Daculsi G, Bouler JM, Guicheux J, Gauthier O macroporous, resorbable and injectable calcium phosphate-based cements (MCPC) for bone repair, augmentation, regeneration and osteoporosis treatment. patent no WO 2005/077049.

75. Bourges X., Baroth S., Goyenvalle E., Cognet R., Moreau F., Pilet P., Aguado E., Daculsi G., Vertebroplasty in goat model using macroporous calcium phosphate cement (MCPC), Key Engineering Materials. 2008 ;361-363: 377-380

76. Daculsi G, Khairoun I, leGeros RZ, Moreau F, Pilet P, Bourges X, Weiss P, Gauthier O. Bone Ingrowth at The Expense of a Novel Macroporous Calcium Phosphate Cement. Key Engineering Material. 2007;330-332:811-4

77. Smucker J.D;, Aggarwal D., Zamora P.O., Atkinson B.L., Bobst J.A., Nepola J.V., Fredericks D.C., B2A peptide on ceramic granules enhance posterolateral spinal fusion in rabbits compared to autograft, *Spine* **33**: 1324-1329 (2008)

78. Smucker J.D;, Aggarwal D., Zamora P.O., Atkinson B.L., Bobst J.A., Nepola J.V., Fredericks D.C., Assessment of B2A2-K-NS coated on an osteoconductive granule in a rabbits posterolateral fusion model, proceedings AAOS, 12-14 February San Diego USA (2007)

BIOCOMPATIBILITY ASPECTS OF INJECTABLE CHEMICALLY BONDED CERAMICS OF THE SYSTEM CaO-Al$_2$O$_3$-P$_2$O$_5$-SiO$_2$

Leif Hermansson[1,2], Adam Faris[1], Gunilla Gomez-Ortega[1] and Jesper Lööf[1,2]

[1]Doxa AB, Axel Johanssons gata 4-6, SE-75451 Uppsala, Sweden
[2]Department of Engineering Sciences, The Angstroem Lab, Uppsala University, Sweden
leif.hermansson@doxa.se

ABSTRACT

The paper will discuss the biocompatibility including bioactivity of the system CaO -Al$_2$O$_3$ - P$_2$O$_5$ - SiO$_2$. The bioactivity was determined *in vitro* by studying the formation of apatite on the biomaterial surface when soaked in simulated body fluid. Complementary *in vivo* biocompatibility and bioactivity studies were conducted in animals (rabbit femur and tibia and sheep vertebrae). The evaluation included histology, X-ray diffraction and high resolution TEM using specific site sample preparation using FIB-technique. General aspects of biocompatibility including cytotoxicity, sensitization, irritation/intracutaneous reactivity, systemic toxicity, sub-acute, sub-chronic and chronic toxicity, genotoxicity, carcinogenicity, and haemocompatibility were also evaluated. It is concluded that the injectable biomaterials in the system studied seem to fulfill the required biocompatibility for the intended clinical use by meeting all the toxicological endpoints referred to in the harmonized standard ISO 10993:2003.

INTRODUCTION

Chemically bonded ceramics (CBC), also known as inorganic cements, can be designed to be used as *in vivo* hardening injectable implants to restore and support damaged hard tissue, both dental and bone tissue. The CBC-materials are based on phosphates, sulphates, aluminates and silicates or combinations of these. The two former are well established as so called Bone Void Fillers with good biocompatibility and resorbability, but also with low strength and inferior physical properties such as shear separation during injection and low radiopacity. This type of injectable material is used to stabilize damaged bone tissue where no intrinsic strength is needed. The material is often put in place via a large open wound operation, which leads to high costs for the operation and a long recovery period for the patient. Apatite phases and P$_2$O$_5$-contaning glasses are well documented as bioactive materials [1].

It is beneficial to perform less invasive procedures with a material that hardens for a relatively short time to achieve a high strength level *in vivo*. The current golden standard material for such operations is resin-based materials within dentistry and poly-methylmethacrylate (PMMA) based materials within orthopaedics. Injectable CBC-materials based on Ca-aluminates as alternative materials to PMMA-based ones for stabilizing vertebral fractures, and Ca-aluminate-based materials as alternatives to various dental applications have been suggested and tested for some time with regard to physical, mechanical, chemical and biological properties [2-3].

This paper focuses on the biocompatibility of bioceramics of the system CaO-Al$_2$O$_3$-SiO$_2$ - P$_2$O$_5$ – H$_2$O (CASPH). It also deals with aspects such as definitions and use of the terms *biocompatibility* and *bioactivity*, further test conditions, and test methods to evaluate the system in an appropriate way.

EXPERIMENTALS

Materials

The whole system CaO-Al$_2$O$_3$- SiO$_2$ - P$_2$O$_5$ –H$_2$O, referred to as the CASPH-system, must be taken into account when discussing the biocompatibility of the Ca-aluminate based materials. This is due to the fact that a SiO$_2$-phase for geometrical stability is incorporated [4] as well as phosphate phases and H$_2$O since the materials often are in contact with body liquid. Reference materials in the testing were PMMA-based materials, Vertebroplastic, DePuy and CMW-1, DePuy, and a resin-based cement of bisphenol a diglycidylether methacrylate (Bis-GMA) with glass-ceramics particles Cortoss^{TM}, Orthovita, and in some cases a Ca-phosphate based reference material such as NORIAN (Synthes). For dental evaluation, conventional resin-based dental filling materials (3M and Vivadent) and dental cements (e.g. Ketac Cem and RelyX Unicem) were used as reference. In all materials tested, including the reference materials, inert biocompatible filler particles such as zirconia, glass particles or glass-ceramics, and Ba-sulphate can be found. For the CASPH-system zirconia or glass particles were added.

A typical composition range for Ca-based materials in this study was as follows; 50-65 % Ca-aluminates, < 5 % μ-silica or Ca-silicate, 20-40 % inert particles (glass particles or zirconia) and < 5 % processing agents (LiCl as an accelerator, polycarboxylate and methyl-cellulose to control rheology). For dental cement an additional ingredient is a glass ionomer to reduce initial pH and to control setting time. More details are described elsewhere [5-7].

Definitions used

The terms biocompatibility and bioactivity are used in different ways by different categories of scientists. Below are presented the definitions used in this paper, mainly agreeing with the definitions discussed in [8].

Biocompatibility: "The ability of a material to perform with an appropriate host response in a specific application".

Bioactivity (bioactive material): "A material which has been designed to induce specific biological activity". Another definition according to [9]: "A bioactive material is one that elicits a specific response at the interface of the material which results in the formation of a bond between the tissues and the material".

Thus a material cannot in itself be classified as biocompatible without being related to the specific application, for which it is intended. Bioactivity from a materials viewpoint is frequently divided into *in vitro* and *in vivo* bioactivity. The *in vitro* bioactivity of a material is however only an indication that it might be bioactive in a specific *in vivo* application. Another aspect of bioactivity is that this term can be adequate only when the biomaterial is in contact with a cellular tissue. However, often a material is claimed to be bioactive if it also reacts with body liquids forming an apatite-phase. *In vitro* bioactivity is tested in phosphate buffer systems similar to that of saliva or body liquid, and apatite formation is the claimed sign of bioactivity. A further aspect of bioactivity and also biocompatibility deals with the different curing times and temperatures at which the observation (testing) is performed. This is important to issues such as initial pH-development, cohesiveness and initial strength.

Finally the biocompatibility and bioactivity can only be confirmed in clinical situations, with the actual implant/biomaterial in the designed amount or content and shape. This is especially important for injectable biomaterials which are formed (hydrated) and cured *in vivo*, and for implants where movements, even micro-movement, can influence the outcome.

Standards and methods

Relating to the definition aspects above, the acceptance of a biomaterial is a crucial issue, and to some extent the question has been solved by relating to the following toxicological

endpoints indicating biocompatibility as referred in the harmonized standard ISO 10993:2003 [10], which comprises the following sections:
Cytotoxicity (ISO10993-5), Sensitization (ISO10993-10), Irritation/Intracutaneous reactivity (ISO10993-10), Systemic toxicity (ISO10993-11), Sub-acute, sub-chronic and chronic toxicity (ISO10993-11), Genotoxicity (ISO10993-3), Implantation (ISO10993-6), Carcinogenicity (ISO10993-3) and Hemocompatibility (ISO10993-4).

This will be the main guideline when presenting the status of the biocompatibility of the CASPH-system, but was complemented by corrosion testing, elementary analysis, pH-change and additional cytotoxicity testing.

The corrosion resistance test – using a water jet impinging technique - was conducted according to EN 29917:1994/ISO 9917:1991,where removal of material is expressed as a height reduction using 0.1 M lactic acid as solution, pH 2.7 [11]. The duration time of the test is 24 h. The test starts after 24 h hydration. The test probe accuracy was 0.01 mm. Values below 0.05 mm per 24 h solution impinging are judged as acid resistant.

Determination of Ca and Al in the solution during the hydration process of the Ca-aluminate based material was performed using atomic absorption spectrometry [12]. Standard solutions of different concentrations of Ca and Al were prepared according to the manual. Samples were prepared with a size of 10mm x 2mm height using a wet-press method, corresponding to a surface area of 224 mm^2. The test pieces were placed in plastic bottles in inorganic saliva solution of pH 7. The amount of liquid was 10 ml in each bottle. The temperature selected was 37 °C. The inorganic saliva solution contained calcium chloride, magnesium chloride, sodium chloride, a phosphate buffer, hydro-carbonate and citric acid. The Ca-content in the saliva solution corresponded to 68 ppm. 1 ml solution was removed at 1, 7 and 28 days for analyses, and saliva was exchanged at 1, 7 and 28 days after every measurement. For the 28 days test additional samples were also taken 1 h after new solution was added.

Measurement of pH development during hydration of the material was conducted using a standard pH-meter. Samples were prepared according to the procedure for atomic absorption described above. The pH-testing was conducted in two separate ways. First the samples were immersed in saliva solution (pH =7) at 37 °C, and pH was measured continually over the whole experiment period (Test 1). 1 ml solution was removed at 1, 7, 14 and 28 days for pH measurement. The second type of measurement comprised immersion in 10 ml saliva solution at 37 °C, where the saliva was exchanged at 1, 7, 14 and 28 days. pH was measured at the time of observation and also after one hour in new saliva (data within brackets), Test 2.

Specimens at different setting stages were subjected to cytotoxicity testing by using primary cultures of human oral fibroblasts. A tissue culture insert retaining tested materials was assembled into a 12-well plate above the fibroblast monolayers. The cytotoxicity was determined by MTT reduction assay after various curing times. The test design is described in detail elsewhere [13]. Specimens were set and hydrated at 37 °C for different periods of time, i.e. 0, 5, 30, 60 min, 24 h and 1 week and were then placed on tissue culture netwell for a cytotoxicity test. Both acute (1 and 24 h) and long term (1 week) in vitro toxicity tests were conducted with MTT assay.

Studies were further complemented by presentation of the chemical reactions and microstructure developed in the actual contact zone between the biomaterial and tissue, i.e the analysis of intact interfaces and calcified tissue at the highest level by transmission electron microscopy (TEM) in combination with focused ion beam microscopy (FIB) for site-specific high accuracy preparation of the TEM-samples, described in detail in [14]. Phase and elemental analyses were conducted using traditional XRD, HRTEM, XPS and STEM with EDX.

RESULTS AND DISCUSSION

Reactions of the injectable materials

Before dealing with the biocompatibility of the CASPH-system in detail, some general features of the biomaterial should be pointed out. The CASPH-materials are in most cases used as injectable materials, i.e. they are introduced as a paste (a mixture of reacting oxides in a hydration process) which cures and matures through six identified integration reactions, recently published [15]. These mechanisms affect the integration differently depending on what type of tissue the biomaterial is in contact with, and in what state the Ca-aluminate (un-hydrated or hydrated) is introduced. The reaction mechanisms are presented in Tables 1-2.

Table 1. Summery of the reactions

Reaction	Feature
Mechanism 1	Main reaction, hydration of Ca-aluminate
Mechanism 2	Apatite formation in presence of phosphate ions in the biomaterial
Mechanism 3	Apatite formation in the contact zone in presence of body liquid
Mechanism 4	Transformation of hydrated Ca-aluminate
Mechanism 5	Biological induced integration and ingrowth
Mechansim 6	Mass increase reaction when un-hydrated Ca-aluminate is present

Mechanisms 1-2 and mechanism 6 occur at all contact areas, between the paste and other biomaterials, and between the paste and tissue. Mechanisms 3-5 occur mainly at interfaces.

Table 2. Features of the chemical reactions

Reaction mechanism	Description	Comments
Mechanism 1. Main reaction	3 (CaO Al$_2$O$_3$) + 12 H$_2$O → 3 Ca^{2+} + 6Al^{3+} + 4(OH)$^-$ → 3 Ca^{2+} + 6 Al(OH)$_4^-$ → Ca$_3$ [Al(OH)$_4$]$_2$(OH)$_4$ (katoite) + 4 Al(OH)$_3$ (gibbsite)	Katoite and gibbsite are formed as nano-size hydrates
Mechanism 2. Complementary reaction with phosphate-containing solution	5Ca^{2+} + 3PO$_4^{3-}$ + OH$^-$ → Ca$_5$(PO$_4$)$_3$OH	Additional phase formed: apatite
Mechanism 3. Contact zone reaction with body liquid in presence of the basic CA-phase	HPO$_4^{2-}$ + OH$^-$ → PO$_4^{3-}$ + H$_2$O Thereafter the apatite-formation reaction occurs as mechanism 2, 5Ca^{2+} + 3PO$_4^{3-}$ + OH$^-$ → Ca$_5$(PO$_4$)$_3$OH.	Apatite formation in the contact zone
Mechanism 4. Transformation reaction of the originally formed phase katoite.	Ca$_3$ · (Al(OH)$_4$)$_2$ · (OH)$_4$ → 2 Ca^{2+} + HPO$_4^{2-}$ + 2 H$_2$PO$_4^-$ → Ca$_5$ · (PO$_4$)$_3$ · (OH) + 2 Al(OH)$_3$ + 5 H$_2$O	Apatite and gibbsite formed in the biomaterial towards tissue
Mechanism 5. Biologically induced integration and ingrowth	Bone ingrowth towards the apatite allows the new bone structure to come into integrated contact with the biomaterial.	New bone formation at the contact
Mechanism 6. Mass increase reaction due to presence of unhydrated CA	3 CaO Al$_2$O$_3$ + 12H$_2$O → 3 CaO Al$_2$O$_3$ 6H$_2$O + 2Al$_2$O$_3$ 3H$_2$O, or C$_3$AH$_6$ + 2 AH$_3$ (in cement abbreviation chemistry)	Mass increase and point welding

The hydration yields precipitated phases in the nano-size range 10-40 nm [16]. Typical nano-microstructure of the contact zone to tissue is presented in Figure 1.

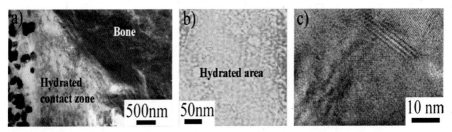

Fig. 1 Microstructure of the Ca-aluminate material, a) the contact zone to bone, black particles to the left are inert zirconia particles, b) the hydrated contact close to tissue, and c) the nano-size structure of the hydrated materials

Biocompatibility evaluation.
  Summarized below are the results from several biocompatibility and bioactivity studies [17-20] where Ca-aluminate is used as a biomaterial in orthopaedic and dental applications. *In vitro* bioactivity studies show apatite formation on the surface of the Ca-aluminate materials exposed to phosphate buffer solution, an example shown in Figure 2 below.

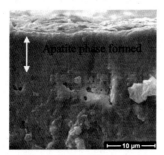

Fig. 2 Cross section of the apatite-containing surface layer formed, SEM [14]

  The evaluation also comprises specific issues related to acid corrosion resistance, release of ions and pH changes during hardening, and stability over time of the Ca-aluminate based materials. The prime use of these is as stable high-strength injectable biomaterials, while biomaterials mainly based on phosphates, sulphates and silicates are classified as slowly resorbable or resorbable materials [21].

Corrosion resistance
  No height reduction at all was observed for two tested Ca-aluminate materials. Thus, according to the acid corrosion test, Ca-aluminate materials are judged as stable materials. The total absence of material loss, measured as height reduction in the acid corrosion test, is related to the general basic nature of the material, with possibility of neutralization of the acid in the contact zone – especially in the earlier stage of the hydration process.

Ion release measured by atomic absorption spectrometry
    The results of Ca and Al determination in the solution during the hydration process of the Ca-aluminate based material are presented in Table 3.

Table 3. Ca and Al dissolution during hardening of the Ca-aluminate material, (The 1h testing at 28 days within brackets)

Ion tested	1h, ppm	24 hrs, ppm	7 days, ppm	28 days, ppm
Ca	66	64	44	50 (70)
Al	11.4	9.3	9.6	8.6 (1.2)

The release of metal ions in water was below 5×10^{-2} ppm/(mm^2 material) for aluminium and below 30×10^{-2} ppm/(mm^2 material) for calcium, whereas somewhat higher aluminium content was measured in artificial saliva. The ion concentrations detected are generally not time-dependent during hydration. After the initial hydration time the ion concentration (molar) is determined by the solubility product of the phases formed (katoite = 5×10^{-26} and gibbsite = 3×10^{-24}). Since the concentration of Ca in saliva is higher than what is obtained in the non-physiological aqueous solution (distilled water), it can be assumed that the filling material releases very limited amounts of Ca or Al once the material has hardened. The presence of Ca in saliva will decrease the solubility tendency of the calcium-aluminate-hydrate phases.
    Based on a search in the literature, the FAO/WHO Joint Expert Committee on Food Additives (JECFA) has provided a provisional figure for tolerable weekly aluminium intake of 7 mg/kg body weight. This corresponds to 1 mg/kg /day. The daily intake of aluminium via digestion/food is approximately 5 mg per day. For calcium the NIH Consensus Development Conference on Optimal Calcium Intake recommended an intake in the range 800 mg/day for young children to 1000 – 1500 mg/day for adults depending on gender and age. For many people there is a need to supply additional calcium in order to stay healthy. The ion concentrations measured and the amounts of Ca and Al released are far below the concentrations of the elements produced from food intake and should therefore not pose any safety concerns at all.

Change in pH during hydration
    The results of the measurement of pH development during hydration of the Ca-aluminate based are shown in Table 4. The initial *in vitro*-pH is 10.5 in saliva. After 1 week, pH after 1 h dissolution time in saliva is approx. 8.

Table 4. Change of pH during initial hydration of Ca-aluminate based materials

Test No.	At start	1h	24 hrs	7 days	14 days	28 days
1	10.5	10.3	10.7	10.3	9.9	9.8
2	10.5	10.2	10.2 (7.7)	9.9 (7.8)	9.5 (8.1)	9.2 (8.1)

    The pH is high during the early stage of the hydration, but decreases with time and approaches neutrality. The reason for the high pH in the beginning is the general basic character of the material and the formation of OH$^-$ during the hydration process. In the clinical situation saliva is produced in a dynamic way, creating an environment capable of buffering surrounding solution to neutrality. In the clinical studies performed so far no adverse reactions have been reported from a possible elevated pH during the early part of the hydration.

Cytotoxicity testing.

The *in vitro* MTT reduction test of the experimental Ca-aluminate dental filling material in human oral fibroblast culture showed no obvious cytotoxicity. The average level of MTT reduction of the experimental dental filling material was close to 100% of the control values. The maximal variation (SD) was less than 30%. Different curing times of the test material did not seem to affect the cytotoxicity test results although one week curing produced the most stable testing results both in the short and long term tests. After a week the material can be considered as fully cured, i.e. stable.

Morphological changes were not observed in any of the test groups at different MTT reduction testing points. As shown in Figure 3, the cell culture was typically fibroblastic with a slender and elongated form in both the control group and the group exposed to the examined material. In the exposed picture B even some precipitated hydrates are seen.

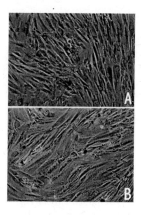

Figure 3. Morphological observation of human oral fibroblasts on an experimental Ca-aluminate based material. A: Normal control. B: After exposure to the experimental filling material for one week.

The standard for cellular biocompatibility in *in vitro* testing has been stated in the International Organization for Standardization (ISO) standards documents [22]. The standard allows for the contact testing of solid dental materials for cytotoxicity with cell lines. Due to several disadvantages of direct contact testing, indirect testing methods have been developed and compared to the direct testing assays [23]. Introduction of a standard cell culture device, i.e. cell culture insert or transwell, provided an opportunity for such cytotoxicity screening of dental materials with indirect contact between material specimens and cell culture monolayer. It is believed that such a testing system more closely mimics the *in vivo* exposure pattern by providing the test of the material in both its solid and dissolved phases at the same time. It has been shown that this testing system has produced the most stable results as compared to other testing systems, such as direct contact test. In a complementary cytotoxicity test using the pulp derived cell response, the experimental material showed no sign of toxicity [24].

Harmonized standard ISO 10993:2003

Further cytotoxicity and other biocompatibility aspects are summarised according to the outline in the harmonized standard ISO 10993:2003 [10]. Xeraspine^{TM}, an orthopaedic Ca-aluminate-based material, was the test material. This material is judged as mildly cytotoxic during the initial curing, and as non-cytotoxic as cured material. See Table 5 below.

Table 5. Cytotoxicity testing of an orthopaedic Ca-aluminate based material, Xeraspine^{TM}

Type of test	Method	Cytotoxicity (scale 0-4 or 100-0%
During curing, undiluted	ISO 10993-5, § 8.2	2 (mild)
During curing, diluted	ISO 10993-5, § 8.2	0-1 (none-slight)
During curing	XTT-test	60 % (slight)
During curing, diluted	XTT-test	> 70 % (none)
Cured, undiluted	ISO 10993-5, § 8.2	0 (none)
Cured, diluted → diluted	XTT-test	> 70% (none)

A sensitization test (ISO 10993-10), Guinea Pig Maximization Test was performed with the orthopaedic Ca-aluminate material during curing. No sensitizing potential was obtained. Additional irritation and delayed hypersensitivity testing according to ISO 10993-10:2002 was conducted with both polar and non-polar extract from cured material, and the results showed no discrepancies after intracutaneous injections in the rabbit compared to the blank injections. The acute systemic toxicity study according to ISO 10993-11 was performed with both polar and non-polar extracts from cured Ca-aluminate material (Xeraspine), and the results showed no signs of acute systemic toxicity. Sub-acute, sub-chronic and/or chronic toxicity studies according to ISO 10993-11 were not conducted explicitly, since data from the two implantation studies in rabbit (see below) were judged to support that no long term toxicity is expressed. From the implantation studies histopathological organ and tissue data is available, and no adverse effects were reported.

According to the outline in "Genotoxicity test according to ISO 10993-3:2003 Biological evaluation of medical devices, Part 3 *Test for genotoxicity, carcinogen toxicity and reproductive toxicity*" three *in vitro* genotoxicity tests were performed on Xeraspine during curing. The findings are presented in Table 6.

Table 6. *In vitro* genotoxicity tests of Xeraspine^{TM}

Type of test	Comment	Result
Reverse mutation assay (AMES test)	+ DMSO vehicle	No bacterial toxicity, No mutagenicity
Chromosomal aberration test		No clastogenicity
Mouse Lymphoma assay	+/- metabolic activation	No mutation inducing capacity

Additionally, in an *in vivo* genotoxicity assay, the mice micronucleus test of bone marrow was used. The extract (Xeraspine during curing and cured material) was administered intraperitoneally twice. The results showed no clastogenic effect.

Three *in vivo* implantation studies based on ISO 10993-6 have been performed. Two studies in rabbit (femur) and one in sheep (vertebrae). *In vivo* implantation studies are judged as the most relevant studies for documentation of safety of a product. In the rabbit implantation studies Xeraspine was compared to the PMMA-material CMW 1, and in the sheep study Xeraspine was compared to the PMMA-material Vertebroplastic, and to the Bis-GMA material Cortoss^{TM}. The results are summarized in Table 7.

Table 7. Implantation studies in femur rabbit, and in vertebrae sheep, details in [25-26]

Implantation studies	Species	Reference material	Result
6-week femur	Rabbit	CMW-1	Minimal inflammation, very few inflammatory cells were present in bone, bone marrow and adipose tissues.
6 (12) -month femur	Rabbit	CMW-1	No Al- accumulation
12-week vertebrae	Sheep	Vertebroplastic Cortoss	No inflammation, no Al-accumulation

The 6 months femur study in rabbits included a 12 months subgroup. The amount of aluminium in blood and selected organs was analysed. The main target organs of the animals (kidney, lung, liver) were histopathologically investigated. Granulomatous inflammation in the cavity, pigmented macrophages and new bone formation were the treatment-related observations at 6- and12-months examination. No difference between Xeraspine and PMMA was detected. There were no signs of aluminium accumulation in the analysed tissues.

In the 12-week study, the histopathology of vertebrae obtained one week after surgery showed the most severe inflammatory reaction to the surgery in the *sham* operated vertebrae. The bone marrow in the vertebrae filled with Xeraspine was not reported to be infiltrated by any inflammatory cells. In vertebrae obtained 12 weeks after surgery no inflammatory reactions were reported, and no obvious differences were observed in the pathological reactions to the surgery (sham) or the filler materials. Overview of the histological contact zone to the Ca-aliminate based material Xeraspine^{TM} is shown in Figure 7.

The analysis of serum samples showed low concentrations of aluminum in comparison to what is normal in humans. Since the concentration of aluminum did not increase after surgery and in some instances was lower after surgery than in the 0-samples, one may regard these concentrations as within the normal variation.

Fig 7. Histology image of Xeraspine (black) in close contact with sheep vertebral bone.

Repeated haemocompatibility studies have been performed to evaluate possible reactions in whole human blood as a result of contact with Ca-aluminate materials [27]. Test items were an experimental Ca-aluminate based material and Xeraspine, Vertebroplastic and Norian (Calcium Phosphate Cement, Synthes Inc). A Chandler loop model was used in which circulating human blood was in contact with the test materials for up to 4 hours. For comparison, loops free from test materials were used. Platelet count (PLT), thrombin-antithrombin (TAT) complex, complement factors C3a and C5b-9 (TCC), and TNF-α were assayed. The degree of haemolysis was assessed by the Drabkin method. Norian (a calcium phosphate based material) invariably induced extensive clotting already after 60 minutes, verified macroscopically and also by significantly reduced PLT in comparison to the Control loops, whereas there was no significant reduction in PLT in the loops with Xeraspine or Vertebroplastic, respectively, neither at 60 nor at 240 minutes. Xeraspine did not induce haemolysis to a greater extent than any of the other materials tested. TCC was activated to a certain degree by Xeraspine, comparable to what is commonly observed for artificial materials. TNF-α generation, indicative of activation of white blood cells, was not enhanced by either Vertobroplastic or Xeraspine.

On-going clinical studies.
Three on-going clinical studies are evaluating the use of Ca-aluminate based materials from safety and handling aspects. These comprise a 1-year evaluation of XeraCem^{TM} as a dental luting cement, and a 2-year evaluation of Xeraspine^{TM} as an orthopaedic material for percutaneous vertebroplasty (PVP) and for kyphoplasty (KVP).

CONCLUSION
The studies presented in this paper can be summarised as follows;

- The biocompatibility studies conducted on the Ca-aluminate based materials, XeraCem^{TM} and Xeraspine^{TM} are judged sufficient for a preclinical safety evaluation, and the results show the materials to be biocompatible and *in vitro* bioactive

- The release of ions, such as calcium and aluminium, from the materials is low and considerably lower than corresponding intake via digestion/food. The cured materials show no cytotoxicity.

- The local tolerance of the Xeraspine was investigated in three implantation studies in two different species and two different anatomic locations (rabbit femur and sheep vertebrae). The results indicate that no distinct differences in general toxicity or in inflammatory response between Xeraspine and the two reference control materials, Vertebroplastic (PMMA) or Cortoss (Bis-GMA) are obtained.

- The Ca-aluminate materials are not degradable and do not induce clotting or haemolysis.

Based on all above mentioned data and generated toxicity data, it is considered that there is no reason to expect that the Ca-aluminate biomaterials when used in accordance with the intended clinical use will create any adverse effects. The Ca-aluminate based materials fulfill the requirements of the harmonized standard ISO 10993:2003.

REFERENCES
[1]L.L. Hench, Biomaterials: a forecast for the future, *Biomaterials*, 19 (1998) 1419-1423

[2]L . Kraft, Calcium Aluminate Based Cement as Dental Restorative Materials, *Ph D Thesis,* Faculty of Science and Technology, Uppsala University, Sweden. 2002

[3]J. Lööf, Calcium Aluminate as Biomaterial – Synthesis, Design and Evaluation, *Ph D Thesis*, Uppsala University, 2008

[4]L. Kraft and L. Hermansson, Hardness and dimensional stability of a bioceramic dental filling material based on calcium aluminate cement, *Am. Ceram. Soc. Advanced ceramics, Materials and structures*, Vol 23B, [4] (2002)

[5]J. Lööf, F. Svahn, T. Jarmar, H. Engqvist and C:H: Pameijer, *Dentals Materials*, Vol 24 (5), 653-659 (2008)

[6]L. Hermansson, H.Engqvist, J. Lööf, G. Gómez-Ortega and K. Björklund, Nano-size biomaterials based on Ca-aluminate Advances in Science and technology, *Key Eng. Mater.* Vol 49 21-26 (2006)

[7]H. Engqvist, J Lööf, S. Uppström, M.V. Phaneuf, J.C. Johnson, L. Hermansson, N-O Ahnfelt, Transmittance of a bioceramic dental restorative material based on calcium aluminate, *J. Biomed Mater. Res.* Part B: Applied Biomaterials, Vol 69 B, 94-98 (2004)

[8]D.D. Williams, Definitions in Biomaterials, Chaper 8, pp 66-71, Ed. Elsevier, New York, 1987

9W. Cao and L.L. Hench, Bioactive materials, Ceramics International, Vol 22 493-507 (1996)

[10]ISO 10993:2003

[11]EN 29917:1994/ISO 9917: 1991

[12]Y. Liu et al, Aspects of Biocompatibility and Chemical Stability of Calcium-Aluminate-Hydrate Based Dental Restorative Material, Paper IX in *Ph D Thesis* by L. Kraft, Uppsala University (2002)

[13]Liu, Y, Cotgreav, I, Atzori L and Grafström, R. The mechanism of Hg2+ toxicity in cultured human oral fibroblasts: the involvement of cellular thiols. *Chem Biol Interact*, 85:69-78, 1992

[14]H. Engqvist, , J-E. Schultz-Walz, J Lööf, G.A. Bottom, D. Mayer, M.W. Phaneuf, N-O. Ahnfelt, L. Hermansson, *Biomaterials* Vol 25, 2781-2787 [2004]

[15]L. Hermansson, J. Lööf and T. Jarmar, Integration mechanisms towards hard tissue of Ca-aluminate based materials, *Key Eng. Mater.* Vol 396-398, 183-186 (2009)

[16]L. Hermansson, J. Lööf and T., Injectable ceramics as biomaterials today and tomorrow, to be publ. in *Proc. ICC 2*, Verona (2008)

[17]H. Engqvist, J. Lööf, L. Kraft, L. Hermansson, Apatite formation on a biomaterial-based dental filling material, *Ceramic Transactions*, Vol 164 , Biomaterials: Materials and Applications V, (2004)

[18]A. Faris, H. Engqvist, J. Lööf, M. Ottosson, L. Hermansson, In vitro bioactivity of injectable ceramic orthopaedic cements, *Key Eng. Mater.* 309-11, 1401-1404 (2006)

[19]H. Engqvist, M. Couillard, G.A. Botton. M.P. Phaneuf, N. Axén, N-O Ahnfelt and L. Hermansson, In vivo bioactivity of a novel mineral based based orthopaedic biocement, *Trends in Biomaterials and Artificial Organs*, Vol 19, 27-32 (2005)

[20]J.Lööf, A. Faris, L. Hermansson, H. Engqvist, In vitro biomechanical testing of two injectable materials for vertebroplasty in different synthetic bone, *Key Eng. Mater.* Vol 361-363, 369-372 (2008)

[21]M. Nilsson, *Ph D Thesis*, Injectable calcium sulphates and calcium phosphates as bone substitutes, 2003, Lund University

[22]ISO 10993-12: 1992 and ISO 10993-5:1992, clause 8.4.1

[23]A.T.H. Tang, J. Li, J. Ekstrand and Y. Liu, Cytotoxicity tests of in situ polymerized resins: Methodological comparisons and introduction of a tissue culture insert as a testing device *J Biomed Mater Res*, 45, 214-222.

[24]G. Schmalz, M.Rietz, M. Federin, K-A Hiller, H Schweikl, Pulp derived cell response to an inorganic direct filling material, Abstact, presented at Cardiff Conference, July 2002

[25]Technical report, Doxa AB 2008

[26]L. Hermansson, U. Höglund, E. Olaisson, P. Thomsen, H. Engqvist, Comparative study of the bevaiour of a novel injectable bioceramic in sheep vertebrae, *Trends in Biomaer. Artif. Organs*, Vol 22, 134-139 (2008)

[27]N. Axén, N-O. Ahnfelt , T. Persson, L. Hermansson,J. Sanchez and R. Larsson, A comparative evaluation of orthopaedic cements in human whole blood, *Proc. 9th Ceramics: Cells and tissue*, Faenza, 2004

ASPECTS OF DENTAL APPLICATIONS BASED ON MATERIALS OF THE SYSTEM CaO-Al$_2$O$_3$-P$_2$O$_5$-H$_2$O

Leif Hermansson[1,2], Adam Faris[1], Gunilla Gomez-Ortega[1], John Kuoppala[1] and Jesper Lööf[1,2]

[1]Doxa AB, Axel Johanssons gata 4-6, SE-75451 Uppsala, Sweden

[2]Department of Engineering Sciences, The Ångström Laboratory, Uppsala University, Sweden

leif.hermansson@doxa.se

ABSTRACT

The chemically bonded bioceramics (CBC), also known as inorganic cements, are based on materials of the system CaO-Al$_2$O$_3$-P$_2$O$_5$-SiO$_2$ (CAPS) comprising phosphates, aluminates and silicates. This paper deals with the property profile of CBC-biomaterials mainly based on hydrates of Ca-aluminates and phosphates (CAPH), identified dental applications, some unique features and challenges. The following product areas have been identified based on experimental material data, pre-clinical studies, pilot studies and on-going clinical studies: dental cement, endodontic products (orthograde and retrograde), sealants, restoratives, underfillings, and pastes for augmentation and dental implant coatings. It is concluded that a potential use of the CAPH-materials for dental applications is based on the following features: nanostructural integration with tissue, possible apatite formation, and a mass increase yielding early point welding between the biomaterial and surrounding tissue. Consequences of nanostructural contact integration of the CAPH-system are reduced risk of secondary caries and restoration failure, and reduced post-operative sensitivity.

INTRODUCTION

The existing dental materials are mainly based on amalgam, resin composites or glass ionomers. Amalgam, originating from the Tang dynasty in China, was introduced in the early 19th century as the first commercial dental material. It is anchored in the tooth cavity by undercuts in the bottom of the cavity to provide mechanical retention of the metal. Although it has excellent mechanical characteristics it is falling out of favor in most dental markets because of health and environmental concerns. One exception is the US in which amalgam still has a large market share.

The second generation material is the resin composites, first introduced in the late 1950s. These are attached to the tooth using powerful bonding agents that glue them to the tooth structure. After technical problems over several decades, these materials today have developed to a level where they work quite well and provide excellent aesthetic results. Despite the improvements, resin composites have some drawbacks related to shrinkage, extra bonding, irritant components, a risk of post-operative sensitivity, and technique sensitivity in that they require dry field treatment in the inherently moist oral cavity. The key problem, due to shrinkage or possible degradation of the material and the bonding, is the margin between the filler material and tooth, which often fails over time leading to invasion of bacteria and secondary caries. Secondary caries is a leading cause of restorative failure and one of the biggest challenges in dentistry today. As a significant number of dental restorations today are replacement of old, failed tooth fillings, it is clear that tackling this problem is a major market need [1]. Secondary caries occurs not only after filling procedures but also following other restorative procedures such as the cementation of crowns and bridges.

Glass ionomers were first introduced in 1972 and today are an established category for certain restorations and cementations. Their main weakness is the relatively low strength and low resistance to abrasion and wear. Various developments have tried to address this, and in the early 1990s resin-modified ionomers were introduced. They have significantly higher flexural and tensile strength and lower modulus of elasticity and are therefore more fracture-resistant. However, in addition to the problems of resin composites highlighted above, wear resistance and strength properties are still inferior to those of the resin composites.

Alternative dental materials and implant materials based on bioceramics are found within all the classical ceramic families: traditional ceramics, special ceramics, glasses, glass-ceramics, coatings and chemically bonded ceramics (CBC) [2]. The CBC-group, also known as inorganic cements, is based on materials in the system CaO-Al$_2$O$_3$-P$_2$O$_5$-SiO$_2$, where phosphates, aluminates, and silicates are found. Depending on *in vivo* chemical and biological stability, the CBC biomaterials can be divided into three groups: stable, slowly resorbable and resorbable. The choice for dental and stable materials is the Ca-aluminate based materials [3]. Slowly resorbable materials are found within Ca-silicates and Ca-phosphates, and fast resorbing materials among Ca-sulphates and some Ca-phosphates. The stable biomaterials are suitable for dental applications, long-term load-bearing implants, and osteoporosis-related applications. For trauma and treatment of younger patients, the preferred biomaterial is the slowly resorbable materials, which can be replaced by new bone tissue [4].

This paper addresses some shortcomings of commercial dental materials, with focus on the resin-based materials, and how the Ca-aluminate based dental materials may meet these. The paper will summarise the property profile of the Ca-aluminate-based materials, and in relation to the setting and curing reactions discuss the relevance of using these materials with regard to different dental applications.

MATERIALS AND METHODS

The Ca-aluminate phase used in the testing was the mono-calcium aluminate (CA), synthesised by the company Doxa AB, Sweden at 1350 °C for 4 hrs. After crushing, the material was jet-milled to obtain fine-grained particles, mean particle size of approximately 4 μm. Inert phases mixed with the CA were glass particles or ZrO$_2$ depending on the applications aimed at. Li-chloride was used as an accelerator agent and polycarboxylate as a dispersing agent. For low viscosity and early hardening of the CA, a complementary glass ionomer was used. In the endodontic study, the fillings were based on a mixture of the Ca-phases CA and CaOx2Al$_2$O$_3$ (CA$_2$) in the molar ratio 1:1.

The paper describes typical features of the CAPH-material (hydrated Ca-aluminate in phosphate-containing water) with regard to technology, biocompatibility, and typical property profile including different mechanical and microstructural aspects. Young´s modulus, flexural and compressive strength, shear strength development, and fracture toughness were evaluated. The test conditions are presented in details in [5, 6].

In vitro and *in vivo* tests (rabbit and human) have been conducted. These studies are described in more detail elsewhere [7-9]. To analyse intact interfaces and calcified tissue at the highest level, transmission electron microscopy (TEM) was used in combination with focused ion beam microscopy (FIB) for site-specific high accuracy preparation of the TEM-samples [9]. Phase and elemental analyses were conducted using XRD, HRTEM, XPS and STEM with EDX.

RESULTS AND DISCUSSION

The results of the testing will be presented with regard to the general property profile and the specific application the material is aimed for. The main reason for this is that for different applications and various tissues, the chemical reactions may be a bit different. This is due to the fact that the CAPH-materials react *in situ*.

Chemical reactions – general aspects

The six different mechanisms involved during hydration and curing of Ca-aluminate-based materials have recently been presented [10]. The CA-material will be in contact with different tissue - enamel, dentine and hard bone tissue and soft tissue - as well as other biomaterials contact surfaces. These six mechanisms affect the integration differently depending on a) what type of tissue the biomaterial is in contact with, b) in what state (un-hydrated or hydrated) the CA is introduced, and c) what type of application is aimed for (cementation, dental fillings, endodontic fillings, sealants, coatings or augmentation products). Both a pure nanostructural, mechanically controlled integration, and a chemically induced integration seem plausible. Table 1 presents a summary of the six mechanisms involved in the integration of CAPH-materials towards tissue and implant surface.

Table 1. Chemical reactions of the CAPH-system in contact with different environment.

Reaction mechanism	Description	Comments
Mechanism 1. Main reaction	$3 (CaO\ Al_2O_3) + 12\ H_2O \rightarrow 3\ Ca^{2+} + 6Al^{3+} + 4(OH)^- \rightarrow 3\ Ca^{2+} + 6\ Al(OH)_4^- \rightarrow Ca_3\ [Al(OH)_4]_2(OH)_4$ (katoite) $+ 4\ Al(OH)_3$ (gibbsite)	Katoite and Gibbsite are formed as nano-size hydrates
Mechanism 2. Complementary reaction with phosphate-containing solution	$5Ca^{2+} + 3PO_4^{3-} + OH^- \rightarrow Ca_5(PO_4)_3OH$	Additional phase formed: Apatite
Mechanism 3. Contact zone reaction with body liquid in presence of the basic CA-phase	$HPO_4^{2-} + OH^- \rightarrow PO_4^{3-} + H_2O$ Thereafter the apatite-formation reaction occurs as mechanism 2, $5Ca^{2+} + 3PO_4^{3-} + OH^- \rightarrow Ca_5(PO_4)_3OH.$	Apatite formation in the contact zone in presence of body liquid
Mechanism 4. Transformation reaction of the originally formed phase Katoite.	$Ca_3 \cdot (Al(OH)_4)_2 \cdot (OH)_4 \rightarrow 2\ Ca^{2+} + HPO_4^{2-} + 2\ H_2PO_4^- \rightarrow Ca_5 \cdot (PO_4)_3 \cdot (OH) + 2\ Al(OH)_3 + 5\ H_2O$	Apatite and Gibbsite formed in the biomaterial towards tissue
Mechanism 5. Biologically induced integration and ingrowth	Bone ingrowth towards the apatite allows the new bone structure to come into integrated contact with the biomaterial.	New bone formation at the contact zone
Mechanism 6. Mass increase reaction due to presence of unhydrated CA	$3\ CaO\ Al_2O_3 + 12H_2O \rightarrow 3\ CaO\ Al_2O_3\ 6H_2O + 2Al_2O_3\ 3H_2O.$	Mass increase and point welding

The actual contact zone developed depends on a combination of the above discussed mechanisms and the tissue. The latter varies from a cellular-free high content apatite tissue in the case of a dental enamel, via dentine to a bone structure with cellular and body liquid contact. Also the material can be in contact with other implant materials as dental crowns, dental screws or coatings on implants. In Table 2, the different reactions with regard to type of tissue and applications are summarized.

Table 2. Reactions occurring depending on dental application and tissue/contact surface

Application	Enamel contact	Dentine contact	Bone contact	Biomaterial contact
Cementation	Mechanisms 1-2	Mechanisms 1-3	-	Mechanism 1
Dental fillings	Mechanisms 1-2	Mechanisms 1-3	-	-
Endo filling: Orthograde	Mechanisms 1-2	Mechanisms 1-2	-	Mechanism 1
Endo filling: Retrograde	Mechanisms 1-2	Mechanisms 1-2	Mechanisms 1-5	-
Augmentation	-	-	Mechanisms 1-5	-
Coatings on implants	-	-	Mechanisms 1-6	Mechanism 1

Property profile aspects

In addition to the main binding phases (the Ca-aluminate and water), filler particles are included to contribute to some general properties of interest when used for different applications. The general contribution of added particles regards the microstructure (homogeneity aspects) and mechanical properties (especially hardness, Young's modulus and strength). For dental applications additives are mainly glass particles. For many dental applications translucency is desired, why inert particles must have a refractive index close to that of the hydrates formed [11]. A preferred high-density oxide for increased radiopacity, for endodontic applications, is zirconia, the latter material also used as a general implant material [12].

The Ca-aluminate cements exhibit an inherent property of importance for high-strength cement materials, namely a huge water uptake capacity of Ca-aluminate cement. The water consumption during hydration and curing (Mechanism 6) is as high as 45 w/o water, as compared to approximately 5 w/o for Ca-phosphates. Practically, this results in high-strength, low-porosity materials, if an appropriate w/c ratio is selected, i.e. a w/c ratio close to that of complete reaction of the Ca-aluminate phase. The total residual porosity is also depending on the amount of solid filler particles added but is often as low as 5-10 %. Figure 1 illustrates the typical nano-size microstructure of the hydrated material with nano-size porosity between precipitated nano-size hydrates. More aspects are presented elsewhere [13, 14].

Fig. 1 Nanostructure of Ca-aluminate hydrates. The pore channels are estimated to be 1-2 nanometers and the hydrates in the interval 10-40 nm. (White bar 10 nanometers)

Due to reduced porosity, the Ca-aluminate material exhibits the highest strength among the chemically bonded ceramics. The inherent flexural strength is above 100 MPa based on measurement of the fracture toughness, which is about 0.7 - 0.8 MPam$^{1/2}$, and a largest distance in the microstructure of approximately 10-15 micrometers. The actual flexural strength is controlled by external defects introduced during handling and injection of the material. Table 3 summarises results from some studies [15].

Table 3. Mean property data of experimental dental Ca-aluminate based materials

Property	Mean value
Hardness (H$_v$ 100 g)	110-130
Young's modulus (GPa)	12-18
Compressive strength (MPa) after 28 days	180 -260
Flexural strength (MPa) after 7 days	50-80
Weibull modulus, m	7-10
Dimensional stability, in linear-%	< 0.3
Thermal conductivity, W/mK	0.7
Thermal expansion, α	9.5 x 10^{-6}

The thermal and electrical properties of Ca-aluminate based materials are close to those of hard tissue, the reason being that Ca-aluminate hydrates chemically belong to the same group as Ca-phosphates, the biomineral phases of bone. Another important property related to Ca-aluminate materials is the possibility to control the dimensional change during hardening. In contrast to the shrinkage behaviour of resins and PMMA-based biomaterials, the Ca-aluminates exhibit a small expansion, 0.1-0.3 linear-% [16]. Biocompatibility and inherent bioactivity are presented in a separate paper [17].

Dental applications

The nature of the mechanisms utilized by CAPH-materials (especially Mechanism 1 above) when integrating and adhering to tooth tissue and other materials makes the CAPH-materials compatible with a range of other dental materials, including resin composite, metal, porcelain, zirconia, glass ionomers and gutta-percha. This expands the range of indications for CAPH-products from not only those involving tooth tissue, e.g. cavity restorations, but also to a range of other indications that involve both tooth tissue and other dental materials. Examples include dental cementation, base and liner and core build-up and endodontic sealer /filler materials, which involve contact with materials such as porcelain, oxides and polymers and metals, and coatings on dental implants such as titanium or zirconia-based materials.

Dental cement.

Long-term success after cementation of indirect restorations depends on retention as well as maintenance of the integrity of the marginal seal. Sealing properties of great importance deal with microleakage resistance, the retention developed between the dental cement and the environment, compressive strength and acid resistance. Data presented below support the CAPH-system to be a relevant dental cement material.

Integration with tooth tissue is a powerful feature of CAPH and the foundation of the CAPH-technology platform. Secondary caries occurs not only after filling procedures but also after other restorative procedures such as the cementation of crowns and bridges. The consequence of the difference in the mechanism of action between CAPH-products and conventional products is illustrated by the study presented in details in [18], illustrated in Figure 2 below. It shows that the micro leakage, measured by dye penetration after thermo cycling, of a leading dental cement (Ketac Cem®, 3M) was significantly higher, both before and after thermo cycling compared to XeraCem™, a CAPH-product recently approved by FDA. This has also recently been verified using techniques for studying actual bacterial leakage. Those results will be presented at the 2009 IADR meeting. The above described nanostructural precipitation upon tissue walls, biomaterials and within the original Ca-aluminate paste is the main reason for this, in addition to a high acid corrosion resistance [19].

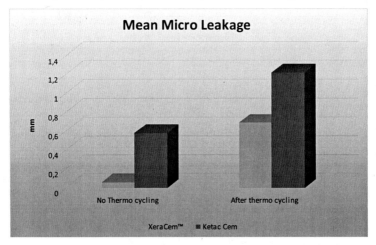

Fig. 2 Micro leakage leakage of XeraCem and Ketac Cem

General properties of the CAPH-system used as dental cement were presented at the IADR-meeting 2008 [20, 21]. A summary of some of the data is presented in Table 4.

Table 4. Selected properties, Test methods according to SO 9917-1

Material	Net setting time, min	Film thickness, μm	Compressive strength, MPa	Crown retention, Kg/Force
Xeracem	4.8	15	196 (at 30days)	38.6
Ketac Cem	-	19	-	26.6
RelyX-Unicem	-	-	157	39.4

A feasibility study of the CAPH-material was performed in 2007. After one year all restorations were intact. An on-going clinical 1-year study comprising 35 cemented crowns at Kornberg School of Dentistry, Temple University started in the beginning of 2008. Six month follow up data and feedback from participating dentists are excellent. To be presented in detail at IADR-2009 [22].

Endodontics

In a review of the biocompatibility of dental materials used in contemporary endodontic therapy [23], amalgam was compared with gutta-percha, zinc oxide-eugenol (ZOE), polymers, glass ionomer cements (GICs), composite resins and mineral trioxide aggregate (MTA). A review [24] of clinical trials of *in vivo* retrograde obturation materials summarized the findings. GIC's appeared to have the same clinical success as amalgam, and orthograde filling with gutta-percha and sealer was more effective than amalgam retrograde filling. Retrograde fillings with composite and Gluma, EBA cement or gold leaf were more effective than amalgam retrograde fillings. However, none of the clinical trials reviewed in [26] included MTA. In a 12 week microleakage study, the MTA performance was questioned compared to that of both amalgam and a composite [25].

The CAPH-based material discussed in this paper belongs to the same material group as MTA, the chemically bonded ceramics [26]. MTA is a calcium silicate (CS) based cement having bismuth oxide as filler material for improved radio-opacity, whereas the Ca-aluminate material consists of Ca-aluminate phases CA and CA$_2$ with zirconia as filler material. MTA is claimed to prevent microleakage, to be biocompatible, to regenerate original tissues when placed in contact with the dental pulp or periradicular tissues, and to be antibacterial. The product profile of MTA describes the material as a water-based product, which makes moisture contamination a non-issue [27]. The CA-cement materials are more acid resistant than the CS-based materials, and in general show higher mechanical strength than the CS materials. A two-year retrospective clinical study of Ca-aluminate based material has been conducted [28]. The study involved patients with diagnosis of either chronic per apical osteitis, chronic per apical destruction, or trauma. Surgery microscope was used in all cases. For orthograde therapy the material was mixed with solvent into appropriate consistency and put into a syringe, injected and condensed with coarse gutta-percha points. Machine burs were employed for root canal resection. For the retrograde root fillings, the conventional surgery procedure was performed [29]. The apex was detected with surgery microscope and rinsed and prepared with an ultrasonic device. Crushed water-filled CA-tablets were then inserted and condensed with dental instruments. The patients' teeth were examined with X-ray, and three questions regarding subjective symptoms were put to patients: 1. Have you had any persistent symptoms? 2. Do you know which tooth was treated? 3. Can you feel any symptoms at the tooth apex?

In 13 of the 17 treated patients the diagnosis was chronic perapical osteitis (c p o). These were treated with retrograde root filling (rf) therapy. Three patients suffered from trauma or chronic perapical destruction, and these patients were treated with orthograde therapy. Out of 17 patients (22 teeth) treated, 16 patients (21 teeth) were examined with follow-up x-ray after treatment and also after two years or more. The additional patient was asked about symptoms. The results of both the clinical examination and the subjective symptoms were graded into different groups related to the success of the therapy. The results are shown in Table 5.

Table 5. Summary of the results (Score 1 and 2 considered successful, score 3 and 4 failure)

	1 Complete healing	2 Incomplete healing	3 Uncertain	4 Failure
Nos. of teeth	18	3	-	1
Percentage	82	14	-	4

Figures 3-4 show examples of the X-ray examination of orthograde and retrograde treatments.

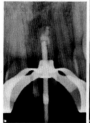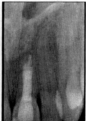

Fig. 3 Tooth 21 (patient14) a) condensing with a Gutta-percha pointer, b) just after treatment and c) at two year control

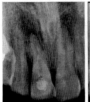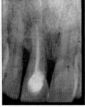

Fig. 4 Tooth 21 (patient 9) a) at treatment and b) at two year control

In summary 21 out of 22 treated teeth have acceptable results being either symptom free or judged healed after clinical examination. The single failure can probably not be attributed to the material, but rather to the difficulty of treating and sealing a multi-channelled tooth. The use of CA's as root canal sealers is indirectly supported in "Introduction to Dental Materials" by van Noor [30], where the following materials characteristics are looked for; biocompatible, dimensionally stable, antibacterial and bioactive. The results in this study can be interpreted as a success in meeting these materials requirements.

Already in the 1970s, Calcium aluminate (CA) was suggested as a biomaterial and tested *in vivo*. Hentrich et al [31] compared CA with alumina and zirconia in an evaluation of how the different ceramics influenced the rate of new bone formation in femurs of rhesus monkeys. Hamner et al [32] presented a study in which 22 CA roots were implanted into fresh natural tooth extraction sites in 10 baboons for periods ranging from 2 weeks to 10 months. In both studies CA successfully met the criteria for tissue adherence and host acceptance.

Dental filling materials

An important feature of the hydration mechanisms of the Ca-aluminate based materials is the nanostructural integration with and the high shear strength developed towards dental tissue. This makes both undercut (retention) technique and bonding techniques redundant. The CAPH approach to dental filling technique is new. With CAPH-technique, the chemical reactions cause integration when the bioceramic material is placed in the oral cavity at body temperature and in a moist treatment field. Figure 5 shows a TEM (transmission electron microscopy) illustration of the interface between the CA-based material and dentine. This establishes a durable seal between bioceramic and tooth. Whereas amalgam attaches to the tooth by mechanical retention and resin-based materials attach by adhesion, using bonding agents, etchants, light-curing or other complementary techniques, the CA-materials integrate with the tooth without any of these, delivering a quicker, simpler and more robust solution.

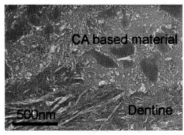

Fig. 5 Nanostructural integration of CAPH-material with dentine (gray particles in the biomaterial are glass particles)

The general aspects of Ca-aluminate based materials have been presented in two Ph D Thesis-publications. Important aspects of Ca-aluminate materials as dental filling materials are dealt with, such as dimensional stability, acid corrosion and wear resistance, and biocompatibility and mechanical properties [3, 33].

Coatings on dental implant and augmentation

For successful implantation of implants in bone tissue, early stabilisation is of great importance [34]. Even small gaps may lead to relative micro-motions between implant and the tissue, which increases the risk of implant loosening over time due to formation of zones of fibrous tissues at the implant-tissue interface. Early loading of implants is of particular interest for dental implants [35]. The use of surface coatings technology is today an established method to reduce the problem with poor interfacial stability for implants. With coatings technology, structural characteristics of the implant (e.g. strength, ductility, low weight or machinability) may be combined with surface properties promoting tissue integration [36]. There are several established coating deposition techniques, e.g. physical vapour deposition (sputtering) and thermal spraying techniques [37, 38]. Coatings based on calcium phosphates are the most used ones.

This paper deals with coatings deposited with established methods, with the aim of improving particularly the early stage anchoring of metal implants to bone tissue by exploring *in vivo* hydration of coatings or pastes based on chemically curing ceramics. The study focuses on calcium aluminate in the form of coatings and paste. Results are presented from an implantation study with flame-sprayed coating on titanium implants and uncoated implants augmented with a calcium aluminate paste in the hind legs of rabbits. Implants were applied with the paste composed of a mixture of CaO·Al$_2$O$_3$ and CaO·2Al$_2$O$_3$. The paste was applied manually as a thin layer on the threaded part of the implant just before implantation. The uncoated and coated implants were sterilised with hot dry air at 180 °C for 2 hrs. Female albino adult New Zealand White rabbits with a body weight around 2.5 kg were used. Each animal received four implants, two in each hind leg. Implants were placed in the distal femoral metaphysis as well as in the proximal tibial metaphysis. Surgery followed standard procedure. The implants were screwed into predrilled and threaded cavities. Necropsy took place after 24 hrs, 2 and 6 weeks [8].

No negative effects of the implants on the general welfare of the animals were observed. The healing progressed in a normal and favourable way. As for the removal torque recordings, all calcium aluminate coatings types provided an improved implant anchoring to bone tissue after *in vivo* hydration, as compared to that of the pure metal implants. Implants on the tibia and femur side of the knee gave similar removal torques. Table 6 provides average values from both tibia and femur sides.

Table 6. Removal torque (Ncm) for dental implants in rabbit hind legs (tibia and femur).

Implant type	24 hrs	(n)	2 weeks	(n)	6 weeks	(n)
Flame spraying	7.0	(8)	7.0	(8)	25	(6)
Paste augmentation	6.6	(8)	15	(6)	13	(4)
Rf-PVD	12	(4)	-	-	10	(4)
Uncoated reference	3.8	(8)	5.7	(6)	14	(4)

24 hrs after implantation, calcium aluminate in-between the implant and tissue increased the removal torque to about double that of the uncoated reference implants, independently of means of application (coatings or paste). This is considered to be attributable to the point-

welding according to integration mechanism 6 above. Two weeks after implantation, implants combined with paste augmentation provide the highest removal torque; flame sprayed coatings also improve the torque relative to the uncoated system. Six weeks after implantation, all systems are relatively similar (considering the uncertainty due to scatter and statistics), apart from the sprayed system which shows significantly higher values.

CONCLUSIONS

Nano-structural integration and apatite formation provide important benefits to both the dentist and patient, notably minimal micro-leakage, perfect seal at the interface between tooth and material and as a result longer-lasting treatment results. These powerful features and benefits are summarized in Table 7, and make the CAPH-materials suitable for a large number of dental applications such as dental cements, dental fillings, endodontic fillers, fissure sealing, and implant coatings, including implants with a CAPH-paste.

Table 7. Features and Benefits of CAPH Technology Platform

• Nano-structural integration & apatite formation • No shrinkage • Integration/stability/strength • No bonding /dry field required • Variable consistency and compatibility to other materials	• Reduced risk of secondary caries • No or limited post-op sensitivity • Longevity/durability • Easy and fast • Broad spectrum of usage with products targeting indication needs

REFERENCES

[1] I. Mjör et al, *International Dental Journal*, Vol 50 [6], 50 (2000)

[2] A. Ravagliolo and A. Krajewski, *Bioceramics*, Chapman and Hall, 1992

[3] L . Kraft, Calcium Aluminate Based Cement as Dental Restorative Materials, *Ph D Thesis,* Faculty of Science and Technology, Uppsala University, Sweden. 2002

[4] M. Nilsson, *Ph D Thesis*, Injectable calcium sulphates and calcium phosphates as bone substitutes, 2003, Lund University

[5] J. Lööf, H. Engqvist, G. Gomez-Ortega, H. Spengler, N-O. Ahnfelt, L.Hermansson, Mechanical property aspects of a biomaterial based dental restorative system, *Key Eng. Mater.* Vols 284-286, 145-148 (2005)

[6] L-Kraft and L. Hermansson, A method for the examination of geometrical changes in cement pastes, *International RILEM* 17, 401-413 (2000)

[7] H. Engqvist, J. Lööf, L. Kraft and L. Hermansson, Apatite formation on a biomaterial-based dental filling material, *Ceramic Transactions* Vol 164, Bioceramics: Materials and Application V (2004)

[8] N. Axén, H. Engqvist, J. Lööf, P. Thomsen and L. Hermansson, *Key Eng. Mater.* Vols. 284-286 831-834 (2005)

[9]H. Engqvist, , J-E. Schultz-Walz, J Lööf, G.A. Bottom, D. Mayer, M.W. Phaneuf, N-O. Ahnfelt, L. Hermansson, Chemical and biological integration of a mouldable bioactive ceramic material capable of forming apatite in vivo inteeth, *Biomaterials* Vol 25, 2781-2787 [2004]

[10]L. Hermansson, J. Lööf and T. Jarmar, Integration mechanisms towards hard tissue of Ca-aluminate based biomaterials, *Key Eng. Mater.* Vols. 396-398 , 183-186 (2009)

[11]H. Engqvist, J. Lööf, S. Uppström, M.W. Phaneuf, J.C. Johnson, L. Hermansson and N-O. Ahnfelt, Transmittance of a bioceramic dental restorative material based on calcium aluminate, *J. Biomed. Mater.* Res. Part B: Applied Biomaterials Vol 69 B, 94-98 (2004)

[12]E. Adolfsson and L. Hermansson, Zirconia-Fluoroapatite materials produced by HIP, *Biomaterials* Vol 20, 1263-1267 (1999)

[13]L. Hermansson, H. Engqvist, J. Lööf, G Gómez-Ortega and K. Björklund, Nano-size biomaterials based on Ca-aluminate, *Advances in Science and Technology*, Vol 49, 21-26 (2006)

[14]F. Bultmark, J. Lööf, L. Hermansson and H. Engqvist, Continuous hydration of low w/c calicium-aluminate cement, *Submitted to J. Mater. Sci* (2008)

[15]L. Hermansson, H. Engqvist, J. Lööf, G. Gomez-Ortega and K. Björklund, Nano-size biomaterials based on Ca-aluminate, *Key Eng. Mater.* Vols. 309-311, 829-832 (2006)

[16]L. Kraft and L. Hermansson, Hardness and dimensional stability of a bioceramic dental filling material based on calcium aluminate cement, *Am. Ceram. Soc. Advanced ceramics, Materials and structures*, Vol 23B, [4] (2002)

[17]L. Hermansson, G. Gomez-Ortega, J. Lööf, Daytoona Beach Conference on Advanced Ceramics, 2009, to be published

[18]C.H. Pameijer, S.R. Jeffries, J. Lööf and L. Hermansson, Microleakage Evaluation of XeraCem in Cemented Crowns, Poster # 3098, *IADR*, 2008

[19]Y. Liu et al, Aspects of Biocompatibility and Chemical Stability of Calcium-Aluminate-Hydrate Based Dental Restorative Material, Paper IX in *Ph D Thesis* by L. Kraft, Uppsala University (2002)

[20]C.H. Pameijer, S.R. Jeffries, J. Lööf and L. Hermansson, Physical properties of XeraCem, Poster # 3099, IADR, 2008

[21]C.H. Pameijer, S.R. Jeffries, J. Lööf and L. Hermansson, A comparative crown retention test using XeraCem, Poster # 3100, IADR, 2008

[22]C. H. Pameijer et al, Clinical study on cemented dental crowns, accepted for IADR-conference, Miami Beach, 2009

[23]Haumann, C.H.J. and R.M. Love, Biocompatibility of dental materials used in contemporary endodontic therapy: a review. Part 2. Root–canal–filling materials. *Int Endo J*, 2003. 36: p. 147-160.

[24]Niederman, R. and J.N. Theodosopoulou, Review: A systematic review of in-vivo retrograde obturation materials. *Int Endo J*, 2003. 36: p. 577-585.

[25]Alamo, H.L., et al., A Comparison of MTA, Super-EBA, composite and amalgam as root-end filling materials using a bacterial microleakage model. *International Endodontic Journal*, 1999. 32: p. 197-203.

[26]Hermanssson, L. and E. Kraft L., H., Chemically Bonded Ceramics as Biomaterials, *Key Engineering materials*, 2003. **247**: p. 437-442

[27]Dentsply, *ProRoot MTA* White Brochure B18b. www.dentsply.co.uk/literature/.pdf, 2003

[28]L. Polhagen, L. Kraft, M. Saksi and L. Hermansson, A two-year retrospective investigation of a calcium-aluminate based material in root canal sealing, in preparation

[29]Rashedi, A., *Treatment of periapical pathology with Retrograde Endodontic Technique.* PM, Institute of Odontology, Karolinska Institutet, Huddinge, Sweden, http://www.ki.se/odont/cariologi_endodonti/T10/Atousa%20Rashedi.pdf, 2001

30Noort, R.v., *Introduction to Dental Materials.* Mosby, 1994

[31]R.L. Heintricht et al, An evaluation of inert and resorbable ceramics for future clinical applications, *J. Biom. Res.* Vol 5 [1], 25-51 (1971)

[32]J.E. Hammer, M. Reed, R.C. Gruelich, Ceramic root implantation in baboons, *J Biom. Res* Vol 6 [4] 1-13 (1972)

[33]J. Lööf, Calcium Aluminate as Biomaterial – Synthesis, Design and Evaluation, *Ph D Thesis*, Uppsala University, 2008

[34]J.E. Ellingsen, S.P. Lyngstadaas (Eds.), *Bio-implant interface, improving biomaterials and tissue reactions*, (2003) CRC Press LLC.

[35]S.Vercaigne et al, Bone healing capacity of titanium plasma-sprayed and hydroxylapatite coated oral implants, *Clin. Oral Implants Res*, 9, 261 (1998).

[36]X. Liu, C. Ding, Z. Wang, Apatite formed on the surface of plasma sprayed Wollastonite coating immersed in simulated body fluid, *Biomaterials*, 22 (2001) 2007-2012.

[37]D.S. Rickerby, A. Matthews (Eds.), *Advanced Surface Coatings*, (1991) USA:Chapman & Hall, New York.

[38]J.M. Schneider, S. Rohde, W. D. Sproul and A. Matthews, Recent developments in plasma assisted physical vapour deposition, *J. Phys. D: Appl. Phys.* 33 (2000) R173-R186.

SYNTHESIS AND CHARACTERIZATION OF BIOACTIVE-GLASS CERAMICS

Saikat Maitra[1*], Ariful Rahaman[2], Ram Pyare[2], Hilmi B. Mukhtar[1] and Binay K. Dutta[1]
1. Dept. of Chemical Engineering, Universiti Teknologi Petronas, 31750, Tronoh, Malaysia
2. Dept. of Ceramic Engineering, Banaras Hindu University, Varanasi, India

ABSTRACT
Synthesis and characterization of bioactive glass ceramics containing lithium oxide and sodium oxide based alkali-lime-silica glasses have been presented in this work. Both powder processing route and bulk glass crystallization route with controlled heat treatment schedule were employed in the synthesis process. P_2O_5 in different proportions was used as the nucleating agent in these compositions. The glass ceramics formed were characterized by measurement of specific gravity, hardness, DTA, XRD, FT-IR spectra, SEM and EDAX study. In-vitro tests with simulated body fluid composition and protein absorption tests with albumin were conducted with the synthesized materials. It was observed that the composition of the glass ceramics affect significantly the mechanical behaviour and bioactivity of the materials.

1. INTRODUCTION

Development, evaluation and application of biomaterials which come into contact with human tissues to meet the specifications for research and practice in biology and medicine have attracted great interest in recent times. As a class of materials bio glass-ceramics have opened a new vista in the field of medicinal and material science. The two important properties of bio glass-ceramics which make them as attractive as a material for research are biocompatibility, which refers to the introduction of the material in the tissues of the human body without rejection or a toxic effect and bioactivity which relates the property of properly growing of the tissues on the biomaterials inside the human body. Bio-active glass chemically attaches to both hard and soft tissues by development of a biologically active apatite layer[1]. Combination of bio-active glass in porous composites encourages bonding to bone and may affect calcification in the artificial soft tissues[2]. In recent time many interesting works were done in the area of bioactive glass and glass ceramics. The more important works are reviewed here. Kobuto[3] made a detailed study on the surface chemistry of bioactive glass ceramics. Medvedev[4] worked on the ceramic and glass ceramic materials for bone implants. Takadama et al[5] studied the mechanism of biomineralization of apatite on a sodium silicate glass by in-vitro TEM –EDX study. Sinha et al[6] reported the development of calcium phosphate based bio-ceramics. Ross et al[7] investigated on the tissue adhesion to bioactive glass coated silicone tubing in a rat model of peritoneal dialysis of catheter and catheter tunnels. Avent et al [8] studied the dissolution of silver-sodium-calcium phosphate glass for the control of urinary tract infections. Fujbayashi et al[9] made a comparative study of in-vivo bone in growth and in-vivo apatite formation on Na_2O-CaO-SiO_2 glasses. Venne et al[10] worked on the surface characterization of silver-based bioactive glass. Zhou et al[11] worked on the preparation and bioactivity of sol-gel macro-porous bioactive glass.

Although, many works have been done on the synthesis and characterization of bioactive glass and glass ceramics, not much information is available regarding the role of compositional variations and processing conditions on the properties of the biomaterials. In the present investigation therefore, bio-glass ceramics were prepared by thermally treating glasses based on the alkali-lime-alumina–silica system with different alkali cations. Two different processing routes, ca. bulk glass forming route and powder pelletization route was followed for the development of the glass ceramics. P_2O_5 was used as a nucleating agent in different proportions in these compositions. The glass ceramics formed were thoroughly characterized by specific gravity, hardness, DTA, XRD and FTIR studies. Bio-activity of the samples was assessed by in-vitro test in simulated body fluid and also by protein absorption test.

From these studies the relationship between the compositional variation and the properties of the bi-ceramic materials was assessed

2. EXPERIMENTAL

For the present investigation, glass compositions from the SiO_2-Al_2O_3-P_2O_5-CaO-R_2O (R^+= Na^+, Li^+) system have been selected. The glasses were prepared from analar grade $CaCO_3$, Na_2CO_3, $NH_4H_2PO_4$, Al_2O_3 and acid washed SiO_2 powder. The composition of the different batches is detailed in Table-1. The batches after thorough mixing were melted in a high alumina crucible at 1400°C for 5 hrs in an electrically heated muffle furnace. The melts were cast in a steel mould and annealed in an electric muffle furnace at 500°C for 2 hrs to remove the strains and then allowed to cool down slowly to room temperature. In the present investigation the following routes were employed to obtain the glass ceramics. (1) Glass ceramics by powder processing route (GCP) and (2) glass ceramics by bulk glass crystallization route (GCB). The heat treatment schedule for the glass ceramic samples is given in table-2. The rate of heating was 4°C/min up to the nucleation temperature and 2°C/minute from the nucleation to growth temperature.

In the first route, the glass samples were crushed in an alumina-lined mortar until the powder passed through 75-micron mesh. The green compacts that were mixed with PVA and pressed uni-axially at 75 KN in a 10 mm diameter steel die. In the second route, the samples were taken in the form of bulk glass.

The differential thermal analysis (DTA) was done by Netzsch (STA-409) thermal analyzer. The tests were performed under nitrogen atmosphere at a heating rate of 10°C/min. XRD analysis of the samples were carried out with Seifert, 2000D, X-ray diffractometer. SEM study of fracture surface polished samples was carried out with FEI-Quanta-Phillips electron microscope. For in-vitro test the simulated body fluid (SBF) was prepared by dissolving analar grade NaCl, KCl, K_2HPO_4,$3H_2O$, $MgCl_2$, $6H_2O$, $CaCl_2$ and Na_2SO_4 in distilled water (ion concentration is given in table-3) followed by buffering at pH 7.36 with tris-hydroxy-methyl amino methane and hydrochloric acid at 36.5°C[12]. 1gm of each sample was immersed in 25ml of SBF for 48 hours. FT-IR spectra of the samples were taken with a Hitachi spectrophotometer (270-30) before and after the treatment with SBF. Protein absorption study of the samples was carried out with albumin solution with a concentration of 5gm/litre[13]. I gm of the samples was treated with albumin solution for 6 hours in vibratory shaker followed by centrifugation and collection of the supernatant liquid which was analyzed with Beckman DU 640B spectrophotometer at 280 nm wavelength.

3. RESULTS AND DISCUSSIONS

Apatite is a major crystalline phase of bioactive glass ceramics that crystallizes evenly throughout the material. Depending on the compositions, the apatite may be hydroxyapatite or fluoroapatite which has the same crystal structure and similar lattice parameters. The constituent of most bioactive glass ceramics are almost the same as those for bioactive glasses. But bioactive glass ceramics contain higher proportion of P_2O_5 and less alkali oxides compared to the bioactive glasses. P_2O_5 acts as a nucleating agent in these materials and it is also an important constituent of the crystalline phase apatite.

Na_2O and Li_2O are effective fluxes making melting, homogenizing, casting and flame spraying of glass easier. Again these ions get partially dissolved in body fluid and maintain a physiological balance of Na^+ and Li^+ modifying the local pH.

3.1: Thermal Analysis

The DTA patterns for all the glasses (Figure-1 and 2) revealed an isotherm in the temperature range of 730°C to 800°C that was associated with glass crystallization. The glass containing Li^+ gave a sharper exothermic peak. It can be related to the higher ionic potential of Li^+ compared to Na^+ by virtue of which it exerts greater ordering effect on surrounding oxygen ions resulting in more crystallization. The DTA of air quenched glass samples are shown in figure 3. Glass-1 gave an exothermic peak at 729°C. Glass-2 and Glass-3 gave an exothermic peak at 730.8°C and 765.8°C respectively. The phase transformation temperature of the glass increased when a part of the alkali oxide was substituted with P_2O_5.

3.2: Specific gravity of the samples:
Specific gravity or true density of the glass ceramics depend on the nature and the amount of different phases formed during the heat treatment of the bulk and the pellet glass. From the specific gravity values given in table-3 it is obvious that lithium glasses have higher specific gravity than sodium glasses. With the increase in the P_2O_5 content the specific gravity value increased in both types of glass samples (Figure-3). Pellet ceramized glasses exhibited higher specific gravity than the bulk ceramized glasses. In compacted masses the densification kinetics increased appreciably as a result of the increased chemical potential gradients of the atoms between the particles

3.3: XRD analysis
The XRD patterns for the glasses (Figure-4) exhibit no peaks at all, indicating an amorphous structure Figure-4a shows the XRD patterns of the heat-treated Na_2O containing glass ceramic powders. It was observed that the XRD results of the GCP and GCB route derived glass ceramics were almost the same. The main phases identified in the XRD diagram were different forms of sodium calcium silicates ($Na_2Ca_3SiO_9$, $Na_4CaSi_3O_9$), sodium calcium aluminium phosphate ($Na_2Ca_3Al_2(PO_4)_2(SiO_4)_2$) present in the glass matrix. In Na_2O containing glass ceramics, prominent peaks were appeared at 2θ values of 22.8°, 26°, 33.5°, 48.2°, 58.4°, and 59.9°, which correspond to (204), (311), (440), (800), (406), and (334) reflections, as listed in the Joint Committee on Powder Diffraction Standards (JCPDS). High intensity lines of two different phases were coinciding with sodium calcium silicates and sodium calcium phosphates phases. Although sodium calcium phosphates were present, but calcium phosphates or sodium phosphates phases were not observed. In the diffraction patterns of Li_2O containing glass ceramics the major phases identified were Li_2CaSiO_4 and $Li_2Ca_4Si_4O_{13}$. A sharp peak at $2\theta = 27.1°$ (300) and several small peaks were observed in the diffractogram corresponding to (110) and (221) planes at 2θ values 24.8° and 34.3° corresponding to these phases.

3.5: SEM analysis
Figure-5 shows SEM images of SP-1, 2 and SB-1, 2 specimens. The main phases were sodium calcium silicates and sodium calcium phosphates, which was indicated in figure 4b. Other phases were sodium calcium aluminium silicates, wollastonite etc. These all phases are bio-active[1]. It has been shown that the magnitude of crystalline phase formation were almost equal in GCP and GCB process. The average crystal sizes in the glass ceramics were calculated using Scherrer's formula and it was observed that in glass ceramic SP-2 with 3 mole% of P_2O_5 crystal of the average size of 0.81 μm was formed, whereas in the glass ceramic SP-1 with 1.5 mole% of P_2O_5 crystals of the average size of 0.98 μm was formed. P_2O_5 addition to the glass samples promoted to a reduction in the average crystallite size and the also the phase separation in the glass system.. Small spherical liquid droplet phase was found on the surface heat-treated glasses because the interface energy between glass phases is lower than that between glass and crystals. Some cracks were also observed in the samples. This arises because different crystals were formed and the volume change, which accompanies crystallization, can result in the generation of

stresses in both the glass phase and crystal phases. P_2O_5 creates phase separation in the glasses and cause reduction in the rate of crystal growth.

3.4: EDAX analysis
From the EDAX (Energy Dispersive X-ray Analysis) of the samples the composition of glass ceramics (Table-6) was verified. The glass ceramics contained Si, O, Al, Ca, Na and P and the EDAX results were matched with batch composition of glass. The results indicated compositional homogeneity in the samples and no change in the composition before and after processing of the samples.

3.5: Hardness of the Samples:
With the increase in the P_2O_5 content the hardness of the glass ceramics increased (table-7). The hardness of the lithium containing glass was found to be more. It can be related to the higher level of phase separation in Li_2O-P_2O_5 system in presence of P_2O_5 and consequent development of finer grain sizes in the microstructure. GCP route derived glass ceramics gave higher hardness due to improved densifiaction.

3.6: FT-IR spectra of the samples before and after the in-vitro test:
The formation of hydroxy carbonate apatite (HCA) layer on the surface of bioactive glasses and glass ceramics involves an ion exchange reaction, which increases the pH and causes cytotoxicity in the surrounding environment. In SBF alkali ions are expected to be removed from the glass by a diffusion controlled process. Figure -6 (a) and (b) represents the FT-IR spectral behaviour of the samples before and after the in-vitro test. O-H stretching vibration was present in the samples at about $3450Cm^{-1}$. Peaks near wave numbers 620, 600, 575, 560 and 530 cm-1 belong to v_4 PO_4 domain and these peaks can be assigned to a calcium phosphate or hydroxyapatite phase [14-16].No splitting of the rounded P-O bending vibration near 575cm-1 was observed and therefore it may be assumed that the hydroxyapatite layer formed were not entirely composed of crystalline hydroxyapatite[17,18]. Si-O stretching vibration was observed at 800-$1000Cm^{-1}$ and at 650-$800Cm^{-1}$. Si-O-Si bending vibration was observed at 430-$460Cm^{-1}$. After soaking in SBF, absorption peaks were shifted to higher side. As the percentage of Na_2O was increased, the intensity at $1000Cm^{-1}$ was decreased. With the increasing proportion of Na_2O, the same band became progressively broader. This may be related to a release of Na^+ and Ca^{2+} into solution[19,20]. During the dissolution of the glass, alkali silicate phase is preferentially attacked by SBF. Glass containing higher P_2O_5 is more soluble due to the bonding of P_2O_5 with CaO. Li-containing glasses are less soluble due to the lower diffusion coefficient of Li compared to sodium.

3.7: Protein adsorption on glasses and glass-ceramics
Albumin was chosen for the protein absorption test since it is the most abundant protein in blood plasma. The protein adsorption on bioglasses and glass ceramics were depicted in figure- 7. It has been found that glass and glass ceramic having higher molar composition of P_2O_5 adsorbed minimum protein and vice versa. This result agrees very well with that from the previous study on glasses. SP-3 adsorbed the minimum protein compared to SP-2 and SP-1. Similarly, SB-3 adsorbed minimum protein compared to SB-2 and SB-1.

4. SUMMARY AND CONCLUSIONS:
Bioactive glass ceramics of different compositions in the system Na_2O/Li_2O-CaO- Al_2O_3-SiO_2 was prepared using P_2O_5 as the nucleating agent by both bulk and palletized forms. The materials were characterized by DTA, XRD, SEM and also by the measurement of specific gravity, hardness, FT-IR spectroscopy after exposure to simulated body fluid and by protein absorption test. From the work the following observations were made:

- P_2O_5 encourages phase separation in the glass and reduction in the rate of crystal growth. Phase transformation in glasses is increased when a part of Na_2O and Li_2O is replaced by P_2O_5.
- X-ray studies revealed the presence of different forms of sodium calcium silicates, sodium calcium phosphates, wollastonite, sodium calcium aluminium silicates and quartz in the glass ceramic samples. In the glass ceramic GCP-2 with 3 mole % of P_2O_5 crystals of the size of 0.81 μm was formed whereas in the glass ceramic GCP-1 with 1.5 mole % of P_2O_5 crystals of the size of 0.98 μm was formed.
- Specific gravity and hardness of the lithium containing glasses are more compared to sodium containing glasses
- Lithium containing glasses are less soluble in SBF. With the increase in the P_2O_5 content the solubility of the glasses in the SBF increases.
- With the increase in P_2O_5 content the protein absorption tendency decreased in both the pellet and the bulk forms. GCP-3 adsorbed minimum protein compared to GCP-2 and GCP-1. Similarly, GCB-3 was adsorbed minimum protein compared to GCB-2 and GCB-1.

ACKNOWLEDGEMENT:
The authors acknowledge the financial support provided by the the the Universiti Teknologi Petronas, Mala for continuing this work and also acknowledge the support provided by the Banaras Hindu University, I for doing some of the experimental part of the work.

5. REFERENCES
[1]L.L. Hench, Bioceramics: from Concepts to Clinic, *J. Am. Ceram. Soc.*,74 1480-510 (1991)
[2]C. Loty, N. Forest, H. Boulekbache, T. Kokubo, and J.M. Sautier, Behaviour of Fetal Rat Chondrocytes Cultured on a Bioactive Glass Ceramics, *J. Biomed. Mater. Res*, 37 137-49 (1997)
[3]T. Kobuto, Surface Chemistry of Bioactive Glass Ceramics, *J. Non-Cryst. Sol.*, 120 138-46 (1990)
[4]E.E. Medbedev, Ceramics and Glass Ceramics materials for Bone Implants, *Glass Ceram.*, 50 81-88 (1993)
[5]H. Takadama, H.M.Kim, T. Kobuto and T. Nakamura, Mechanism of Biomineralization of Apatite on a Sodium Silicate Glass: TEM-EDX Study in Vitro, *Chem. Mater.* 13 1108-13 (2001)
[6]A. Sinha, A. Ingle, K.R. Munim, S.N.Vaidya, B.P.Sharma and A.N.Bhisey, Development if Calcium Phosphate Based Bio-ceramics, *Bull. Mater. Sci*, 24 653-659 (2001)
[7]E.A.Ross, C.D.Batich, W.L.ClappJ.E.Sallustio and N.C.Lee, Tissue adhesion to Bioactive Silicone Tubing in a rat Model of Peritoneal Dialysis Catheter and Catheter Tunnels, *Kid. Intl.* 63 702-10 (2003)
[8]A.G.Avent, C.N.Carpenter, J.D.Smith, D.M.Healey and T.Gilchrist, The dissolution of Silver-Sodium –Calcium Phosphate Glasses for The Control of Urinary Tract Infection, *J. Non-Cryst Sol*, 328 31-8 (2003)
[9]F. Fujbayashi, M. Neo, H.M.Kin, T.Kokubo and T. Nakamura, A Comparative in Vivo Bone in Growth and in-Vivo Apatite Formation on Na_2O-CaO-SiO_2 Glasses, *Biomater*, 24 1349-57 (2003)
[10]E. Venne, S. Dinunzio, M.Bosetti, P. Appendino and C. Vitale Brovarone, Surface Characterization of Silver Doped Bioactive Glass, *Biomater.*, 26 5111-19 (2005)
[11]Z. Zhou, J. Quan, J. Zou and Z. Zhou, Preparation and Bioactivity of Sol-Gel Macroporous Bioactive Glass, *J. Univ Sci. Technol. Beijing, Miner., Metal, Mater* , 15 290-96 (2008)
[12]M. Renke Gluszko and M. E I Fray, The Effect of Simulated Body Fluid on The Mechanical Properties of Multiblockpoly(Aliphatic/Aromatic Ester) Copolymers, *Biomater.*, 25 5191-98 (2004)
[13]P.A.A.P. Marques, S.C.P. Cachinho, M.C.F. Magalhaes, R.N. Correia and M.H.V.Fernandes, Mineralisation of Bioceramics in Simulated Body Plasma with Physiological CO_2/HCO_3^- Buffer and Albumin, *J. Mater. Chem*, 14 1861-66 (2004)

[14]C.Rey, H.M. Kim, L. Gerstenfeld and M.J. Glimcher, Characterization of The Apatite Crystals of Bone And Their Maturation in Osteoblast Cell Culture: Comparison With Native Bone Crystals, *Connect Tissues Res.*, **35** 343-49(1996)

[15]I. Reehman and W. Bonfield, Characterization of Hydroxyapatite and Carbonated Apatite By Photo Acoustic FTIR Spectroscopy, *J. Mater Sci.: Mater Med* **8** 1-9 (1997)

[16]H.M. Kim, C. Rey and M.J. Glimcher, Isolation Of Calcium Phosphate Crystals Of Bone By Non Aqueous Methods At Low Temperatures, *J Bone Min Res*, **10** 1589-95(1995)

[17]J.R. Asplin, N.S.Mandel and F.L. Coe, Evidence For Calcium Phosphate Supersaturation In The Loop of Henle", *Am. J. Physiol.*, **270** F604-10(1996)

[18]S. Radin, P. Ducheyne, B. Rothman and A. Conti., The Effect of In Vitro Modeling Conditions on The Surface Reactions of Bioactive Glass, *J. Biomed Mater Res*, **37** 363-69(1997)

[19]L.L.Hench, and D.E..Clarke, Physical Chemistry of Glass Surfaces, *J. Noncryst Solids*, **59**, 62-68 (1978)

[20]K.D.Lobel and L.L. Hench, In Vitro Adsorption and Activity of Enzymes on Reaction Layers of Bioactive Glass Surfaces, *J. Biomed Mater Res.*, **39**, 575-83(1998)

Table 1:Glass compositions (wt.%)

Glass	Li_2O	Na_2O	CaO	Al_2O_3	SiO_2	P_2O_5
Glass-1	-	20.5	25	5	48	1.5
Glass-2	-	19	25	5	48	3
Glass-3	-	18	25	5	48	4
Glass-4	20.5	-	25	5	48	1.5
Glass-5	19	-	25	5	48	3
Glass-6	18	-	25	5	48	4

Table-2: Codes of the Glass-Ceramic Compositions

Glass Ceramic System	Code
From Glass-1 in bulk form	SB-1
From Glass- 1 in pellet form	SP-1
From Glass-2 in bulk form	SB-2
From Glass-2 in Pellet form	SP-2
From Glass-3 in Bulk form	SB-3
From Glass-3 in Pellet form	SP-3
From Glass-4 in bulk form	LB-1
From Glass- 4 in pellet form	LP-1
From Glass-5 in bulk form	LB-2
From Glass-5 in Pellet form	LP-2
From Glass-6 in Bulk form	LB-3
From Glass-6 in Pellet form	LP-3

Table-3: Heat Treatment Schedule

Glass Ceramic	Nucleation		Growth	
	Temperature(°C)	Time(h)	Temperature(°C)	Time(h)
SB-1	550	2	730	2
SP-1	550	2	730	2
SB-2	560	2	750	2
SP-2	560	2	750	2
SB-3	600	2	780	2
SP-3	600	2	780	2
LB-1	560	2	780	2
LP-1	560	2	780	2
LB-2	550	2	750	2
LP-2	550	2	750	2
LB-3	530	2	720	2
LP-3	530	2	720	2

Table-4: Specific gravity of the glasses

Batch	Specific Gravity
Glass-1	2.6273
Glass-2	2.6332
Glass-3	2.6552
Glass-4	2.6298
Glass-5	2.6334
Glass-6	2.6668

Table-5: Specific gravity of the crystallized glasses

Crystallized Glass	Specific Gravity
Bulk	
SB-1	2.7168
SB-2	2.7665
SB-3	2.8234
LB-1	2.7751
LB-2	2.7950
LB-3	2.8213
Pellet	
SP-1	2.9530
SP-2	2.9648
SP-3	2.9901
LP-1	2.9835
LP-2	2.9947
LP-3	3.0153

Table-6: EDAX Analysis of the Samples

Sample	Atomic wt (%)					
	Na	Ca	Al	Si	P	O
SP-1	15.20	14.28	2.64	30.4	0.65	36.83
SP-2	14.09	14.28	2.64	30.4	1.30	37.29
SP-3	13.35	14.28	2.64	30.4	1.75	37.58

Table-7: Hardness of the Glass Ceramics

Sample	Hardness (kg/mm^2)	Sample	Hardness (kg/mm^2)
SB-1	600	LB-1	624
SB-2	620	LB-2	657
SB-3	682	LB-3	695
SP-1	634	LP-1	668
SP-2	678	LP-2	712
SP-3	715	LP-3	737

Table-8: Ion Concentration of Simulated Body Fluid (SBF)

Ions	Conc. in SBF (ppm)
Na^+	129.4
K^+	8.5
Mg^{2+}	5.2
Ca^{2+}	6.5
Cl^-	10.0
HCO_3^-	14.6
HPO_4^-	7.1
SO_4^{2-}	4.6

Table-9: Wave Number of Various Peaks of FT-IR before and After Soaking in SBF

Peak	Wave Number (Cm^{-1})			
	SP-1 (Before)	SP-1 (after)	LP-1 (before)	LP-1 (after)
1	3487.6	3553.1	3497.3	3368.0
2	2928.2	1637.7	1633.7	1637.7
3	1743.8	1429.4	1444.8	1419.7
4	1483.4	1074.5	1265.4	1078.3
5	1113.0	781.2	1026.2	902.8
6	1030.1	463.0	966.4	563.3
7	918.2	-	735.0	453.3
8	698.3	-	704.1	-
9	617.3	-	603.8	-
10	449.5	-	459.1	-
11	-	-	403.2	-

Figure-1: DTA of Glass 1, 2 and 3 (From bottom to top)

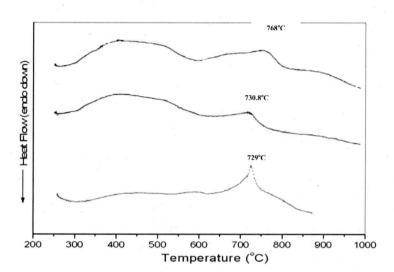

Figure-2: DTA of Glass 4, 5 and 6 (From top to bottom)

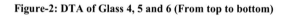

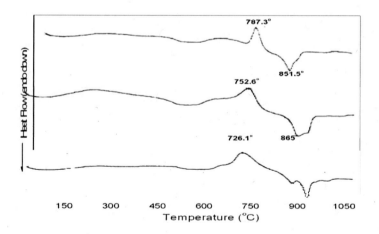

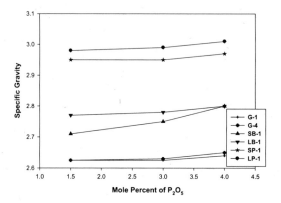

Figure- 3: Variation of Specific gravity with P_2O_5 Content in Sodium and Lithium Containing Glass adn Glass Ceramics

Figure4–XRD graphs of as received glass-1 and glass ceramics (a) (GCB-1 and GCP-1) (b) (GCB-2 and GCP-2)

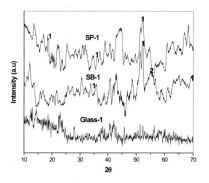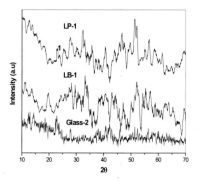

Figure-5: (i), (ii) SEM images of SP-1, LP-1 (ii), (iv) SEM images of SP-2, LP-2 (v), (vi) SEM images of fracture sample SP-2, LP-2(vii), (viii) SEM images of SB-1, LB-1 (ix), (x) SEM images of SB-2, LB-2

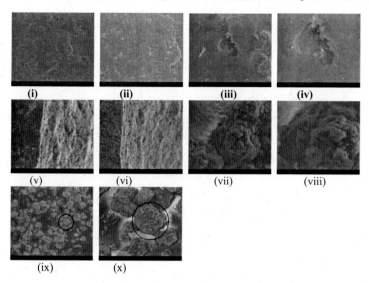

Figure-6(a): FT-IR Spectra of SP-1 Glass ceramics before and after soaking in SBF

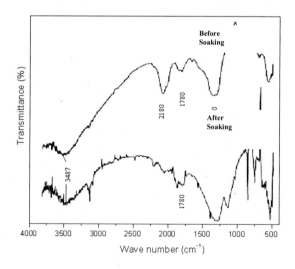

Figure-6(b): FT-IR spectra of LP-1 glass ceramics before and after soaking in SBF

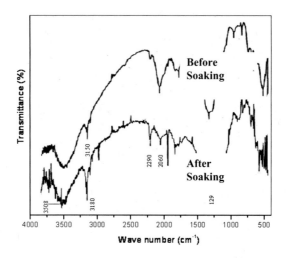

Figure -7. Protein adsorptions of Bio-glass and Bio-glass ceramics (Glass, GCP and GCB) with variation of P$_2$O$_5$ content

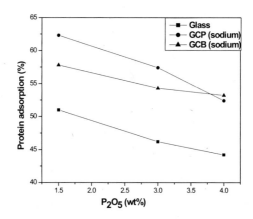

EVALUATION OF A PDLLA/45S5 BIOGLASS COMPOSITE: MECHANICAL AND BIOLOGICAL PROPERTIES

Ginsac Nathalie[1], Chevalier Jérôme[1], Chenal Jean Marc[1], Meille Sylvain [1], Hartmann Daniel[2], Zenati Rachid[3].
1. Université de Lyon, INSA-Lyon, Laboratoire Matériaux Ingénierie et Sciences (MATEIS), UMR CNRS 5510
Villeurbanne, France
2. Université de Lyon, Faculté de Pharmacie, Rockefeller, Laboratoire des Biomatériaux, Unité Réparation Tissulaire, Interactions Biologiques et Biomatériaux" (RTI2B)
Lyon, France
3. Noraker, Adresse, Villeurbanne, France

ABSTRACT
The present study aims to characterize a composite made of poly-L,DL-lactic acid (P(L,DL)LA, Mv: 120 KDa) containing 30 wt% of 45S5 bioactive glass particles. Glass transition (around 52°C) of the polymer was assessed by differential scanning calorimetry (DSC). The mechanical properties of the neat polymer and the composite were evaluated by tensile and compressive tests. From these tests, it was confirmed that the addition of bioglass into the polymer matrix leads to a slight decrease in tensile strength and an increase in elastic modulus. In vitro bioactivity of this composite was evaluated by immersion in a simulated body fluid (SBF) at 37°C for different durations. Formation of hydroxyapatite crystals on the surface of the composite was recorded by scanning electron microscopy and confirmed by X-ray diffraction. Many hydroxyapatite crystals covered the surface of the composite after 14 days of immersion in simulated body fluid. Osteoblast cells MG-63 (human osteosarcoma cell line) were cultured in direct contact with the polymer and the composite. Cells morphology and attachment were analysed using SEM and MTT viability test. Scanning electron microscopy analysis showed the presence of cells at the surface of the composite. These results confirmed the biocompatibility of the composite and the positive effect of the bioglass on the osteoblast cells adhesion and proliferation on the composite.

INTRODUCTION
The need of innovation in bone reconstruction is a major requirement. Currently the ultimate approach in bone reconstruction is the use of donated bone allograft, but it presents infections risks. Autograft bone does not transmit pathologies and does not create immune response, but the quantity of bone available is low. Synthetic biomaterials could offer an attractive approach. Bioabsorbable devices are the best alternative for internal fixation. Indeed during healing process, bioabsorbable devices maintain the fixation, decompose gradually, and the stresses are transferred gradually to the healing tissue so that no stress shielding will occur. Furthermore, bioabsorbable surgical devices do not require a removal operation, and so may reduce the total cost when compared to metallic devices. The copolymer P(L,DL)LA offers an attractive choice for different applications due to its wide range of properties. Since the discovery of bioglass by Hench[1], the bioactivity of a variety of glass compositions has been well established. Bioglass, 45S5, is a particular commercial bioactive glass containing 45% SiO_2, 24,5% Na_2O, 24,5% CaO et 6% P_2O_5 (%wt). The lack of in situ mould-ability and the relative brittleness of bioactive glasses limit the range of their applications. A composite could combine the osteoconductive properties of bioactive glass with the processability of biodegradable polymer and overcome the brittleness of the bioglass. Bioactive glasses react in vitro in many solutions and in vivo in body fluids. Due to ion dissolution, a silica gel layer is formed on their surface during the first hours, and is followed by calcium phosphate precipitation and crystallisation, creating an apatite layer at the bioglass surface. Composites made of biodegradable polymers and bioactive glasses have been studied

for several medical applications that included non-load bearing applications[2,3,4]. The aim of the current study is to evaluate the mechanical properties of the polymer matrix and of the composite made of 30(wt%) bioglass and also to confirm its biocompatibility and bone bonding ability.

EXPERIMENTAL

Materials

P(L,DL)LA 70:30 copolymer (inherent viscosity = 5.5-6.5 $l.mol^{-1}$, density = 1.27 g/cm^3, Boehringer, Ingelheim, Germany) was used. This copolymer is filled with 45S5 bioglass particles. The composition of bioglass (quoted hereafter BG) is 45.0% SiO_2, 24.5% Na_2O, 24.5% CaO, and 6.0% P_2O_5 by weight. The matrix and the composite with 30% (by weight) BG were prepared by a process protected by Noraker (patent n°WO2008116984 (A2)). Briefly, polymer was first dissolved in a solvent and mixed with ceramic powder (average diameter 3 μm), and then the solvent was evaporated. Plates, cylinders and dumbbell specimens of the matrix and the composite were obtained by injection molding.

Thermal characterization

Differential scanning calorimetry (DSC) thermograms were recorded with a Pyris Diamond apparatus. Approximately 6 mg of the matrix and the composite were placed in aluminium pans. Samples were cooled to 20°C and held for 3 min prior heating from 20 to 160°C at a heating rate of 2°C/min. Then samples were cooled from 160 to 20°C at 10°C/min, prior a second similar heating run. Thermograms were analyzed to determine thermal transitions of the materials.

Bioactivity study

The development of a surface Ca-P layer on the P(L,DL)LA/BG composite was evaluated in a simulated body fluid (SBF) with solution ion concentrations similar to those of blood plasma. Preparation of SBF solution was realized according to ISO23317 Standard. P(L,DL)LA and P(L,DL)LA/BG rectangular specimens (10x10x2 mm^3) were incubated at 37°C in SBF for 7, 14 and 28 days. After each immersion period, samples were rinsed with distilled water and dried in a desiccator. The formation of Ca-P layer on the surface of the samples was characterized using X-ray diffraction (Rigaku) and scanning electron microscopy (JEOL 840 A LGS).

Mechanical characterization

The following mechanical properties were measured on both the polymer matrix and the composite.

Dog bone-shape specimens having a length of 35 mm and a rectangular cross-section (2 x 5 mm^2) were tested in tension with a MTS machine coupled with video extensometer according to NF EN ISO 527-2 standard. The tests were conducted using a cross-head speed of 0.35 mm/min (0.1% of the length). Tensile strength (St) and elastic modulus (Et) were calculated. Rational stress and strain were also calculated to account for the section modification during the test.

Cylinders with a diameter of 6 mm and a length of 11 mm were tested in compression with an INSTRON 8502 machine, using a cross-head speed of 1 mm/min according to ISO 604:2002 standard. Compressive strength (Sc), elastic modulus (Ec) and rational stress and strain were calculated.

Biocompatibility in vitro

In vitro biocompatibility study was carried out on the neat polymer and the composite by cells culture. Rectangular specimens (10x10x2 mm^3) were used for the culture cell. Materials were first soaked in cell culture medium for 8 days. MG63 cells were cultured in Dulbecco's Modified Eagle's Medium (DMEM), supplemented with 10% foetal calf serum, penicillin and streptomycin. MG 63 cells were seeded onto materials at a density of 10^5 cells.ml^{-1}. The same number of cells was seeded on 6 control wells, without material. The plates were further incubated at 37°C in humidified air with 5% CO_2, for a period of 7 and 14 days. The culture medium was changed every two days. In order to evaluate the biocompatibility of the polymer and the composite, MTT test was carried out at 7 and 14 days.

In the cell adhesion study, viability was measured according to MTT-based colorimetric assays. This assay relies on the ability of living cells to reduce a tetrazolium salt into soluble coloured formazan product, thanks to succinate deshydrogenase mitochondrial activity present in living cells. This enzyme converts the yellow MTT in blue formazan crystals. After 7 and 14 days of incubation, MTT test was carried out on the materials. Formazan measurement was realized by spectrophotometry at 570 nm. Results are determined by measuring the optical density, and reported as a percentage of cells on the materials relative to cells without material.

Osteoblast morphology was examined using scanning electron microscopy (SEM) after 14 days of culture on the materials. Two wells, containing polymer and composite specimens without cells, were used for SEM negative control. For the SEM observation, cells were fixed with paraformaldehyde during minimum 24h at 4°C. Cells were then dehydrated through a series of alcohol concentrations (30°, 50°, 70°, 95°, 100°), and by air drying. The day after, cells were gold coated and examined under the SEM at an accelerating voltage of 5KeV.

RESULTS AND DISCUSSION

Thermal characterization

Only the second DSC heating run was analyzed, the first one being used to erase the thermal history of the polymer. The P(L,DL)LA copolymer is amorphous, with a glass transition temperature (Tg) between 52 and 53°C. Addition of bioglass particles on the polymer matrix has no effect on the glass transition temperature.

Bioactivity study

SEM observation of the composite was realized after different immersion times and compared with observations of the neat polymer (Fig. 1 to 3). Crystals appear on the edges of the composite after 7 days of immersion in SBF (Fig. 1 c and d), and on its surface after 14 days of immersion (Fig. 2b). The XRD analysis of materials after different immersion times in SBF evidenced the existence of the main characteristic diffraction peaks of HA after 14 days soaking in SBF. The peak intensity increased after 28 days soaking in SBF. This was confirmed by the comparison of the XRD spectra with the standard pattern of HA (JCPDS 00-01-1008) (Fig. 4). Others minor peaks (not identified) appear on the diffractograms, and could be attributed to salt crystallization from the SBF solution. XRD and SEM results showed that after 14 days in SBF, an apatite layer appears on the composite surface. This result confirms the bioactive character of the composite. In the literature, a similar result has been observed but with a different kinetic. Y. Shikinami et al.[5] have studied bioactivity of the composite PLLA with HA particles. They have shown HA crystals between 3 and 6 days of immersion in SBF, and a totally covered surface after 7 days.

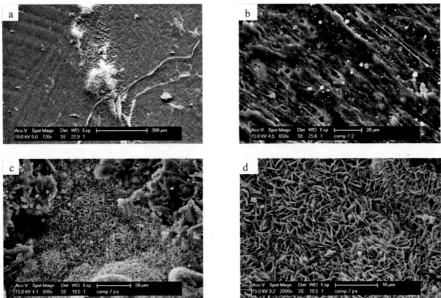

Figure 1. SEM micrographs after 7 days of immersion in SBF, (a) of the neat polymer x 100, (b) of the surface composite x 650, and the edges of the composite (c) x 800, (d) x 2000.

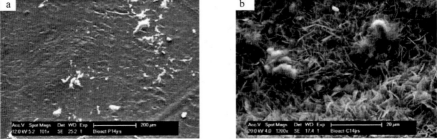

Figure 2. SEM micrographs after 14 days of immersion in SBF, (a) of the neat polymer x 100, (b) of the composite x 1200.

Figure 3. SEM micrographs after 28 days of immersion in SBF, (a) of the neat polymer x 100, (b) of the composite x 100, (c) of the composite x 800, (d) of the composite x 1200.

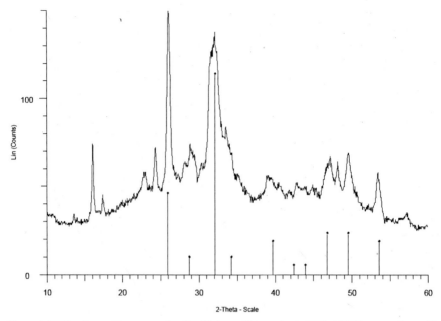

Figure 4. DRX pattern of the composite after 28 days of immersion in SBF at 37°C, comparison with HA standard pattern (JCPDS 00-001-1008).

Mechanical characterization

The results of tensile tests for the polymer matrix and the composite are presented in Table I. The addition of ceramic particles decreases significantly the tensile strength, but as expected increases the Young modulus. Stress – strain curves for the polymer and the composite show a first linear part followed by a plastic domain (the strength decreases and reaches a plateau). Extensive cavitation was observed on the polymer specimens at the fracture section. The composite shows a more brittle fracture. Maximal tensile stress obtained with the materials are lower than human cortical bone one (between 130 and 150 MPa). Few materials of this kind have been tested in previous works. Compared with those works, the tensile stress is similar to that obtained by Verheyen[6], but lower compared to that obtained by Shikinami[5]. The Young modulus obtained here is higher than those observed in the literature[5]. It is to note however that the materials studied by both authors (PLLA/HA composites) are different from the current composite (P(L,DL)LA/BG), and the comparison is difficult.

Table I. Tensile strength and Young modulus of the polymer matrix and the composite.

	Tensile strength (MPa)	Young modulus (GPa)
Polymer	46 ± 11	3.2 ± 0.2
Composite	31 ± 2	4.4 ± 0.1

The compressive tests results for the polymer and the composite are presented in Table II. As expected, and as already reported in previous works[5,6], the addition of ceramic particles to the polymer matrix improves the Young modulus. The polymer presents slightly higher values of yield stress and maximal strength. Our results are low compared to the literature, but quite close to those obtained by Shikinami et al.[5], who worked on a PLLA/HA composite. Composite compressive strength is very low compared to cortical human one (130-200 MPa), but remains acceptable.

Table II. Compressive strength and Young modulus of the polymer matrix and the composite.

Materials	Compressive maximal strength (MPa)	Yield stress (MPa)	Young modulus (GPa)
Polymer	101 ± 7	74 ± 1	3.5 ± 0.1
Composite	96 ± 2	62 ± 2	4.1 ± 0.3

Biocompatibility in vitro
    Figure 5 shows the percentage of cell viability on polymer and composite materials, for MG63 cells. After 7 days of cell incubation, 88% of cell viability was obtained on the neat P(L,DL)LA, while only 56% was obtained on the composite. But after 14 days of cell incubation, 93% and 88% were obtained on the polymer and the composite respectively. These results show that materials are not toxic. After 14 days of culture, cell viability on the composite returns to normality. The apparent decrease in cell viability observed at 7 days, can be due to ions leaching from the bioglass. It might be limited or suppressed if pre-soaking duration was increased. SEM micrographs realized on the materials with cells, show that MG63 cells adhere on both materials after 14 days (Fig. 6).

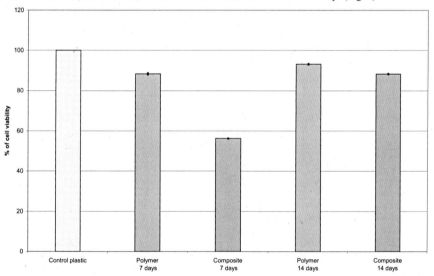

Figure 5. Cell viability on the polymer and the composite after 7 and 14 days of incubation.

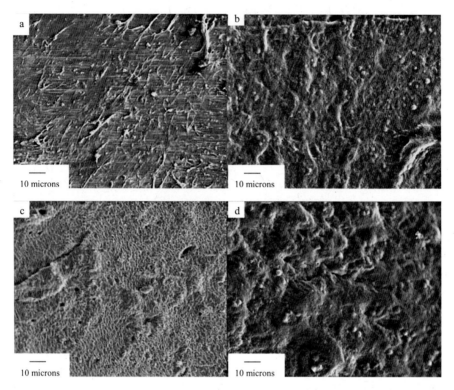

Figure 6. (a) Electron micrograph of the polymer without cell, 1000 x, (b) of the polymer with cells, 1000 x, (c) of the composite without cell, 1000 x, (d) of the composite with cells, 1000 x.

CONCLUSION

The addition of 30% in weight of BG particles in the P(L,DL)LA matrix causes a decrease in tensile strength and a slight decrease in compressive strength. The Young modulus of the composite is however increased. Further mechanical tests have to be carried out to better understand the mechanical behaviour of the composite: precise elastic modulus values can be obtained by ultrasonic waves propagation. Bioactivity study has confirmed bioactivity ability of the composite. A complementary study has been carried out on the polymer, the composite and the bioglass with a more complete kinetic. First results tend to show the apparition of HA crystals even after 3 days by DRX. Complementary studies will be conducted on cell cultures, in order to better characterize the extracellular matrix, in particular collagen I and osteocalcin assays. Moreover, osteoblast differentiation will be determined by alkaline phosphatase assay. An in vitro degradation study will be realized, to simulate the degradation of the composite and the evolution of its properties. Biocompatibility in vivo study is planned, in order to check the biocompatibility of the composite. Composites with amounts of bioglass fillers (20 and 50% wt) will also be studied. The final aim is to choose the best composition that presents optimal mechanical, chemical and biological properties for bone fixation devices. The first study seems to prove that this material is a good target for medical devices conception, and in particular bone fixation devices, like plates, screws, pins … for orthopaedic, oral-maxillo facial surgeries. This composite is a dense material, easy to shape, biocompatible and bioresorbable.

REFERENCES
[1]L. L. Hench, The story of Bioglass, *J Mater Sci : Mater Med*, **17**, 967-978 (2006).
[2]H. H. Lu, A. Tang, S. C. Oh, J. P. Spalazzi, and K. Dionisio, Compositional effects on the formation of a calcium phosphate layer and the response of osteoblast-like cells on polymer-bioactive glass composites, *Biomaterials*, **26**, 6323-6334 (2005).
[3]J. Rich, T. Jaakkola, T. Tirri, T. Narhi, A. Yli-Urpo, and J. Seppala, In vitro evaluation of poly(ε-caprolactone-co-DL-lactide)/ bioactive glass composites, *Biomaterials*, **23**, 2143-2150 (2002).
[4]H. Wilda, and J. E. Gough, In vitro studies of annulus fibrosus disc cell attachment, differentiation and matrix production on PDLLA/45S bioglass composite films, *Biomaterials*, **27**, 5220-5229 (2006).
[5]Y. Shikinami, and M. Okuno, Bioresorbable devices made of forged composites of hydroxyapatite (HA) particles and poly-L-lactide (PLLA): Part I. Basic characteristics, *Biomaterials*, **20**, 859-877 (1999).
[6]C.C.P.M. Verheyen, J.R. de Wijn, C.A. van Blitterswijk, and K. de Groot, Evaluation of hydroxylapatite/poly(L-lactide) composites : Mechanical behavior, *Journal of Biomedical Materials Research*, **26**, 1277-1296 (1992).

SYNTHESIS AND CHARACTERIZATION OF WET CHEMICALLY DERIVED MAGNETITE-HAP HYBRID NANOPARTICLES

S. Hayakawa, K. Tsuru, A. Matsumoto, A. Osaka
Graduate School of Natural Science and Technology
Research Center for Biomedical Engineering, Okayama University
3-1-1 Tsushima, Okayama, 700-8530

E. Fujii, K. Kawabata
Industrial Technology Center of Okayama Prefecture
5301 Haga, Okayama, 701-1296, Japan

ABSTRACT
        A new type of hyperthermia or magnetic resonance imaging materials with bone-bonding ability was explored within the framework of magnetic-bone mineral composite ceramics. That is, hydroxyapatite (HAp) nanoparticles, hybridized with ferrous (Fe^{2+}) and ferric (Fe^{3+}) ions (Fe(II) and Fe(III)), were synthesized through the wet chemical procedure, and characterized in terms of crystal structures, magnetic properties and protein adsorption properties. The as-synthesized particles derived from precursor solutions with $FeCl_2$ consisted of hydroxyapatite (JCPDS 09-0432) (20-30 nm in size) and magnetite (JCPDS 19-0629) (2-5 nm in size). They showed super-paramagnetic behavior, yet their saturation magnetization increased with the content of Fe(II) in the solutions up to 4.3 emu/g. From TEM observations, the HAp particles were rod-like, by which the magnetite particles that seemed spherical rather than showing cubic morphology, were surrounded. The particles from the solutions with Fe(III) gave only HAp, and hence Fe(III) was considered to form an amorphous phase. Moreover, Fe(III) incorporation suppressed HAp crystal growth.

INTRODUCTION

        Apatite, or hydroxyapatite (HAp) having the stoichiometric formula $Ca_5(PO_4)_3OH$, is the major inorganic component which constitutes the human bone framework.   It accommodates most of the inorganic cations and anions by replacing the constituent calcium, phosphate, and hydroxyl ions: they include monovalent or divalent metallic cations[1-3], and the multivalent oxoanions SiO_4^{4-} [6] and BO_2^{2-} [7,8] as well as CO_3^{2-} [9] in the bone mineral apatite.   Such ion substitution not only modifies the lattice structures but also affects the physical and chemical properties of apatite.   For example, Zn(II) and Mg(II) control protein adsorption behavior of apatite particles, while Si(IV) introduction favors their dissolution in aqueous media[4,5].   In protein adsorption, Zn(II) in the lattice is active probably due to the 3d orbitals.   Thus, cations with incomplete 3d ortbitals like ferrous or ferric ions are likely to be more active.   Moreover, nano- or micrometer particles of a composite structure having a magnetite core with an apatite shell are applicable to drug-delivery systems, hyperthermia clinics or magnetic beads for DNA separation and protein isolation, taking advantage of good protein affinity due to the apatite shell and heat-evolution under electromagnetic wave irradiation due to the magnetite core.   In addition, it is a common inorganic chemistry topic that ferrous and ferric ions in highly basic conditions yield magnetite particles: Fe(II) + Fe(III) + OH^- → Fe_3O_4.[10]   That is, magnetite particles will be yielded as Fe(II) and Fe(III) are added into calcium hydroxide suspension.   Then, that suspension is mixed with a phosphate solution, whose pH is adjusted in the alkali range, and apatite is nucleated to grow on the magnetite particle surface through heterogeneous nucleation steps, and apatite is separately precipitated under the wet chemical process:

$$5Ca^{2+} + 3PO_4^{3-} + OH^- \rightarrow Ca_5(PO_4)_3OH \qquad (1)$$

Under the precipitation process, some of the Fe ions, except for those involved in the magnetite lattice, might get involved in the apatite lattice to form Fe-substituted apatite, while the others remain in the form of hydrated iron oxide. The hydrated iron oxide may play the role of an adhesive layer between the magnetite core and the HAp shell. In addition, the HAp shell may contribute to the stability and dispersion of magnetite-HAp hybrid nanoparticles in liquid media.

In this study, magnetite-HAp hybrid nanoparticles were wet chemically synthesized, and their microstructures were characterized by transmission electron micrographs, ^{1}H, ^{31}P MAS and CP-MAS NMR spectroscopy, X-ray diffractometry, and other techniques. Their magnetic properties were examined with a vibrating sample magnetometer (VSM). The adsorption of proteins was also examined, using bovine serum albumin (BSA) and lysozyme (LZM) as examples of acidic protein and basic protein, respectively.

MATERIALS AND METHODS
Reagent-grade calcium hydroxide and diammonium hydrogen phosphate were employed as well as iron (II) chloride tetrahydrate and iron (III) chloride hexahydrate. The phosphate was dissolved in ultra-pure water to prepare 0.4 M diammonium hydrogen phosphate solution, while 0.67 M calcium hydroxide solution was also prepared, to which appropriate amounts of ferrous and ferric chlorides were added so that the molar ratio Fe/Ca was in the range of $0 \sim 0.5$. Following previous studies[1-3], 150 mL of both the calcium and phosphate solutions were added dropwisely at a rate of 3 ml/min to 100ml of distilled water held in a three-necked flask kept at 60°C under an N_2 atmosphere. The pH of the suspension was adjusted to 10 by adding a 28 mass% NH_4OH aqueous solution during the addition. After completion of the addition, the resulted suspension was further stirred at 60°C for 24 h. The precipitates were washed with distilled water three times, and finally dried *in vacuo* at 105°C for 24 h. The dry cakes were milled and sieved to obtain Fe-containing apatite particles under 150 μm in size. The Fe, Ca, P content of the samples were analyzed by inductively coupled plasma emission spectroscopy (ICP, iCAP6000, ThermoFisher). The crystalline phases were identified by an X-ray diffractometer (XRD, CuKα, RINT2500, RIGAKU; 40 kV-200 mA), and the crystallite diameters were derived from their XRD peaks following the Scherer equation. The specific surface area (SSA) was measured by the Brunauer-Emmett-Teller (BET) N_2 adsorption method. The morphology of those sample particles was observed under a transmission electron microscope (TEM, JEM-2100, JEOL). The zeta potential of the samples in the physiological saline (0.142 mol/L NaCl aqueous solution, pH7.4) was measured (ZETASIZER, 3000HSA, MALVERN). Magic-angle spinning solid-state ^{31}P nuclear magnetic resonance (NMR) was performed on a Fourier transform NMR spectrometer (^{UNITY}INOVA300, Varian) at room temperature. Inversion recovery experiments were performed using a standard π-τ-π/2 radio frequency pulse sequence, where τ is the time (sec) of variable delay to measure the relaxation time, T' of all samples. Magnetic characterization was performed with a vibrating sample magnetometer (VSM, VSM-15, TOEI INDUSTRY). BSA and LZM adsorption was examined in the same way as in our previous study[11]. Briefly, the saline solution (4 mL) of BSA (4.0 mg/mL) and LZM (0.1mg/mL), held in glass test tubes, whose pH was kept at pH 7.4 with TRIS, were in contact with each sample (0.05g) at 36.5°C under shaking for 6h. Then, the amount of the proteins left in the supernatant was measured spectroscopically (UV-3100, SHIMADZU) using the optical absorption density at about 280nm.

RESULTS
Figure 1 shows the XRD profiles of HAp and Fe-containing HAp synthesized at various Fe concentrations. Here, the valence of iron ions in the precursor calcium hydroxide suspensions was indicated as Fe(II) and Fe(III). All diffraction peaks were assigned to hydroxyapatite (closed circles; JCPDS 09-0432) except one at $\sim 35.5°$(closed diamond). Only one diffraction peak is ever

insufficient for definite identification of any crystalline phase. Yet, as Fig. 2 indicates that the samples with those XRD peaks, dispersed in distilled water, were completely attracted toward a magnet. Therefore, it is reasonable to assign the 35.5° diffraction to the 311 diffraction of magnetite (JCPDS 19-0629). Note that only Fe-HAp's derived from the starting solutions with Fe(II) showed the presence of magnetite, and those from the Fe(III) containing solutions gave only diffractions for apatite: The Fe(III)-HAp samples were not attracted by the magnet, indicating they were paramagnetic. In addition, the XRD profile decreased in sharpness or was broadened with increase in the Fe(II) and Fe(III) content. This broadening is commonly attributed to decrease in size or crystallinity of the crystallites.

Fig. 3 shows the TEM images of two samples. Comparing a) with b) indicates that Fe(III)-HAp particles were smaller than Fe(II)-HAp and the addition of Fe(III) led to

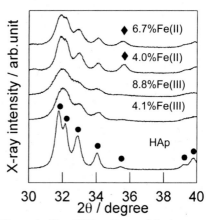

Figure 1. The XRD patterns of hydroxyapatite (HAp) and Fe-containing HAp nanoparticles. The analyzed Fe content is indicated. ♦: the 311 diffraction of magnetite; •: HAp.

Figure 2. Images of a) 6.7%Fe(II) nanoparticles dispersed homogeneously in water: b) all 6.7%Fe(II) nanoparticles were attracted toward a permanent magnet.

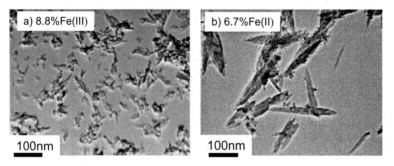

Figure 3. TEM photographs of Fe-HAp nanoparticles.   Total analyzed Fe content and Fe resource presented were both responsible for the change in particle morphology (see text).

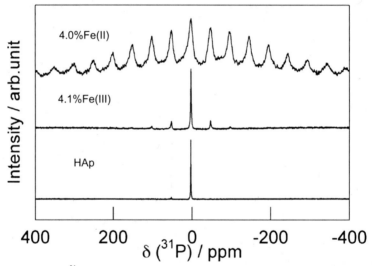

Figure 4. ^{31}Phosphorous NMR spectra of HAp, 4.0%Fe(II) and 4.1%Fe(III).

finer particles, and at the same time, the aspect ratio decreased.   Similar results were obtained from other Fe(III)-HAp particles with the lower Fe content. This result is in agreement with the broadening in XRD profile of Fe(III)-HAp.   One of the most important features in Fig. 3b) was that the darker spots, rich in Fe, were only seen in the apatite particles or they were only associated with the apatites, but were not present in independent particles.   That is, the HAp particles were rod-like, by which spherical magnetite particles were surrounded.

The observation above was closely related to the distribution of the iron ions.   The magnetic moments of the Fe(II) and Fe(III) affect the nuclear magnetic moments of adjacent atoms, hence ^{31}P

MAS NMR spectra were taken and some of them are presented in Fig. 4: HAp, 4.0%Fe(II) and 4.1%Fe(III). The ^{31}P spectrum of 4.1%Fe(III) had four small spinning sidebands in addition to the P

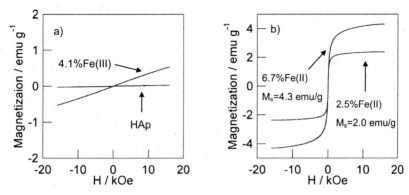

Figure 5. VSM curves of hydroxyapatite (HAp) and Fe-containing hydroxyapatite nanoparticles.

(orthophosphate) at 2.8 ppm in $\delta(^{31}P)$ as that of HAp. The empirical data agreed well with the isotropic ^{31}P chemical shift in the literature, 2.8 ppm[9]. In contrast, the Fe(II) samples showed more complex sideband structures, i.e., many stronger and wider sidebands, than the Fe(III) sample. Moreover, 4.0%Fe(II) yielded more distinct sideband structure. Such differences in the MAS NMR spectral profile indicate the ^{31}P-electron spin of Fe interactions, or the distribution of Fe in the particles.

As magnetite was precipitated, Fe(II)-HAp should exhibit ferromagnetic behavior. Fig. 5 shows the magnetization curves of (a) the HAp and 4.1%Fe(III)-HAp nanoparticles and (b) 2.5% Fe(II)-HAp and 6.7%Fe(II)-HAp ones at room temperature, respectively. In Fig. 5(a) the 4.1%Fe(III)

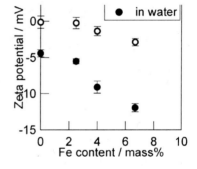

Figure 6. Zeta potential of HAp and Fe(II)-HAp nanoparticles as a function of the analyzed Fe content in the apatite particle.

nanoparticles showed a magnetization curve characteristic of paramagnetic materials, and HAp showed a typical curve for diamagnetic materials. In contrast, in Fig.5(b) the Fe(II)-HAp particles behaved ferromagnetically and, what is more, exhibited a superparamagnetic hysteresis loop. It is reasonable that 6.7%Fe(II)-HAp had greater saturation magnetization than 2.5%Fe(II)-HAp. From the saturation magnetization of 73.8 emu/g for commercial magnetite particles (Kojundo Chemical Co.), the 2.5%Fe(II)- and 6.7%Fe(II)-HAp nanoparticles contained magnetite particles to the extent of ca. 2.7 mass% and 5.8 mass%, respectively.

From the viewpoint of application to drug-delivery systems, hyperthermia clinics or magnetic beads for DNA separation and protein isolation, the Fe(II)-HAp particles were characterized in terms of their protein adsorption properties and zeta potential. Fig. 6 shows the zeta potential in saline (pH7.4) for Fe(II) -HAp as a function of Fe content. Their zeta potential decreased with Fe content.

Figure 7 shows the amount of BSA and LZM adsorbed per unit area as a function of Fe content. The amount of adsorbed protein per unit area is useful as a measure of the intrinsic adsorption ability of the adsorbent. As the Fe content increased, the amount of BSA adsorbed per unit area decreased from $1.63 mg/m^2$ to 1.43 mg/m^2 while that of LZM increased from 6.7 $\mu g/m^2$ to $16.5 \mu g/m^2$.

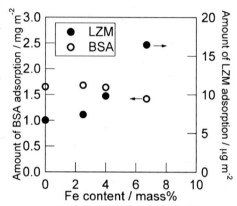

Figure 7. BSA and LZM adsorption on HAp and Fe(II)--HAp as a function of the Fe content.

DISCUSSION

Hydroxyapatite (HAp) nanoparticles, hybridized with ferrous (Fe^{2+}) and ferric (Fe^{3+}) ions (Fe(II) and Fe(III)) were synthesized through the wet chemical procedure. From XRD and TEM observation, the particles from the solutions with Fe(III) gave only HAp, and hence Fe(III) was considered to form an amorphous phase. One can suggest the microstructure model for a) Fe(III)-HAp and b) Fe(II)-HAp, as shown in Fig.8. In Fe(III)-HAp, the HAp particles were covered by a thick layer of amorphous iron hydroxide oxide, while in Fe(II)-HAp, the magnetite particles were covered by HAp particles.

Another important issue is whether Fe ions are present inside or outside HAp lattice structure.

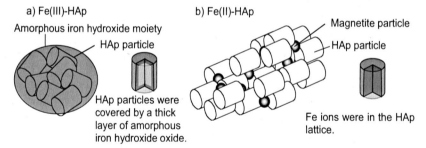

Paramagnetic ions such as Fe incorporated into the HAp lattice enhance the relaxation of a ^{31}P nucleus.

Fig. 8. Microstructure model of a) Fe(III)-HAp and b) Fe(II)-HAp.

After Schroeder and Pruett,[12] the ^{31}P NMR signal is completely eliminated by paramagnetic ions that are close (<1nm) enough to ^{31}P nuclei, which contributes to reduction in the integrated intensity, while paramagnetic ions that are >1nm from ^{31}P nuclei reduced but did not eliminate the ^{31}P NMR signal. In order to get insight into Fe ion substitution into the HAp lattice structure or the distribution of Fe atoms, the relaxation time, estimated relaxation time T' of ^{31}P is used below as a measure of the distance between Fe and P atom due to the interaction between strong electron spin and the ^{31}P nuclear magnetic moment. A short relaxation time, T', of a ^{31}P NMR signal must lead to the appearance of multiple spinning side bands. In Fig. 4, 4.1%Fe(III) had four small spinning side bands in addition to the center peak, indicating that Fe ions were not incorporated into the HAp lattice structure. This was explained by considering that HAp particles were covered by the thick layer of amorphous iron hydroxide oxide, as shown in Fig. 8. On the other hand, observation of the multiple spinning sidebands for 4.0%Fe(II) agreed well with Sutter *et al.*[13] Moreover, Fe(II) ions were substituted for the Ca position in the HAp lattice.[14,15] Indeed, the enhanced relaxation of ^{31}P nuclei of 4.0%Fe(II) indicated that Fe(II) ions were close to ^{31}P nuclei or distributed in the HAp particles. Therefore, considering the results of XRD analysis, Fe(II) ions were present inside and outside of the HAp lattice structure and in magnetite.

Since BSA (isoelectric point ca. 5) has a negative charge and LZM (isoelectric point ca. 11) has a positive charge in the saline (pH 7.4), they will be attracted by substance with an opposite charge. The zeta-potential of Fe-HAp in saline, as shown in Fig. 6, was decreased negatively by the incorporation of Fe ions. The negative zeta potential of Fe(II)-HAp reduced BSA adsorption and increased LZM adsorption (Fig. 7).

CONCLUSION

We synthesized hydroxyapatite (HAp) nanoparticles hybridized with ferrous (Fe^{2+}) and ferric (Fe^{3+}) ions (Fe(II) and Fe(III)) by the wet chemical procedure. The as-synthesized particles from the precursor solutions with Fe(III) gave only nanocrystalline HAp with an amorphous phase of iron hydroxide oxide. On the other hand, the particles derived from the solutions with FeCl$_2$ consisted of HAp and magnetite, where the spherical magnetite particles were surrounded by the rod-like HAp particles. They showed a super-paramagnetic behavior and their saturation magnetization increased with the content of Fe(II). It was found that the incorporation of Fe ions into HAp particles promoted the adsorption of LZM but inhibited the adsorption of BSA in a physiological saline solution (pH7.4). This indicated that the electrostatic interaction between Fe(II)-HAp and BSA or LZM is the dominant factor for their adsorption behavior.

REFERENCES
[1]S. Hayakawa, K. Ando, K. Tsuru, A. Osaka, E. Fujii, K. Kawabata, C. Bonhomme, and F. Babonneau, Structural Characterization and Protein Adsorption Property of Hydroxyapatite Particles Modified With Zinc Ions, *J. Am. Ceram. Soc.*, **90**, 565-569 (2008).
[2]S. Hayakawa, A. Osaka, K. Tsuru, E. Fujii, K. Kawabata, K. Ando, C. Bonhomme, and F. Babonneau, Synthesis and Characterization of Mg-Containing Nano-Apatite, *Key Eng. Mat.*, **361-363**, 47-50 (2008).
[3]E. Fujii, M. Ohkubo, K. Tsuru, S. Hayakawa, A. Osaka, K. Kawabata, C. Bonhomme and F. Babonneau, Selective protein adsorption property and characterization of nano-crystalline zinc-containing hydroxyapatite, *Acta Biomaterialia*, **2**, 69-74 (2006).
[4]I.R. Gibson, S.M. Best, W. Bonfield, Chemical characterization of silicon-substituted hydroxyapatite, *J. Biomed. Mat. Res.*, **44**, 422-428 (1999).
[5]A.E. Porter, N. Patel, J. N. Skepper, S. M. Best and W. Bonfield, Comparison of in vivo dissolution processes in hydroxyapatite and silicon-substituted hydroxyapatite bioceramics, *Biomaterials*, **24**, 4609-4620 (2003).
[6]G. Gasqueres, C. Bonhomme, J. Maquet, F. Babonneau, S. Hayakawa, T. Kanaya, A. Osaka, Revisiting silicate substituted hydroxyapatite by solid state NMR, *Magn. Reson. Chem.*, **46**, 342-346 (2008).
[7]A. Ito, H. Aoki, M. Akao, N. Miura, R. Otsuka, and S. Tsutsumi, *J. Ceram. Soc. Japan*, **96**, 305-309 (1988).
[8]S. Hayakawa, A. Sakai, K. Tsuru, A. Osaka, E. Fujii, K. Kawabata, C. Jäger, Preparation and

Characterization of Boron-containing Hydroxyapatite, *Key Eng. Mat.*, **361-363**, 191-194 (2008).

[9]Y-T. Wu, M. J. Glimcher, C. Rey and J. L. Ackerman, A Unique Protonated Phosphate Group in Bone Mineral Not Present in Synthetic Calcium Phosphates : Identification by Phosphorus-31 Solid State NMR Spectroscopy, *J. Mol. Biol*, **244**, 423-425 (1994).

[10]P.H. Refait, J.M.R. Génin, The oxidation of ferrous hydroxide in chloride-containing aqueous media and pourbaix diagrams of green rust one, *Corros. Sci.*, **34**, 797-817 (1993).

[11]A. Matsumoto, K. Tsuru, S. Hayakawa, A. Osaka, E. Fujii, K. Muraoka, K. Kawabata, Synthesis and protein adsorption property of ferric ion-containing hydroxyapatite, *Arch. BioCeram. Res.*, **7**, 171-174(2007).

[12]P.A. Schroeder and R.J. Pruett, Fe ordering in kaolinite; insights from ^{29}Si and ^{27}Al MAS NMR spectroscopy, *Am. Mineral*, **81**, 26-38 (1996).

[13]B. Sutter, R.E. Taylor, L. R. Hossner, and D. W. Ming, Solid State 31Phosphorus Nuclear Magnetic Resonance of Iron-, Manganese-, and Copper-Containing Synthetic Hydroxyapatites, *Soil. Sci. Soc. Am. J.* **66**, 455-463 (2002)

[14]M. Jiang, J. Terra, A. M. Rossi, M. A. Morales, E. M. Baggio Saitovitch, and D. E. Ellis, Fe^{2+}/Fe^{3+} substitution in hydroxyapatite: Theory and experiment" *Phys. Rev. B*, **66**, 224107 (2002)

[15]I. Gutowska, Z. Machoy, B. Machalinski, The role of bivalent metals in hydroxyapatite structures as revealed by molecular modeling with the HyperChem software, *J. Biomed. Mater. Res.*, **75A**, 788-793 (2005)

LOW TEMPERATURE CONSOLIDATION OF NANOCRYSTALLINE APATITES TOWARD A NEW GENERATION OF CALCIUM PHOSPHATE CERAMICS

D. Grossin[1], M. Banu[1], S. Sarda[1], S. Martinet-Rollin[1,3], C. Drouet[1], C. Estournès[2], E. Champion[3], F. Rossignol[3], C. Combes[1], C. Rey[1].
[1]CIRIMAT, University of Toulouse, INPT, CNRS, Toulouse, France
[2]PNF2/CIRIMAT, University of Toulouse, UPS, CNRS, Toulouse, France
[3]SPCTS, University of Limoges, Limoges, France

ABSTRACT

Biomimetic nanocrystalline apatites analogous to bone mineral can be prepared by different ways. These non-stoichiometric compounds possess a high reactivity related to the presence of a metastable hydrated layer on the surface of the nanocrystals. The processing of such unstable phases by conventional techniques at high temperature strongly alters their physico-chemical and biological properties. Therefore, several low temperature routes have been investigated taking advantage of the structural characteristics of these compounds. Self-setting, injectable cements leading to nanocrystalline apatites have been developed. Solid mesoporous ceramic-like materials can also be obtained at low temperature by drying aqueous suspensions of nanocrystalline apatites. Among the most promising routes, however, are pressure sintering and spark plasma sintering at temperatures lower than 300°C. These techniques produce ceramics retaining most of the characteristics of the nanocrystals. The consolidation mechanism is thought to implicate the high mobility of ions within the hydrated layer. This work should be helpful for the preparation of a new generation of resorbable highly-reactive bioceramics.

INTRODUCTION

Biomimetic nanocrystalline apatites (NCA) analogous to bone mineral are involved in several medical devices due to their unique biological properties. They have been shown to form at the interface between most bioactive materials and bone tissue and this process is considered as essential in the biointegration of implants[1-3]. The ability of a material to nucleate a nanocrystalline apatite layer appears so important that it has become a widespread measure of the biological performance of biomaterials, promoted especially by T. Kokubo, using simulated body fluid (SBF)[4]. Although nanocrystalline apatites are used as coatings of metallic orthopaedic prosthesis, in bioabsorbable mineral cements and in bioabsorbable mineral-polymer composites[5-7], they are rarely proposed as bioceramics due to their thermal instability and the lack of low temperature processing technique preserving their surface and bulk properties.

Biomimetic NCA can be prepared by many different ways. These non-stoichiometric compounds possess a high reactivity related to the presence of a metastable hydrated layer on the surface of the nanocrystals[8-9]. The processing of such unstable phases by conventional techniques, at high temperature, strongly alters their physico-chemical and biological properties. Therefore, several low temperature routes have been investigated taking advantage of the structural characteristics of these compounds. Several processes will be compared involving gel-drying, cements, pressure sintering and Spark Plasma Sintering (SPS).

MATERIALS AND METHODS

Apatite nanocrystals

NCA were prepared by precipitation in aqueous media from calcium and phosphate (and possibly carbonate) solutions at room temperature and physiological pH as described elsewhere[10].

Briefly, a calcium solution is rapidly poured in a phosphate (and in some experiments, carbonate) solution. An excess of phosphate (and carbonate) ions in solution insured pH buffering. Some preparations were aged in the mother solution for different periods of time. The samples were then filtered on Büchner funnel, washed with deionized water and lyophilized or used in the wet state.

Gel drying method

The wet, gel-like cake was agitated until it became liquid (thixotropy) and was cast (50 ml) in a polytetrafluorethylene beaker (diameter 64 mm) for solidification and drying at room temperature. The shrinkage during drying was considerable and sometimes cracks appeared. A slow drying rate was maintained (typically 3 weeks for a 50 cm^3 original volume) to avoid such problems. This method allows easy associations of the nanocrystals with organic molecules. The association with albumin and casein was made by adsorption on NCA: after filtering, the solid was re-suspended in the protein solution (15 g/1000 ml) containing 0.1 % of NaN3 to prevent any development of bacteria during the slow drying stage. Gelatin was simply dissolved in the aqueous PCA gel (2% of wet gel weight). The suspensions were filtered and washed and treated as previously described. Small blocks (a few centimeters in size) free of cracks, despite an important shrinkage, were obtained.

Consolidation at mild temperature

Two processing techniques were investigated: uniaxial pressing and spark plasma sintering (SPS).

The uni-axial pressing of powders was performed by associating a Hounsfield press (model H25K-S) with a custom-built setup. A 8-mm diameter mold was used. Heating was provided simultaneously by a custom-built low-temperature regulated oven (maximum operational temperature: 300°C), and the actual temperature was determined by a thermocouple located near the sample.

Spark plasma sintering (SPS) experiments were performed on an SPS 2080 Sumitomo Coal Mining equipment. The temperature was measured by means of a thermocouple close to the sample. Powder samples were placed in a graphite mold (8 mm). A mechanical force of 2.5 kN was applied on the sample and the temperature was kept constant until completion of the sintering process. Fast cooling (~100°C/min) and a release of the mechanical force were then operated.

Physical-chemical characterization

The samples were characterized by X-ray diffraction (Inel CP 120, Cobalt anticathod, Kα radiation), FTIR (Nicolet 5700), chemical analyses and electron microscopy observations (SEM; LEO 435 VP and cryo-FEG-SEM; JEOL). The crystal size was determined by the Scherrer method from XRD peaks width (002 and 310 peaks). The mechanical characteristics were determined by different techniques depending on the samples: ultrasonic measurements (Young Modulus of gel dried samples), compressive tests on cylinders (uniaxial pressing) and diametral compressive tests on disks (SPS), using a Hounsfield press (H25K-S).

RESULTS

1-Gel drying at low temperature.

This process can be applied directly to precipitate samples after filtration. The gel-like cake contains about 10% of apatite nanocrystals. Drying was performed in a narrow range of temperature from 4°C to 37°C.

Characteristics of the nanocrystalline apatites

In the study presented here, the blocks were obtained using wet carbonate-containing nanocrystalline apatites aged in solution for different periods of time (maturation) and associations of these with biological macromolecules[11].

The chemical composition and crystal characteristics of the matured apatites and of the composites are reported in Table I.

Table I: Chemical Composition and Crystal Size of the Nanocrystals (after lyophilisation)

Sample	Maturation time (days)	Ca/P (atomic) ($\pm$ 0.02)	Ca/(P+C) (atomic) ($\pm$ 0.03)	C/P (atomic) ($\pm$ 0.01)	OM (% weight) ($\pm$ 0.5)	L_{002} (nm) ($\pm$ 0.3)	L_{310} (nm) ($\pm$ 0.3)
without proteins	0	1.42	1.35	0.05		14.6	3.9
	3	1.52	1.39	0.10		16.1	4.1
	10	1.57	1.40	0.12		17.3	4.5
	30	1.60	1.40	0.14		20.3	5.2
alb	0	1.50	1.42	0.06	34.6	15.1	3.5
	3	1.51	1.37	0.11	32.0	18.0	4.2
gel	0	1.52	1.43	0.06	24.5	13.4	3.5
	3	1.53	1.41	0.08	21.1	17.6	4.0
cas	0	1.53	1.46	0.05	35.9	14.6	3.1
	3	1.52	1.39	0.08	38.5	16.5	3.5

C: carbonate; OM: organic matter, L002: apatite crystal length (002 peak); L310 apatite width-thickness (310 peak)

The ceramic-like materials obtained by gels exhibit a variable density related to maturation stage of the crystals (Figure 1). Aged crystals with larger dimensions gave ceramics with a high porosity comparable to the Ca-P cements whereas freshly precipitated gels give the most densified ceramics.

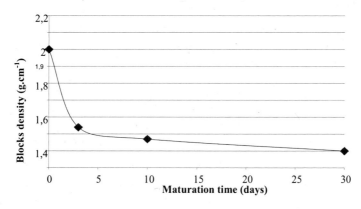

Figure 1: Apparent density of blocks obtained by gel drying as a function of the maturation time.

The Young modulus appeared to vary almost linearly with the density for blocks obtained from apatites at different maturation time, without proteins (Figure 2). The incorporation of proteins in these compounds is very easy (see materials and method section) and results in an increase of the Young modulus, compared to protein-free blocks obtained from apatites at the same maturation time.

The compressive strength of such blocks (for maturation time 0) was found to be close to 56.3 ± 7.1 MPa and reached 79.3 ± 4.6 MPa in the presence of casein.

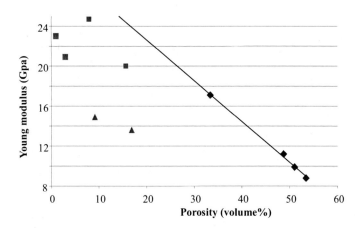

Figure 2: Variation of Young modulus of blocks obtained by gel drying as a function of overall porosity for nanocrystalline apatites at different maturation stages (blue diamond) and with different molecules (albumin: red squares; gelatin: green squares; and casein : brown triangles).

During hardening an evolution of the nanocrystals was observed compared to the initial lyophilized precipitates. Hardening involved a decrease of the carbonate content. XRD revealed an increase of the crystal size and FTIR spectra indicated an increase of the OH^- and HPO_4^{2-} contents for the hardened samples.

2- Blocks obtained by uniaxial compression at mild temperatures
These experiments were made with lyophilized non-carbonated nanocrystalline apatites. The lyophilized powders exhibited the characteristics reported in Table II.

Table II: Characteristics of nanocrystalline apatites used in pressure sintering

Maturation time	Atomic Ca/P ratio (± 0.02)	HPO_4 content (% of P; ± 0.5)	L_{002} (nm ; ± 0.3)	L_{310} (nm ; ± 0.3)
0	1.38	29.0	16.6	4.8

After several trials the samples were produced according to the conditions reported in Figure 3.

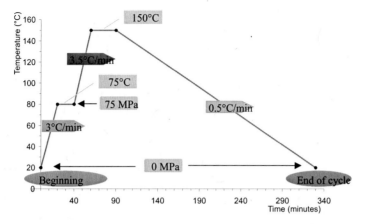

Figure 3: Sintering cycle for lyophilized apatite nanocrystals.

The temperature was first raised to 80 ($\pm$5)°C at a rate of 3°C per minute. This temperature was maintained for 20 minutes. At the end of this period a pressure of 75 MPa was applied. Then the temperature was raised to 150 ($\pm$5)°C at a rate of 3.5 °C per minute, under pressure. The temperature and pressure were maintained for 30 minutes. The mold was then cooled at a rate 0.5°C per minute and the sample was withdrawn from the mold at room temperature.

Fully dense ceramics could not be obtained by this method and apparent densification rates around 65% were measured. Nevertheless, the compacts obtained using non matured apatites, exhibited a high compressive strength in the range 100-150 MPa and a Young modulus of ca. 10 GPa. Very interestingly, despite a poor densification, the values of compressive strength compare well with those obtained after sintering of stoichiometric HA at much higher temperatures (1100-1250°C).

SEM examination of the fractured samples showed a continuous matrix with crystalline inclusions identified as anhydrous dicalcium phosphate (DCPA or monetite) (Figure 4).

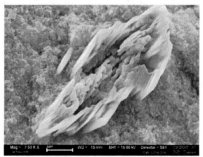

Figure 4: Micrograph of a fractured sample showing the DCPA inclusions

XRD analyses confirmed the presence of well crystallized DCPA and nanocrystalline apatite with larger crystal dimensions (22.4 nm length and 5.5 nm width-thickness) (Figure 5). No specific orientations of the nanocrystals were observed. FTIR spectra showed a chemical evolution of the samples with an increase of the amount of OH⁻ ions in the apatite lattice (Figure 6).

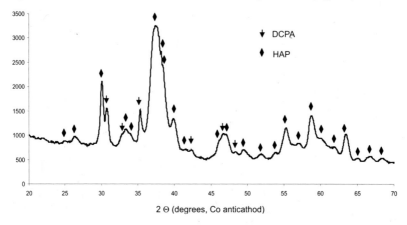

Figure 5 : XRD diagram of a pellet obtained by pressure sintering

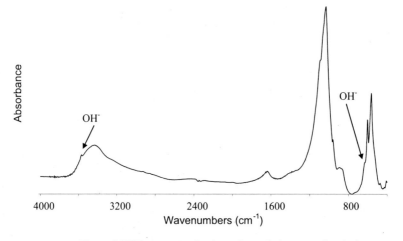

Figure 6: FTIR spectrum of a sintered sample (pressure sintering)

These findings suggest an alteration of the nanocrystals and a partial decomposition involving the hydrated surface layer of the nanocrystals and the apatite core. The process involved here leads to a loss of most of the water molecules associated with the nanocrystals.

SPS with its fast heating rates appeared then as an interesting technique prone to preserve the hydrated surface layer and the associated mobility of the surface ions.

3- Blocks obtained by SPS

These experiments were also performed with lyophilized non-carbonated nanocrystalline apatites, matured for 24 hours. The characteristics of these samples are reported in Table III.

Table III: Characteristics of nanocrystalline apatites used in pressure sintering

Maturation time	Atomic Ca/P ratio ($\pm$ 0.02)	HPO$_4$ content (% of P; $\pm$ 0.5)	L$_{002}$ (nm ; $\pm$ 0.3)	L$_{310}$ (nm ; $\pm$ 0.3)
1 day	1.43	18.9	23.6	6.5

The sintering conditions are schematized in figure 7. The total duration of an experiment is 13 minutes. The sintering temperature was: 150°C, and the pressure: 100MPa.

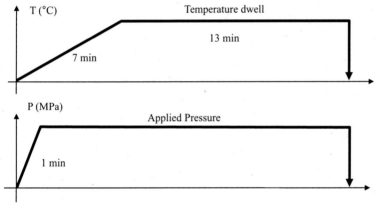

Figure 7: Sintering conditions of the apatite nanocrystals (SPS)

The spontaneous densification of the powder began at 100°C and ended at 150-190°C (Figure 8). The samples described in this report correspond to the best conditions investigated: T°=150°C; pressure=100MPa. Due to the small size of the cylinders obtained, the mechanical testing could not be made by classical compression. The tests were carried out using diametral compression (Brazilian test). The tensile strengths thus measured were in the range 18 to 25 MPa. Interestingly, these values are close to those obtained with stoichiometric HA despite extremely short heating times.

XRD analyses showed that the nanocrystals were preserved in the process and no other crystalline phase was detected. Depending on the processing conditions a texturation of the samples was noticed corresponding to a common orientation of the c axis of the hexagonal unit-cell perpendicular to the applied force (Figure 9). Comparison with the pattern of the initial powder indicated an increase of the degree of crystallinity after processing. No formation of monetite was observed by SPS for dwell temperatures under 200°C. Above this temperature however DCPA and calcium pyrophosphate were observed

FTIR analysis showed a partial loss of water and the increase of the OH‾ bands (Figure 10). The FEG-SEM micrograph (Figure 11) shows the nanocrystals, no inclusion of foreign phase was observed unlike for low temperature regular pressure sintering.

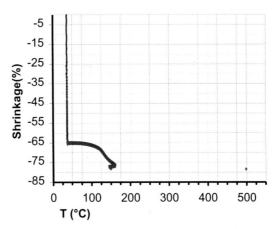

Figure 8: Shrinkage of the nanocrystalline apatite samples (SPS).

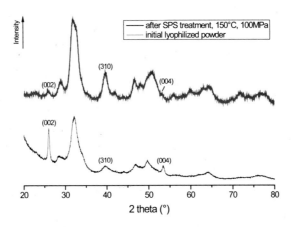

Figure 9: Comparison of the XRD diagrams of the lyophilized powder (bottom, black) and the sintered pellet (top, red). The relative intensity of 002 and 004 peaks has considerably decreased indicating a texturation of the nanocrystalline material.

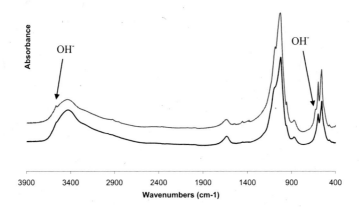

Figure 10: FTIR spectra of the lyophilized powder (bottom, black) and the sintered sample (red, top).

Figure 11: FEG-SEM micrograph of SPS pellets obtained from apatite nanocrystals.

DISCUSSION

Several reports have been published on low temperature processing of apatites leading to solid bodies. Most of them concern calcium phosphate mineral cements. Commercial cements exhibit generally compressive strengths between 12 and 80 MPa. Using low liquid to solid ratio and pressure Martin and Brown[12] reached tensile strengths of 18 MPa and compressive strength of 174 MPa with a Young's modulus of 6 GPa. The direct consolidation of pre-formed apatites nanocrystals at low temperatures has however been rarely attempted. Tadic and Epple[13] report the formation of monoliths from apatite nanocrystals by cold isostatic pressing (4kbar, 25 °C) with compressive strengths in the range 20-50 MPa. The present data obtained by SPS on biomimetic apatite nanocrystals clearly indicate the existence of a narrow range of temperature (100-200°C) for which a characteristic and significant shrinkage is observed, which corresponds to a new phenomenon.

Biomimetic nanocrystalline apatites are characterized by the existence on the surface of the crystals of an hydrated layer containing mainly, but not exclusively, bivalent ions (calcium, HPO_4^{2-} and CO_3^{2-}) as schematized in Figure 11[9, 14].

It has been shown that this hydrated layer progressively decreases on ageing in solution as the more stable apatite domains slowly grow. Analyses of wet samples by spectroscopic techniques (FTIR and solid-state NMR) have demonstrated that this hydrated layer is structured[15]. On drying however the loosely bound water molecules are removed and the surface structure is lost leaving the surface mineral ions in amorphous-like environments. However even after lyophilization an important amount of water molecules still remains (10 to 15 weight %) which might enhance the surface mobility of the mineral ions.

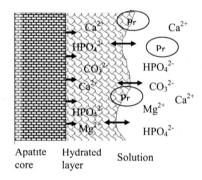

Apatite   Hydrated   Solution
core   layer

Figure 11: Schema of the surface hydrated layer on apatite nanocrystals. The structured hydrated layer constitutes a pool of loosely bound ions which can facilitate intercrystalline interactions. These ions nourish the growth of the more stable apatite domains.(Pr: proteins).

An important point concerning apatite nanocrystals is the wide range of compositions and properties exhibited[16]. All nanocrystalline apatites used in the different studies reported in this paper can be qualified as biomimetic due to their conditions of formation at low temperature and physiologic pH and their general physical-chemical characteristics (non-stoichiometry, crystal size and presence of non-apatitic species). However for the simplicity and understanding of apatite modifications, low temperature sintering (uniaxial mild temperature-pressing and SPS) were performed with non-carbonated samples. Undoubtedly, carbonate ions could considerably modify the behavior of these apatites in the different processing methods investigated here, but they would also have added more complexity to the system.

The chemical composition of biomimetic apatites is generally approximated by the chemical formulae describing non-stoichiometric apatites[16,17]:

$$Ca_{10-x} (PO_4)_{6-x}(HPO_4; CO_3^{2-})_x (OH)_{2-x} \quad (I)$$

in which the loss of a negative charge due to the replacement of a trivalent PO_4^{3-} anion by a bivalent one (HPO_4^{2-} or CO_3^{2-}) is compensated for by the combined loss of a bivalent cation Ca^{2+} and a monovalent anion OH⁻. Thus at short maturation times in solution the apatite nanocrystals contain a very high amount of HPO_4^{2-} close to the maximum whereas at long maturation times the amount of HPO_4^{2-} decreases and the amount of OH⁻ increases. Another equivalent description, from the point of view of the chemical composition, considers, as proposed by W. Brown, that non-stoichiometric apatites result of the interlayering of Octacalcium phosphate (OCP) and stoichiometric hydroxyapatite (sHA):

$$(1-\alpha) \quad Ca_{10}(PO_4)_6(OH)_2 \quad sHA \quad (II)$$
$$\alpha \quad Ca_8(PO_4)_4(HPO_4)_2(H_2O)_5 \quad OCP \quad (III)$$

where α represents the molar fraction of OCP. These representations of non-stoichiometric apatites lead to the same results (with $x=\alpha/2$). However they do not describe correctly the wide range of compositions observed for nanocrystalline apatites. Also, the presence of a hydrated layer, with a still undetermined composition, cannot be neglected.

A common evolution has been observed in the different processing techniques used (wet or dry) corresponding to an increase of the OH⁻ content and crystal dimensions during the ceramization processes. As the mineral content does not change, this phenomenon can be associated with a hydrolysis of the PO_4^{3-} ions by residual water molecules:

$$H_2O + PO_4^{3-} \rightarrow HPO_4^{2-} + OH^- \quad (reaction \ 1)$$

This reaction has been described by Heughebaert[18] in the hydrolysis of amorphous calcium phosphate into apatite and is believed to occur also in OCP[19]. As suggested by Heughebaert the formation of monovalent anions could favor the structuration and growth of apatite domains and could be associated with the increase of the apatite domains size. It shall be noticed that the evolution of nanocrystalline apatites on ageing in solution or in the different processes studied here are quite similar indicating a common stabilization mechanism except for the change of the Ca/P ratio, increasing during maturation and necessarily constant in the different processing methods.

Similarly a decrease of the amount of non-apatitic environments of the mineral ions has been observed, especially in the case of gel-drying, which supports the assumption that, at these low temperatures, the growth of the apatite domains involves essentially the mobilization of the mineral ions of the hydrated layer.

These reactions located within a nanocrystal illustrate their high reactivity; however they do not explain the consolidation process.

Unlike mineral cement setting reactions due to recrystallization in an aqueous media, and crystals interdigitation, the evolution of crystal size is negligible in these processes, and recrystallization has been excluded[9]. Thus the formation of a solid body is essentially related to intercrystal interactions. Such interactions may be favored by different mechanisms depending on the processing technique.

In the case of gel drying, the evaporation of water and the shrinkage of the gel could bring the crystals in close proximity, favoring their surface interaction possibly related to hydrogen bonding of adsorbed water molecules. However as drying goes on, the gel becomes more rigid and strains accumulate leading sometimes to cracks in the solid body. The higher density achieved by the less matured apatites suggests that their nanocrystals (due to their small sizes and developed hydrated surface layer) can establish multiple interactions and possibly re-orient themselves. As the apatite nanocrystals mature however, the crystals dimensions increase and reorientations are more difficult. In addition, the surface reactivity is partly lost and intercrystalline interactions are weaker. These phenomena result in less dense ceramics with lower Young's modulus. The presence of proteins seems to change the drying mode and the interactions between crystals. For immature crystals, albumin and gelatin produce an increase of Young's modulus, probably by filling the gaps between crystals, thus increasing the rigidity of the associations, the blocks appear then translucent. In the case of casein which interacts very strongly with apatite, however, the density is much lower indicating that crystals are farther apart. The reasons for this behavior may be found in the complex association of casein micelles and apatite crystals[20] and/or in possible changes of surface properties of the crystals (surface tension, zeta potential, ...). In every case involving organic-mineral composites, however, the alterations due to maturation are weaker than those observed for the blocks obtained without proteins. The decrease of density with maturation is faint or even not observed when proteins are present and the hardening process seems altered. Despite the increase in crystal dimensions, the porosity does not seem to increase when proteins are present. Proteins prevented Young's modulus variations related to maturation and maintained its values in a narrow range depending on the nature of the proteins. Globally the gel-drying process does not give however ceramics with very high compressive strength because of the important shrinkage and resulting strains in the solid body.

In the case of pressure sintering and SPS, the increase of the pressure favors the intercrystalline interactions, however pressure alone does not seem sufficient to consolidate the pellets. Temperature plays a key role in mineral ions mobility in the partly hydrated surface layer. Simultaneously the removal of water facilitates direct ionic interaction but reduces considerably the ion mobility. The rapidity of SPS sintering appears then as a considerable advantage.

An interesting phenomenon is the decomposition of the nanocrystals in pressure sintering. This reaction does not occur when the nanocrystals are heated without pressure. The decomposition of the nanocrystals resembles that of octacalcium phosphate as described by Brown [Brown 1957, 1962][21]. This author has shown that OCP decomposes into hydroxyapatite and dicalcium phosphate dihydrate (DCPD) or dicalcium phosphate anhydrous depending on the heating conditions:

$$2\ Ca_8\ (PO_4)_4\ (HPO_4)_2,\ 5H_2O \rightarrow\ 6\ CaHPO_4 + Ca_{10}(PO_4)_6(OH)_2 + 8H_2O$$

The reaction implies a hydrolysis of PO_4^{3-} ions analogous to reaction I observed in our samples, however several differences appear. For OCP this reaction occurs without pressure; in addition, it has been shown that the nanocrystals simply heated (without pressure) show also an increase of the OH^- content with an increase of apatite crystal size but no formation of DCPA was observed. This observation suggests that the hydrolysis reaction and the formation of DCPA are not related. Moreover, the amount of DCPA observed is always very low (a few percent) and does not correspond to a possible decomposition of the non-stoichiometric apatite domains, which would have given more DCPA considering the initial HPO_4^{2-} content of the samples and the concomitant hydrolysis reaction of phosphate ions. A plausible hypothesis is that under the conjunction of pressure and temperature, the mobile ions of the hydrated layer, mainly bivalent, could be excluded of the intercrystalline space and migrate then crystallizing separately into DCPA. In the case of SPS the sintering process seems too fast to allow this process to take place under 200°C.

The advantage and disadvantages of different consolidation processes of apatite nanocrystals are summarized in table IV.

Table IV: Summary of the Different Characteristics of Low Temperature Processing of Nanocrystalline

Table IV: Summary of the Different Characteristics of Low Temperature Processing of Nanocrystalline Apatites

Method	Associations with proteins	Porosity	Compressive Strength (MPa)	Decomposition	OH⁻ formation and crystal dimensions
Cements	+/-	≈ 50%	12-80	No	Slow No change
Gel-drying	+	35 %	40-70	No	Slow Small increase, or no change
Pressure sintering	-	40-50%	90-120	DCPA impurities	Fast Increase
SPS	-	30-40 %	-	DCPA impurities (above 200°C)	Fast Increase

CONCLUSION

The low temperature consolidation processes described here, preserve the original mineral structure with nanometric crystal dimensions and the non-stoichiometry of the apatite phase. The gel drying process seems to preserve best the original characteristics of the nanocrystals but does not give solid bodies with interesting mechanical properties. The processes involving pressure sintering and SPS at heating temperatures between 150 and 200°C give samples with potentially better mechanical properties. Future developments will include the formation of porous bodies and tough composite materials.

REFERENCES:
[1]L. Hench, J.W. Hench, D. Greenspan, Bioglass a short history and bibliography, *J. Aust.Ceram.Soc.*, **40**,1-42 (2004).
[2]T. Kokubo, H.M. Kim, M. Kawashita, Novel bioactive materials with different mechanichal properties, *Proceedings of Biomaterials 24*, 2161-2175 (2004).
[3]P. Li, X. Ye, I. Kangasniemi, J.M. de Blieck-Hogervorst, C.P. Klein, K. de Groot, In vivo calcium phosphate formation induced by sol-gel prepared silica. *J. Biomed. Mater. Res.*, **29**, 325-328 (1995).
[4]T. Kokubo, H Takadama, How useful is SBF in predicting in vivo bone bioactivity, *Biomaterials*, **27**, 2907-2915 (2006).
[5]D. Knaack D., M.E.P. Goad, M. Ailova, C. Rey, A. Tofighi, P. Chakravarthy, D. Lee, Resorbable calcium phosphate bone substitute. *J. Biomed. Mat. Res.*, **43**, 399-409 (1998).
[6]P. Habibovic, F. Barrère, C.A. van Blitterswicjk, K. de Groot, P. Layrolle, Biomimetic hydroxyapatite coating on metal implants. *J. Am. Ceram. Soc.*, **85**, 517-522 (2002).
[7]A. Tampieri, G. Celotti, E. Landi, M. Sandri, N. Roveri, G. Falini, Biologically inspired synthesis of bone-like composite: self–assembled collagen fibers/hydroxyapatite nanocrystals, *J. Biomed. Mater. Res.*, **67**, 618-625 (2003).
[8]C. Jäger, T. Welzel, W. Meyer-Zaika, M. Epple, A solid state NMR investigation of the structure of nanocrystalline hydroxyapatite, *Magn. Reson. Chem.*, **44**, 573-580 (2006).
[9]C. Rey, C. Combes, C. Drouet, H. Sfihi, A. Barroug, Physico-chemical properties of nanocrystalline apatites: Implications for biominerals and biomaterials, *Materials Science and Engineering C*, **27**, 198-205 (2007)..
[10]C. Rey, A. Hina, A. Tofighi, M.J. Glimcher, Maturation of poorly crystalline apatites: chemical and structural aspects in vivo and in vitro, *Cells and Mater.*, **5**, 345-356 (1995).
[11]S. Sarda, A. Tofighi, M.C. Hobatho, D. Lee, C. Rey, Associations of low temperature apatites

ceramics and proteins, *Phosphorus Res. Bull.*, **10**, 208-213 (1999).

[12]R.I. Martin, P.W. Brown, Mechanical properties of hydroxyapatites formed at physiological temperature, *J. Mater. Sci. Mater. Med.,* **6**, 138-143 (1995).

[13]D. Tadic, M. Epple, Mechanically stable implants of synthetic boe mineral by cold isostatic pressing, *Biomaterials*, **24**, 4565-71 (2003).

[14]D. Eichert, H. Sfihi, C. Combes, C. Rey, Specific characteristics of wet nanocrystalline apatites. Consequences on biomaterials and bone tissue, *Key Engineering Materials*, **254-256**, 927-930 (2004).

[15]D. Eichert, C. Combes, C. Drouet, C. Rey, Formation and evolution of hydrated surface layers of apatites, *Key Engineering Materials* **284-286** 3-6 (2005).

[16]D. Eichert, C. Drouet, H. Sfihi, C. Rey, C. Combes, In *"Biomaterials Research Advances"*, ed. JB Kendall, Nova Science Publishers, p. 93-143 (2007).

[17]J.C. Elliott, *Structure and Chemistry of the Apatites and Other calcium Orthophosphates*, Amsterdam, Elsevier, 1994.

[18]J.C. Heughebaert, G. Montel, Conversion of amorphous tricalcium phosphate into apatitic tricalcium phosphate, *Calcif. Tiss. Int.*, **34**, S103-S108 (1982).

[19]M. Mathew, W.E. Brown, L.W. Schroeder, Crystal structure of octacalcium bis-(hydrogenphosphate)tetrakis(phosphate)-pentahydrate, $Ca_8(HPO_4)_2(PO_4)_4$, $5 H_2O$. *J. Cryst. Spect. Res.,* **18**, 235-250 (1988).

[20]C. Holt, An equilibrium thermodynamic model of the sequestration of calcium phosphate by casein micelles and its application to the calculation of partition salt in milk, *Eur. Biophys. J.,* **33**, 421-34 (2004).

[21]W.E. Brown, J.P. Smith, J.R. Lehr, A.W. Frazier, Octacalcium phosphate and hydroxyapatite: Crystallographic and chemical relations between octacalcium phosphate and hydroxyapatite; *Nature*, **196**, 1050-1055 (1962).

SINTERING BEHAVIOR OF HYDROXYAPATITE CERAMICS PREPARED BY DIFFERENT ROUTES

Tan Chou Yong, Ramesh Singh, Aw Khai Liang, Yeo Wei Hong
University Tenaga Nasional
Kajang, Selangor, Malaysia.

ABSTRACT

The sintering behaviour of three different HA, i.e. a commercial HA(C) and synthesized HA by wet precipitation, HA(W) and mechanochemical method, HA(M) were investigated over the temperature range of 1000°C to 1350°C. In the present research, a wet chemical precipitation reaction was successfully employed to synthesize highly crystalline, high purity and single phase stoichiometric HA powder that is highly sinteractive particularly at low temperatures below 1100°C. It has been revealed that the sinterability and mechanical properties of the synthesized HA by this method was significantly higher than that of the commercial material and HA which was synthesized by mechanochemical method. The optimum sintering temperature for the synthesized HA(W) was 1100°C with the following properties being recorded: 99.8% relative density, Vickers hardness of 7.04 GPa and fracture toughness of 1.22 MPam$^{1/2}$. In contrast, the optimum sintering temperature for the commercial HA(C) and synthesized HA(M) was 1300°C with relative density of 98% and 95.5%, Vickers hardness of 5.47 GPa and 4.73 GPa, fracture toughness of 0.75 MPam$^{1/2}$ and 0.82 MPam$^{1/2}$ being measured, respectively.

INTRODUCTION

Hydroxyapatite, $Ca_{10}(PO_4)_6(OH)_2$ (HA) material has been clinically applied in many areas of dentistry and orthopedics because of its excellent osteoconductive and bioactive properties which is due to its chemical similarity with the mineral portion of hard tissues[1]. Bulk material, available in dense and porous forms, is used for alveolar ridge augmentation, immediate tooth replacement and maxillofacial reconstruction[2]. Nevertheless, the brittle nature and the low fracture toughness (< 1 MPam$^{1/2}$) of HA constraint its scope as a biomaterial in clinical orthopaedic and dental applications[3]. Hence, the development of an improved toughness HA material is required. As a result, various studies have been carried out to improve the mechanical properties of sintered HA[4].

The success of HA ceramic in biomedical application is largely dependent on the availability of a high quality, sintered HA that is characterized having refined microstructure and improved mechanical properties[4]. Intensive research in HA involving a wide range of powder processing techniques, composition and experimental conditions have been investigated with the aim of determining the most effective synthesis method and conditions to produce well-defined particle morphology[1-3]. Among the more prominent methods used to synthesize HA are wet precipitation method, mechanochemical technique, sol-gel technique and hydrothermal. Although numerous studies on HA synthesized via wet precipitation technique and mechanochemical method are carried out, nevertheless, reports on the sinterability of HA synthesized through these technique are rather scarce. Therefore, the primary objective of the present work was to synthesize a well-defined, crystalline, pure hydroxyapatite (HA) phase using two techniques, i.e. wet precipitation technique and mechanochemical technique. The sinterability of both synthesized hydroxyapatite (HA) was evaluated and compared with a commercially available HA (Merck, Germany).

METHODS AND MATERIALS

In the current work, the HA powder used was prepared according to a novel wet chemical method, hereafter named as HA(W), comprising precipitation from aqueous medium by slow addition of orthophosphoric acid (H_3PO_4) solution to a calcium hydroxide ($Ca(OH)_2$)[5]. The HA powder synthesized by mechanochemical method used in the present work, labeled as HA(M), was prepared according to the method reported by Rhee[6]. The starting precursors used were commercially available calcium pyrophosphate, $Ca_2P_2O_7$, and calcium carbonate, $CaCO_3$. In order to evaluate the sinterability and performance of both the synthesized HA, a commercially available stoichiometric HA powder manufacture by Merck, Germany was also studied, hereafter is known as HA(C). The green samples were uniaxial compacted at about 1.3 MPa to 2.5 MPa The green compacts were subsequently cold isostatically pressed at 200 MPa (Riken Seiki, Japan). This was followed by consolidation of the particles by pressureless sintering performed in air using a rapid heating furnace over the temperature range of 1000°C to 1350°C, with ramp rate of 2°C/min. (heating and cooling) and soaking time of 2 hours for each firing. All sintered samples were then polished to a 1 μm finish prior to testing.

The calcium and phosphorus content in the synthesized HA powder were determined by using the Inductively Coupled Plasma-Atomic Emission Spectrometry (ICP-AES) technique. The particle size distributions of the HA powders was determined using a standard Micromeritics® SediGraph 5100 X-ray particle size analyzer. The specific surface area of the powder was measured by the Brunauer-Emmett-Teller (BET) method. The morphology of the starting powder was examined using a Philips ESEM model XL30 scanning electron microscope. The phases present in the powders and sintered samples were determined using X-Ray diffraction (XRD) (Geiger-Flex, Rigaku Japan). The bulk densities of the compacts were determined by the water immersion technique (Mettler Toledo, Switzerland). The relative density was calculated by taking the theoretical density of HA as 3.156 Mgm^{-3}. The microhardness (H_v) of the samples was determined using the Vickers indentation method (Matsuzawa, Japan). The indentation fracture toughness (K_{Ic}) was determined from the equation derived by Niihara[7].

RESULTS AND DISCUSSIONS

Within the accuracy of the analysis, the results show that the Ca/P ratio of all the powders studied was within the stoichiometric range of 1.67. Ozeki et al.[8] have emphasized the importance of the Ca/P ratio since any deviations from the stoichiometric value would have an adverse effect on the sintered properties of the hydroxyapatite body.

The HA(W) powder consists of a mixture of fine powder particles ranging from 1-3 μm diameter and larger particles of 5-10 μm diameter. The larger particles appear to be large agglomerates of loosely packed smaller particles, resulting in a rough surface as shown in Figure 1. The drying of the filter cake of synthesized HA could have resulted in less compaction of the precipitate and, although the dried filter cake was ground and sieved, this probably resulted in the formation of soft agglomerates which was found to break easily using a very low pressing pressure of 1.3-2.5 MPa during powder compaction to form the green body. Due to the soft nature of the powders, attempts to use higher pressures during uniaxial pressing the samples proved to be futile as powders lamination on the die surface was observed and in some extreme cases, a layer of compacted powders separated immediately from the green body upon ejection from the mould.

Figure. 1. SEM micrographs of HA(W) revealing the presences of loosely packed particles.

In contrast, although the particle size for powder synthesized by mechanochemical method ranges from 0.5-4 μm diameters, it should be highlighted that the powder consists of hard agglomerates as typically shown in Figure 2. Additionally, "neck" formation could be observed between smaller particles as a result of the heat treatment process carried out on this powder during the synthesis stage.

Figure 2. SEM micrographs of synthesized HA(M) revealing neck formation between particles and presences of hard particles.

On the other hand, the HA(C) powder consisted of a mixture of small and large particles as shown in Figure 3. The presences of soft agglomerates could not be observed in the commercial powder but instead the particles appeared to be large, up to 10 μm, and seemed to be more compacted when compared to those observed for the HA powder synthesized by the wet chemical method.

Figure 3. SEM micrographs of commercial HA(C) powders.

X-ray diffraction (XRD) analysis of the synthesized HA(W) and HA(M) powders produced only peaks which corresponded to the standard JCPDS card no: 74-566 for stoichiometric HA as shown in Figure 4 and Figure 5, respectively. The only difference in the XRD patterns of HA(W) and HA(M) powders before sintering was in the degree of crystallinity. The HA(W) XRD patterns indicated the powder was poorly crystalline as shown by the broad diffraction peaks, which is a characteristic of HA prepared by an aqueous precipitation route. This observation is in agreement with that reported by Gibson et al.[9] who found that calcination at higher temperature, as in the present case for HA(M) powders, would exhibit a narrower diffraction peaks and not a broad one as observed for the HA(W) powder in the present work.

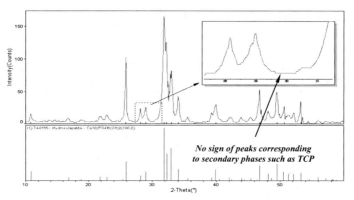

Figure 4. Comparison of XRD patterns of synthesized HA(W) with the standard JCPDS card for stoichiometric HA.

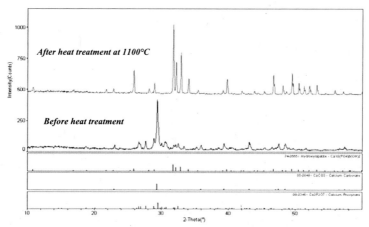

Figure 5. Comparison of XRD patterns of synthesized HA(M) powder before and after heat treatment at 1100°C.

In contrast, powders prepared by mechanochemical process produced a diffraction pattern that indicated a highly crystalline material, with narrow diffraction peaks as a result of heat treatment as typically shown in Figure 5. It should be noted that HA could only be obtained in this powder upon heat treatment at 1100°C. This is shown clearly in Figure 5 where the XRD trace of the prepared powder before heat treatment produced peaks that corresponded to the starting precursors (calcium carbonate and calcium pyrophosphate). Similarly, X-ray diffraction analysis of the commercial powder, HA(C), produced only peaks which corresponded to the standard JCPDS card no: 74-566 for stoichiometric HA.

After sintering in air atmosphere, the commercial HA compacts were observed to have a distinct colour change, i.e. from white (as-received powder) to blue (as-sintered). The intensity of the blue colour was also observed to increase with increasing sintering temperature, i.e. from light blue (1000°C) to dark blue (1350°C). It has been reported by Yubao et al.[10] that most commercial HA powders contained small additions of impurities and the origin of the apatite blue colour was due to the presence of Mn^{5+} or MnO_4^{3-} ions at the PO_4^{3-} sites in the apatite crystal structure. According to these authors, sintering at high temperature not only increases the intensity of oxidation in the oxidizing atmosphere, but also provides enough energy for the oxidized manganese ion (Mn^{2+} to Mn^{5+}) to migrate within the crystal lattice. This colour change, however was not observed in both the synthesized HA(W) and HA(M) compacts. These materials remained white regardless of sintering temperature.

The effect of elemental impurities on the sinterability of the powders could not be confirmed by this study alone. The change in colour in commercial HA was found to have negligible effect on the HA phase stability as confirmed by XRD phase analysis of the sintered HA in the present work. The sintering of the synthesized HA(W) compacts revealed the present of only HA phase as shown in Figure 6. Similar results were observed for HA(C).

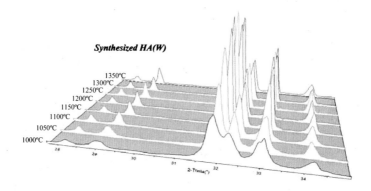

Figure 6. XRD patterns of synthesized HA(W) sintered at various temperatures. All the peaks corresponded to that of stoichiometric HA.

The formation of secondary phases such as tricalcium phosphate (TCP), tetracalcium phosphate (TTCP) and calcium oxide (CaO) was not detected throughout the sintering regime employed. The result shows that the phase stability of HA was not disrupted by the initial pressing conditions, sintering schedule and temperature employed. Similarly, Liao et al.[11] described that HA ceramics would only start to decompose into secondary phases upon sintering beyond 1350°C.

The present results however are not in agreement with other workers who found that decomposition of HA synthesized by wet precipitation method starts at about 1300°C [12]. Kothapalli et al.[13] have reported that sintering at 1200°C would caused the HA to decomposed into α-TCP {$Ca_3(PO_4)_2$}, β–TCP {$Ca_3(P_2O_8)$} and calcium oxide (CaO) according to Equa.1.

$$Ca_{10} (PO_4)_6 (OH)_2 \Rightarrow Ca_3(PO_4)_2 + Ca_3(P_2O_8) + CaO + H_2O \qquad (1)$$

In the present work, decomposition of HA(W) was not observed throughout the sintering regime employed. Additionally, no attempt was made to control the sintering atmosphere and the sintering atmosphere was just plain air (not moisturized). The high local humid atmosphere (i.e. the mean monthly relative humidity falls within 70% to 90% all year around) could have played a role in hindering dehydroxylation in the HA matrix even at 1350°C. In addition, as the HA was produced by wet chemical route and were not calcined prior to sintering, a significant amount of absorbed water would probably remained in the structure. However, it is not clear if the loss of water during sintering plays a role in suppressing dehydroxylation.

The XRD traces of hydroxyapatite synthesized via the mechanochemical technique also revealed the present of only HA phase as indicated in Figure 7. The present results obtained for the sintered HA(M) samples contradicted the findings of Mostafa[14]. In general, the author reported that the powder which was synthesized using the same technique as the present HA(M) transformed partially into β-TCP upon sintering at 1100°C. In another work, Yeong et al.[15] has also confirmed that the HA powder synthesized using mechanochemical method transformed partially into TCP when sintered beyond 1200°C. The difference in the results reported in the literatures as compared to the present work could be attributed to the milling medium used, the milling time and probably due to the high purity of the starting powders. The milling medium used in the current work was water with a milling time of 8 h. Rhee[6] has emphasized the importance of having 100% water content in the milling medium during the mechanochemical synthesis process. Yeong et al.[15] have synthesized their HA powder using ethanol as the milling medium and this could be the reason for obtaining secondary phases upon sintering due to insufficient H_2O present in the medium to suppress decomposition activity in the HA structure. Additionally, the milling time employed is another important factor that could influence the sintering behaviour of HA. Kim et al.[16] synthesized HA powder by mechanochemical method using water as the milling medium and the milling time was set at 60 minutes. Nevertheless, the authors observed secondary phases in their HA matrix upon sintering.

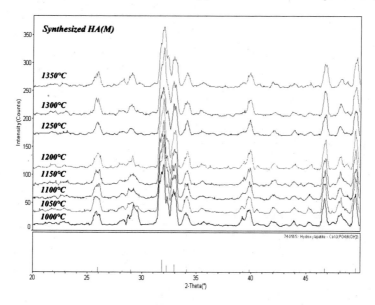

Figure 7. XRD patterns of synthesized HA(M) sintered at various temperatures. All the peaks corresponded to that of stoichiometric HA.

The effects of sintering temperature on the sintered densities of the three HA compacts are shown in Figure 8. All samples were heated to the chosen sintering temperature at 2°C/min. and, after a dwell time of 2 h, cooled to room temperature at 2°C/min..

In general, the bulk density increases with increasing sintering temperature regardless of the type of powder studied. A small increase in density is observed before the onset of densification and this corresponds to the first stage of sintering, where necks are formed between powder particles. The second stage of sintering corresponds to densification (onset of densification) and the removal of most of the porosity. The onset of densification, indicated by a sharp increase in the sintered density, for HA(M) and HA(C) was between 1100°C and 1200°C. In the case for HA(W) samples, sintering were carried out from 700°C so as to determine the onset densification temperature. As shown in Figure 8, this temperature was found to be between 900°C and 1000°C. Generally, sintering above this range resulted in very small increased in density which is associated with the final stages of sintering where small levels of porosity are removed and grain growth begins.

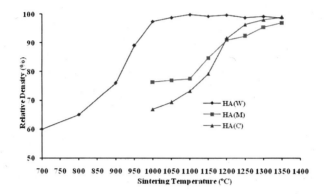

Figure 8. Effect of sintering temperature on the relative densities of HA(W), HA(M) and HA(C) samples.

The synthesized HA by wet precipitation method achieved a final sintered density of 97-99% of theoretical density at 1050-1100°C, whereas the commercial HA required a sintering temperature of 1250-1300°C to attain a similar density. In general, these results indicate that HA synthesized by wet precipitation method are more sinteractive than the commercial HA. The fact that HA(W) shows improved relative density as compared to HA(C) could be attributed to the physical basis of the Herring law of sintering that suggest sintering rate at a given temperature is inversely proportional to the square of the powder particle size. In the present study, the particle size measured using particle size analyzer for HA(W) and HA(C) are 1.78 ± 0.22 μm and 3.26 ± 1.53 μm respectively. Thus, the smaller the particle size, the easier would be for the powder to achieve high density when sintered at the same temperature.

As for HA(M), the highest relative density of 98% could only be attained when the material was sintered at 1350°C. In general, the densification behaviour of HA(W) was more superior to that of HA(M). This could be attributed to the higher surface area of 60.74 m^2/g for HA(W) which promoted a large driving force for densification as compared to a low 2.68 m^2/g for HA(M). Furthermore, the spherical nano particles of HA(W) as observed within the larger agglomerates (Figure 1) would have a higher heat transfer area and thus accelerate densification upon sintering. In contrast, HA(M), which consists of hard agglomerates (Figure 2), could have led to a lower packing density in the green state and hence poor sintering behaviour.

The effect of sintering temperature on the Vickers hardness of the three HA is shown in Figure 9. Throughout the sintering temperature range studied, it can be noted that HA synthesized by the wet chemical precipitation method displayed superior hardness as compared to HA synthesized by mechanochemical method and commercial HA. The hardness for HA(W) increased as the sintering temperature was increased and reached a maximum value of about 7.24 GPa when sintered at 1150°C. Further sintering above 1150°C resulted in deterioration in the hardness. On the other hand, the hardness of HA(C) started to increased sharply between 1100-1250°C and attained a maximum of 5.47 GPa at 1300°C as shown in Figure 9. This trend is similar for HA(M) which achieved a maximum hardness of 4.73 GPa when sintered at 1300°C. Nevertheless, a decreased of hardness was observed for both HA(C) and HA(M) samples when sintered beyond 1300°C.

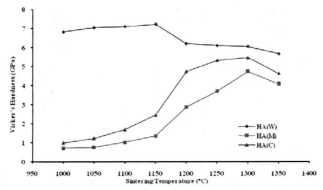

Figure 9. Effect of sintering temperature on the Vickers hardness of hydroxyapatite.

The effects of sintering temperature on the fracture toughness of compacts of three different HA powders are shown in Figure 10. In general, HA(W) exhibited better fracture toughness values as compared to both HA(M) and HA(C) throughout the sintering temperature range studied. The synthesized HA(W) by wet precipitation method achieved maximum K_{Ic} value of 1.22 MPam$^{1/2}$ when sintered at 1100°C. Further sintering at higher temperature would cause a decreased in the toughness as shown in Figure 10. In comparison, both HA(M) and HA(C) could only achieved maximum K_{Ic} of 0.82 MPam$^{1/2}$ and 0.77 MPam$^{1/2}$, respectively when sintered at 1300°C.

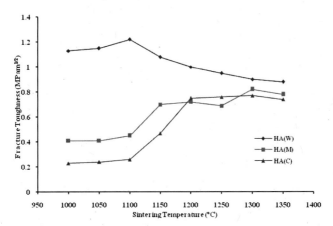

Figure 10. Effect of sintering temperature on the fracture toughness of HA.

In general, the maximum K_{Ic} values for most HA reported in the literature varied between 0.96-1.0 $MPam^{1/2}$ [17-19]. Thus, the high value of 1.22 $MPam^{1/2}$ obtained in the present work for HA(W) when sintered at 1100°C is very encouraging. It is believed that this improvement in fracture toughness in the present HA(W) could be attributed to the improved properties of the starting synthesized HA powder.

Generally, although the chemical and phase compositions of the powders studied were similar, the powder morphology and particle size distribution are very different. It is believed that the different powder characteristics are largely a result of the different powder processing methods used and these differences are evident in the sintering characteristics of the powders.

CONCLUSION

The sinterability of a high purity, single-phase HA powder produced in this study by a wet chemical precipitation reaction, was significantly greater than for powder synthesized via mecahnochemical technique and a commercial HA powder. The X-ray diffraction signatures of both the synthesized HA studied compared very favorably with that of a stoichiometric HA. Decomposition of HA phase to form tri-calcium phosphate, tetra tri-calcium phosphate and calcium oxide was not observed in the present work for all the powders. It has been revealed that the sinterability in terms of density, Vickers hardness and fracture toughness of the synthesized HA via precipitation method was significantly greater than that of the synthesized HA using the mechanochemical method and the commercial HA.

ACKNOWLEDGEMENTS

The authors would like to thank the Ministry of Science, Technology and Innovation of Malaysia for providing the financial support under the grant no. 03-02-03-SF0073. In addition, the support provided by UNITEN and SIRIM Berhad in carrying out this research is gratefully acknowledged.

REFERENCES

[1]H. Oguchi, K. Ishikawa, K. Mizoue, K. Seto, and G. Eguchi, Long-Term Histological Evaluation of Hydroxyapatite Ceramics in Humans, *Biomaterials*, **16(1)**, 33-38 (1995).
[2]L.L. Hench, *Biomaterials*, **19**, 1419-1423 (1998).
[3]W. Suchanek and M. Yoshimura, *J. Mater. Res.*, **13**, 94-117 (1998).
[4]M. Afshar, N. Ghorbani, M. R. Ehsani and C. C. Sorrell: *Materials and Design*, **24**, 197-202 (2003).
[5]S. Ramesh, A Method for Manufacturing Hydroxyapatite Bioceramic. *Malaysia Patent*, **PI. 20043325**, (2004).
[6]S.H. Rhee, Synthesis of Hydroxyapatite Via Mechanochemical Treatment, *Biomaterials*, **23**, 1147-1152 (2002).
[7]K. Niihara, R. Morena and D.P.H, Hasselman, Evaluation of K_{Ic} of Brittle Solids by The Indentation Method With Low Crack-to-Indent Ratios, *J. Mater. Sci. Letts.*, **1**, 13-16 (1982).
[8]K. Ozeki, Y. Fukui and H. Aoki, Influence of The Calcium Phosphate Content of The Target on The Phase Composition and Deposition Rate of Sputtered Films, *Applied Surface Science*, **253**, 5040-5044 (2007).
[9]I.R. Gibson, S. Ke, S.M. Best and W. Bonfield, Effect of Powder Characteristics on The Sinterability of Hydroxyapatite Powders, *J. Mater. Sci.: Mater. Med.*, **12**, 163-171 (2001).
[10]L. Yubao, C.P.A.T., Klein, Z. Xindong and K. DeGroot, Relationship Between The Color Change of Hydroxyapatite and The Trace Element Manganese, *Biomaterials*, **14**, 969-972 (1993).
[11]C.J. Liao, F.H. Lin, K.S. Chen and J.S. Sun, Thermal Decomposition and Reconstitution of Hydroxyapatite in Air Atmosphere, *Biomaterials*, **20**, 1807-1813 (1999).

[12]C.K. Wang, C. P. Ju and J.H. Chern Lin, Effect of Doped Bioactive Glass on Structure and Properties of Sintered Hydroxyapatite, *Materials Chemistry and Physics*, **53**, 138-149 (1998).

[13]C. Kothapalli, M. Wei, A. Vasiliev and M.T. Shaw, Influence of Temperature and Concentration on The Sintering Behavior and Mechanical Properties of Hydroxyapatite, *Acta Materialia*, **52**, 5655-5663 (2004).

[14]N.Y. Mostafa, Characterization, Thermal Stability and Sintering of Hydroxyapatite Powders Prepared by Different Routes, *Materials Chemistry and Physics,* **94**, 333-341 (2005).

[15]K.C.B. Yeong, J. Wang and S.C. Ng, Mechanochemical Synthesis of Nanocrystalline Hydroxyapatite from CaO and $CaHPO_4$, *Biomaterials*, **22**, 2705-2712 (2001).

[16]W. Kim, Q. Zhang and M. Saito, Mechanochemical Synthesis of Hydroxyapatite from $Ca(OH)_2$-P_2O_5 and CaO-$Ca(OH)_2$-P_2O_5 Mixtures, *J. Mater. Sci.*, **35**, 5401-5405 (2000).

[17]W. De With, J.A. Van Dijk, H.N. Hattu and K. Prijs, Preparation, Microstructure and Mechanical Properties of Dense Polycrystalline Hydroxyapatite, *J. Mater. Sci.,* **16**, 1592-1598 (1981).

[18]P. Van Landuyt, F. Li, J.P. Keustermans, J.M. Streydio, F. Delannay and E. Munting, The Influence of High Sintering Temperatures on The Mechanical Properties of Hydroxylapatite, *J. Mater. Sci.: Mater. Med.*, **6**, 8-13 (1995).

[19]G. Muralithran and S. Ramesh, The Effects of Sintering Temperature on The Properties of Hydroxyapatite, *Ceram. Inter.*, **26**, 221-230 (2000).

VATERITE BIOCERAMICS: MONODISPERSE CaCO$_3$ BICONVEX MICROPILLS FORMING AT 70°C IN AQUEOUS CaCl$_2$-GELATIN-UREA SOLUTIONS

A. Cuneyt Tas
Department of Biomedical Engineering, Yeditepe University
Istanbul 34755, Turkey

ABSTRACT
    Calcium carbonate microparticles with a unique biconvex pill shape were produced by simply ageing the prerefrigerated (4°C, 24 h) CaCl$_2$-gelatin-urea solutions at 70°C for 24 h in glass media bottles. Gelatin is known to be the denatured collagen. Thermal decomposition of dissolved urea was exploited to provide the Ca^{2+} ion and gelatin-containing solutions with aqueous carbonate ions. Monodisperse CaCO$_3$ micropills formed in solution had the mean particle size of 4±2.5 µm. CaCO$_3$ micropills were typically biphasic in nature and consisted of 93% vaterite and 7% calcite. Identical solutions used without prerefrigeration yielded only trigonal prismatic calcite crystals upon ageing at 70°C for 24 h. Prerefrigeration of CaCl$_2$-gelatin-urea solutions was thus shown to have an unusual effect on the particle morphology. Samples were characterized by scanning electron microscopy (SEM), Fourier-transform infrared spectroscopy (FTIR) and powder X-ray diffraction (XRD).

INTRODUCTION
    CaCO$_3$ (calcium carbonate) is an important material of marine and geological biomineralization processes. CaCO$_3$ powders are widely used in rubber, plastic, paper making, printing ink, cosmetics, toothpaste, and food industries. Calcium carbonate has three anhydrous polymorphs; calcite, aragonite and vaterite. Amorphous calcium carbonate (ACC) may also be added to the polymorph list as the fourth component [1]. Calcium carbonate monohydrate and calcium carbonate hexahydrate may be regarded as the fifth and sixth CaCO$_3$ polymorphs [2]. At the ambient temperature and pressure, calcite is the most stable and abundant polymorph of calcium carbonate, while vaterite (μ-CaCO$_3$), named after Heinrich Vater [3], is known to be the least stable among the anhydrous polymorphs.
    Vaterite has a higher aqueous solubility than calcite and aragonite [4]. The log(K_S) values for calcite, aragonite and vaterite were experimentally determined by De Visscher and Vanderdeelen [5]. Vaterite is rare in nature, perhaps owing to its instability, as it would readily convert into one of the more stable calcium carbonate phases [6-8]. However, Grasby [9] discovered micron-sized spheres of vaterite at a supraglacial location in the Canadian High Arctic at very low temperatures. Vaterite was known to be a mineralization product in the egg-shells of some gastropodia [10], the spicules of certain sea squirts [11], and the skeletons of woodlice [12].
    By using the method of CO$_2$ gas bubbling through an aqueous solution of Ca-chloride (or Ca-nitrate), either well-crystallized rhombohedra of calcite or spheres of vaterite could be produced [13-15]. Han et al. [15] reported that the higher the concentration of CO$_3^{2-}$, the higher will be the tendency for formation of vaterite rather than its dissolution and gradual transformation into calcite. Using dissolved sodium carbonate (either Na$_2$CO$_3$ or NaHCO$_3$) as the CO$_3^{2-}$ source, in place of CO$_2$ gas bubbling, was another practical option to produce vaterite or calcite crystals [16-18]. CaCO$_3$ spheres were also grown in a desiccator via slow diffusion of CO$_2$ released by the decomposition of (NH$_4$)$_2$CO$_3$ crystals placed at the bottom of the same desiccator, which also contained a glass dish with CaCl$_2$ solution [19].
    Urea (NH$_2$CONH$_2$) was used (instead of CO$_2$ gas bubbling or Na$_2$CO$_3$, NaHCO$_3$ and (NH$_4$)$_2$CO$_3$ additions) to produce calcium carbonate powders [20-26]. Wang et al. [21] synthesized non-agglomerated calcite (trigonal), vaterite (spherical) and aragonite (needle-like) particles by using the decomposition of urea in CaCl$_2$-containing aqueous solutions (50 to 90°C).

Wakayama *et al.* [27] immersed chitosan-coated glass slides into a solution of Ca-acetate and polyacrylic acid (PAA) in the presence of supercritical CO_2 at 50°C and 76.5 kg/cm² (7.5 MPa) and observed the formation of heavily agglomerated but "pill-like" particles of vaterite deposited on the chitosan-coated glass slides.

Attempts to crystallize $CaCO_3$ in the presence of gelatin (or collagen) were found to be rather limited [28-36]. Moreover, none of these studies utilized aqueous Ca^{2+}-gelatin "solutions" as their calcium carbonate synthesis media.

To the best of our knowledge, there was no study in the literature on the *in situ* hydrothermal synthesis of $CaCO_3$ in Ca^{2+} ion-containing aqueous solutions which simultaneously have gelatin and urea.

We have discovered that prerefrigerated $CaCl_2$-gelatin-urea solutions, when simply aged at 70°C in sealed glass bottles, produced monodisperse, biphasic vaterite-calcite biconvex micropills or micropills with a unique morphology not seen and reported before. $CaCO_3$ is used in significant amounts in the powder formulations of new orthopedic and dental cements [37, 38] designed for skeletal repair, and our interest in $CaCO_3$ stemmed from such clinical applications. This manuscript reports, for the first time, the synthesis of micropills of $CaCO_3$.

EXPERIMENTAL PROCEDURE
Preparation of $CaCl_2$-gelatin-urea solutions
Ca-containing gelatin-urea solutions were prepared as follows. 200 mL of deionized water was placed into a 250 mL glass beaker and 11.761 g of $CaCl_2 \cdot 2H_2O$ (>99%, Cat. No. C79-500, Fisher Scientific, Fairlawn, NJ) was added to it, followed by stirring on a hot-plate, with a Teflon®-coated magnetic stir bar, at room temperature (RT: 21±1°C). This solution thus contained 0.4 M Ca^{2+}. 0.30 g of gelatin powder (>99%, Cat. No. G7-500, Fisher Scientific) was then dissolved, by stirring at RT, in the above solution. Finally, 6.00 g of urea powder (>99%, NH_2CONH_2, Cat. No. U15-500, Fisher Scientific) was added to the above Ca-gelatin solution, and the solution was stirred at RT for a minute to dissolve the urea. The transparent solution, which contained 0.4 M Ca^{2+}, 0.5 M urea and 0.3 g gelatin, was then transferred into a 250 mL-capacity Pyrex® media bottle (Cat. No. 06-423-3B, Fisher Scientific). Since these solutions contained urea, and since urea starts going through a very slow decomposition process even at RT, such solutions were not stored at RT for long times; therefore, these solutions must be prepared freshly prior to each synthesis experiment.

These solutions were then used to produce $CaCO_3$ particles with two different morphologies. "As-prepared solutions" and "prerefrigerated solutions" resulted in two different particle morphologies.
Synthesis of trigonal prismatic $CaCO_3$ (calcite) crystals
Only freshly prepared $CaCl_2$-gelatin-urea solutions (prepared in the way described above) were used in this compartment of this study. 200 mL of solution was first placed into a 250 mL-capacity Pyrex® media bottle. Then, one piece of microscope cover glass (Cat. No. 12-542B, 22x22x0.15 mm, Fisher Scientific) was dropped into the bottle and made sure that it laid flat at the bottom of the bottle. The bottle was tightly capped and placed into a microprocessor-controlled oven pre-heated to 70°C, and kept there undisturbed for 24 h. At the end of 24 h, the bottle was opened; the white-coated cover glass was removed, and washed with an ample supply of deionized water, followed by rinsing with ethanol (95%, denatured, Cat. No. S73985, Fisher Scientific). The cover glass was dried in an oven at 37°C, overnight in air.
Synthesis of $CaCO_3$ micropills
A freshly prepared portion (200 mL) of $CaCl_2$-gelatin-urea solution was placed in a 250 mL-capacity Pyrex media bottle, tightly capped and then refrigerated (at 4°C) for 24 h. The pH of the refrigerated solution was measured to be 6.5 (at 6°C). One piece of microscope cover glass was dropped into the bottle and made sure that it laid flat at the bottom of the bottle. The bottle was capped

and placed into a microprocessor-controlled oven pre-heated to 70°C, and kept there undisturbed for 24 hours. At the end of 24 h (solution pH was 7.5 at 68-69°C), the bottle was opened and the white-coated cover glass was removed, and washed with an ample supply of deionized water, followed by rinsing with ethanol. The cover glass was dried in an air atmosphere oven at 37°C, overnight. For further analyses, the white powdery material coating the cover glass was gently scraped off by using a clean and sharp razor blade. The bottom of the glass bottle was also coated with the same material.

Sample Characterization

Samples were characterized by powder X-ray diffraction (XRD; Model XDS 2000, Scintag, Sunnyvale, CA), scanning electron microscopy (SEM; Model S-4700, Hitachi, Tokyo, Japan), and Fourier-transform infrared spectroscopy (FTIR; Nicolet 550, Thermo-Nicolet, Woburn, MA). Powder samples for SEM and XRD analyses (scraped off of the coated cover glasses) were first gently ground in an agate mortar by using an agate pestle and then sprinkled onto ethanol-damped single-crystal quartz sample holders to form a thin layer, followed by tapping to remove the excess of powder. The X-ray diffractometer was operated at 40 kV and 30 mA with monochromated Cu K_α radiation. XRD data (over the typical range of 20 to 50° 2θ) were collected with a step size of 0.03° and a preset time of 1 sec at each step. FTIR samples were first ground in a mortar, in a manner similar to that used in the preparation of XRD and SEM samples, then mixed with KBr powder in a ratio of 1:100, followed by forming a pellet by using a uniaxial cold press. 128 scans were performed at a resolution of 3 cm^{-1}. Coated glass covers examined with the scanning electron microscope (SEM) were sputter-coated with a thin Au layer, to impart surface conductivity to the samples.

RESULTS AND DISCUSSION

Heating of as-prepared and prerefrigerated (24 h at 4°C) CaCl₂-gelatin-urea solutions at 70°C for 24 h resulted in the nucleation of calcium carbonate crystals with two different morphologies. While the as-prepared solutions were nucleating trigonal prismatic crystals, the prerefrigerated solutions produced monodisperse micropills.

The comparative SEM photomicrographs of Figure 1 depicted this drastic change in morphology upon prerefrigeration. Figures 1a-1b, 1c-1d and 1e-1f possessed identical magnifications. Figures 1a, 1c and 1e showed the calcium carbonate particles produced when the freshly prepared CaCl₂-gelatin-urea solutions were directly heated at 70°C for 24 h. On the other hand, Figures 1b, 1d, and 1f exhibited the monodisperse calcium carbonate biconvex micropills obtained when prerefrigerated (24 h at 4°C) CaCl₂-gelatin-urea solutions were heated at 70°C for 24 h. A biconvex tablet or pill geometrically looks like a cylinder of a given height and radius with two hemispheres glued to it at both ends. Biconvex pills (or tablets) have symmetrical top and bottom surfaces. Particle sizes were determined by using the linear intercept method directly on the SEM photomicrographs. The average particle size in powders obtained from the as-prepared solutions was 7±1.5 μm (Figs. 1a, 1c, 1e), whereas that obtained from the prerefrigerated solutions was 4±2.5 μm. The values reported here were the averages of 15 individual particle measurements along 6 lines drawn across each photomicrograph.

Some of the trigonal prismatic calcium carbonate crystals showed very flat surfaces (as shown in Figs. 1a and 1e), and these flat surfaces were considered to be created in direct contact with the glass surfaces, on which the initial phase separation occurred. Such flat surfaces on calcite crystals were also observed by Didymus et al. [39]. In the case of micropills forming in prerefrigerated solutions (Figures 1b, 1d, 1f), the simultaneous observation of small (1 μm in diameter) and large (5 μm) pills indicated the presence of several different nucleation events/waves in progress.

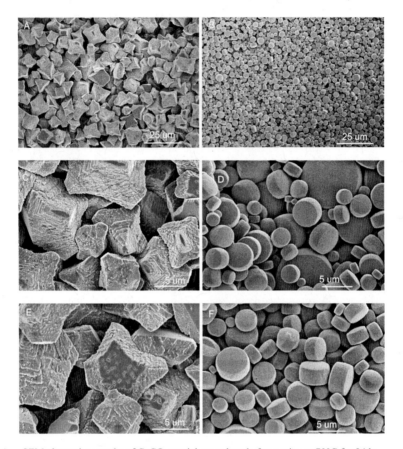

Fig. 1   SEM photomicrographs of CaCO₃ particles produced after ageing at 70°C for 24 h;
(a), (c) and (e): from "as-prepared" CaCl₂-gelatin-urea solutions
(b), (d) and (f): from "prerefrigerated" (4°C, 24 h) CaCl₂-gelatin-urea solutions

The powder XRD traces of samples obtained from both the as-prepared and prerefrigerated solutions were shown in Figure 2. As-prepared solutions, upon ageing at 70°C for 24 h, produced single-phase trigonal prismatic calcite crystals, conforming to the ICDD PDF 5-0586 [40]. Prerefrigerated solutions, on the other hand, produced vaterite [41] micropills contaminated with a minor amount of the calcite phase. The experimental XRD data of the vaterite micropills conformed well to that given in ICDD PDF 72-0506. The only calcite peak appeared in the XRD spectrum of vaterite micropills was indicated by the letter C in the Figure 2b trace. That peak corresponded to the strongest reflection of the calcite phase, i.e., (104).

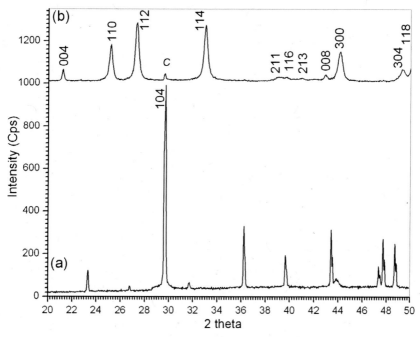

Fig. 2   XRD data of $CaCO_3$ particles produced after ageing at 70°C for 24 h;
        (a) from "as-prepared" $CaCl_2$-gelatin-urea solutions (single-phase calcite)
        (b) from "prerefrigerated" (4°C, 24 h) $CaCl_2$-gelatin-urea solutions (biphasic vaterite-calcite, the only calcite peak was indicated by *C*)

       The FTIR spectra of both samples (trigonal prismatic calcite and vaterite micropills) are depicted in Figure 3. The trigonal prismatic calcite particles obtained from the as-prepared solutions contained some surface adsorbed water (at least at the moment of IR data collection) and this was indicated by the broad water band extending over the range of 3600 and 3100 cm^{-1} (Figure 2a). The H-O-H band observed at 1650 cm^{-1} in Figure 3a was also pinpointing to this fact. The band observed at 1080 cm^{-1}, in both Figures 3a and 3b, was assigned to the symmetric stretching, v_1, and lattice mode vibration. The strong carbonate band seen at 873 cm^{-1} (out-of-plane bending, v_2) was common to both calcite and vaterite. However, based on the IR spectra, it is quite an easy task to distinguish between vaterite and calcite polymorphs. The absorption band at 713 cm^{-1} is characteristic for calcite, whereas in vaterite the same band (in-plane bending, v_4) is characteristically shifted to 744 cm^{-1} [42]. Moreover, in vaterite, the main carbonate band (i.e., asymmetric stretching, v_3) is split into two at 1450 and 1407 cm^{-1} (indicated by an arrow in Figure 3b). This carbonate band splitting was not seen in phase-pure calcite, and the asymmetric stretching band for calcite was observed at 1405 cm^{-1}.

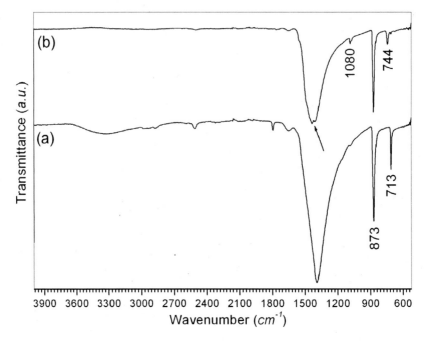

Fig. 3 FTIR spectra of CaCO₃ particles produced after ageing at 70°C for 24 h;
(a) from "as-prepared" CaCl₂-gelatin-urea solutions
(b) from "prerefrigerated" (4°C, 24 h) CaCl₂-gelatin-urea solutions (arrow indicates the characteristic splitting for vaterite)

The amount of calcite phase present in the monodisperse micropills was determined by using both the XRD and the FTIR data according to the methods suggested by Rao [13, 43] and Andersen and Kralj [44], respectively, and the calcite phase was present at about 7±1%. Therefore, the monodisperse micropills were biphasic in nature, i.e., 93% vaterite-7% calcite.

The readers who would be more enthusiastic in learning the decomposition kinetics of urea, in aqueous solutions containing metal ions, are hereby referred to the detailed works of Willard and Tang [45] and Mavis and Akinc [46], which also gave the stepwise decomposition reactions written in full. The ageing temperature was deliberately maintained low at 70°C in this study (in contrast to the use of more common 90°C [46]) to avoid the instantaneous and rapid decomposition of urea, and to provide a much slower supply of HCO₃⁻ ions to the Ca-gelatin solutions.

The experimental amino acid compositions of mammalian (i.e., seal, whale, porcine and bovine) and cod-skin gelatins were reported by Arnesen and Gildberg [47]. The native conformation of collagen molecules is a triple helix; however gelatin, as denatured collagen, is water soluble and forms random coils in solution [48]. Yoshioka *et al.* [49] experimentally determined that in the case of gelatin-water system the conformational coil-helix transition of the protein chains was responsible for the gel formation, and the helix formation was enhanced by lowering the temperature to about 5°C.

Guo *et al.* [48] also observed that upon cooling pure gelatin below its melting temperature (where the melting point of bovine gelatin is 36°C [47]), ordered structures of the gelatin molecules would be re-formed. In other words, gelatin molecules may partially revert to the ordered triple helical collagen-like sequences upon cooling [50].

Joly-Duhamel *et al.* [51] experimentally determined the random coil-to-refolded triple helix transformation percentage in a number of gelatin samples (including bovine gelatin). When the gelatin sols were cooled to around 5°C, the helix amount was found to increase (from zero at 35°C) to about 65% [51, 52]. An annealing time (at 5°C) of at least 6 h was reported to be necessary to achieve the above-mentioned coil-to-helix transformation [51]. Joly-Duhamel *et al.* [51] also stated that renaturation (achieved by the cooling of gelatin sols) was essentially a "nonreversible" process, and the triple helical sequences were stable (stabilized by the hydrogen bonds) in aqueous solutions. This would mean that upon reheating the refrigerated gelatin sols to temperatures above its melting point not all of the triple helices formed would decompose into random coils [53].

The interaction of gelatin with urea, in aqueous solutions, has been a hardly studied topic, however, the article of Jana and Moulik [54] provided a valuable insight into the process disclosed here. The dissociation of amino acids in aqueous solution produces H^+ ions and urea is known to bind hydrogen ion to form Urea-H^+ adduct. Jana and Moulik [54] reported the experimental H^+ ion concentrations generated from a series of individual amino acid solutions (such as, Gly, Pro, Val, Gln, Ser, His, Trp, Arg and Asp) to decrease with an increase in urea concentration. Dissolved urea competes with water for the H^+ ion forming uranium ion (UH^+).

The FTIR spectra shown in Fig. 4 depicted the effect of gelatin concentration used in synthesis solutions on the phase composition of the formed micropills. The samples shown in Fig. 4 were produced by using the same Ca^{2+} and urea concentrations with those presented in Figs. 1 through 3. The gelatin concentration was slowly decreased in the synthesis solutions to produce the samples of Fig. 4. The solution concentrations were given in the figure caption of Fig. 4. 0.1 g gelatin-containing

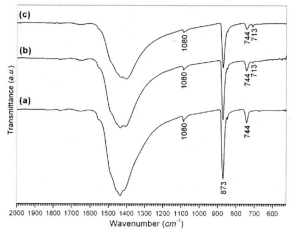

Fig. 4 Influence of gelatin concentration on the phase composition of CaCO₃ micropills
(a) 0.4 M Ca^{2+}, 0.5 M urea, 0.10 g gelatin, prerefrigerate at 4°C, 70°C, 24 h
(b) 0.4 M Ca^{2+}, 0.5 M urea, 0.15 g gelatin, prerefrigerate at 4°C, 70°C, 24 h
(c) 0.4 M Ca^{2+}, 0.5 M urea, 0.25 g gelatin, prerefrigerate at 4°C, 70°C, 24 h

resulted in much purer vaterite powders. The SEM photomicrographs given in Fig. 5 depicted the morphology of biconvex vaterite micropills obtained by using a solution of 0.4 M Ca^{2+}, 0.5 M urea and 0.1 g gelatin (in a total water volume of 200 mL).

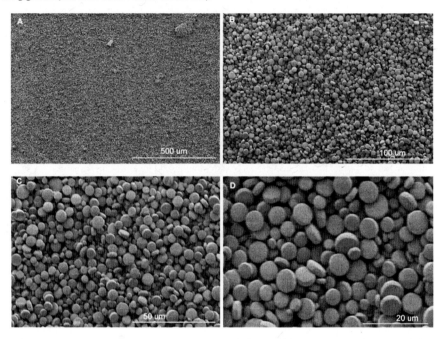

Fig. 5 SEM photomicrographs of vaterite biconvex pills (at increasing magnifications) formed in a solution containing 0.4 M Ca^{2+}, 0.5 M urea and 0.1 g gelatin, following prerefrigeration at 4°C and heating at 70°C for 24 h

It was previously reported that it would be possible to synthesize aragonite rods in CaCl₂-urea solutions free of gelatin [21, 55]. In this study, we have also observed the formation of needle-like aragonite particles, containing rectangular-prismatic calcite crystals, when we simply eliminated gelatin from the synthesis solutions which produced the above-mentioned micropills. These solutions were again heated at 70°C for 24 h. Obviously, prerefrigeration of these solutions had no effect on the observed morphology of the aragonite rods *in situ* formed. Fig. 6 depicted the characteristic FTIR spectrum of the aragonite-calcite biphasic powders of this study, whereas the SEM photomicrographs of the same aragonite-calcite biphasics were reproduced in Fig. 7.

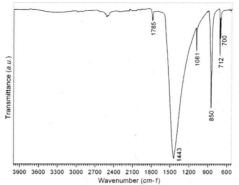

Fig. 6 FTIR spectra of aragonite-calcite biphasics formed at 70°C in 0.4 M Ca^{2+}, 0.5 M urea sols

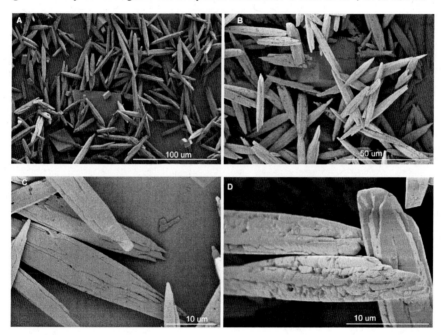

Fig. 7 SEM photomicrographs of aragonite-calcite biphasic powders (at increasing magnifications) formed in a solution containing 0.4 M Ca^{2+} and 0.5 M urea, following prerefrigeration at 4°C and heating at 70°C for 24 h

Would the extent of renaturation of gelatin from "random coils"-to-"triple helix" conformation (by prerefrigeration at 4°C for 24 h) and the thermal stability (while ageing the solutions at 70°C for 24 h) of the formed helices be enhanced by the presence of urea? If so, would this place a light on the formation of vaterite micropills? How does the ratio of random coil to triple helical conformation affect the carbonyl environments in gelatin? These could be the topics of future research.

Monodisperse $CaCO_3$ micropills presented here, besides forming a practical example for *in vitro* biomineralization processes in urea-, gelatin- and Ca^{2+} ion-containing matrices, may also find a number of applications in biomedical, pharmaceutical, cosmetics, polymer, rubber, paper and ink industries.

Birefringence refers to the ability of a mineral crystal to split an incident beam of linearly polarized light into two beams of unequal velocities (corresponding to two different refractive indices of the crystal) which subsequently recombine to form a beam of light that is no longer linearly polarized. The extreme birefringence of $CaCO_3$ makes its crystals appear to light up or glow when viewed through crossed polarizers. For technical applications which would fully exploit the birefringence properties of $CaCO_3$, the changes to be easily obtained in the particle morphology from spherical to biconvex tablets/pills or even to trigonal prisms are, therefore, extremely important.

CONCLUSIONS

(1) $CaCl_2$-gelatin-urea solutions were prepared at room temperature. These solutions nucleated trigonal prismatic calcite particles upon ageing at 70°C, in glass media bottles, for 24 h.

(2) The same $CaCl_2$-gelatin-urea solutions were first refrigerated at 4°C for 24 h and then aged in glass media bottles at 70°C for 24 h. Such solutions nucleated monodisperse, biphasic vaterite-calcite micropills. Such a biconvex pill morphology for $CaCO_3$ was not reported before.

(3) $CaCl_2$-urea solutions free of gelatin formed biphasic mixtures of needle-like aragonite and rectangular-prismatic calcite crystals following ageing at 70°C for 24 h.

NOTES

The names and models of certain commercial equipment, instruments or materials are identified in this paper to enhance understanding. Such identification does not imply any recommendation or endorsement by the author, nor does it imply that the equipment or materials identified are necessarily the best available for the purpose.

REFERENCES
[1]Y. Kojima, A. Kawanobe, T. Yasue, and Y. Arai, Synthesis of Amorphous Calcium Carbonate and Its Crystallization, *J. Ceram. Soc. Jpn.*, **101**, 1145-52 (1993).

[2]S. Mann, "Biomineralization: Principles and Concepts in Bioinorganic Materials Chemistry," Oxford University Press, Oxford, 2001.

[3]H. Vater, Ueber den Einfluss der Loesungsgenossen auf die Krystallisation des Calciumcarbonates, *Z. Kristallogr. Mineral.*, **27**, 477-512 (1897).

[4]A. G. Turnbull, A Thermodynamic Study of Vaterite, *Geochim. Cosmochim. Acta*, **37**, 1593-1601 (1973).

[5]A. De Visscher and J. Vanderdeelen, Estimation of the Solubility Constant of Calcite, Aragonite, and Vaterite at 25°C based on Primary Data using the Pitzer Ion Interaction Approach, *Monatsh. Chem.*, **134**, 769-75 (2003).

[6]S. Mann, B. R. Heywood, S. Rajam, and J. B. A. Walker, Structural and Stereochemical Relationships between Langmuir Monolayers and Calcium Carbonate Nucleation, *J. Phys. D. Appl. Phys.*, **24**, 154-64 (1991).

[7]F. C. Meldrum and S. T. Hyde, Morphological Influence of Magnesium and Organic Additives on the Precipitation of Calcite, *J. Cryst. Growth*, **231**, 544-58 (2001).

[8]D. Rautaray, K. Sinha, S. S. Shankar, S. D. Adyanthaya, and M. Sastry, Aqueous Foams as Templates for the Synthesis of Calcite Crystal Assemblies of Spherical Morphology, *Chem. Mater.*, **16**, 1356-61 (2004).

[9]S. E. Grasby, Naturally Precipitating Vaterite (μ-CaCO$_3$) Spheres: Unusual Carbonates Formed in an Extreme Environment, *Geochim. Cosmochim. Acta*, **67**, 1659-66 (2003).

[10]A. Hall and J. D. Taylor, The Occurrence of Vaterite in Gastropod Egg-Shells, *Mineral. Mag.*, **38**, 521-22 (1971).

[11]H. A. Lowenstam and D. P. Abbott, Vaterite: A Mineralization Product of the Hard Tissue of a Marine Organism (*Ascidiacea*), *Science*, **188**, 363-65 (1975).

[12]A. Becker, U. Bismayer, M. Epple, H. Fabritius, B. Hasse, J. Shi, and A. Ziegler, Structural Characterization of X-ray Amorphous Calcium Carbonate (ACC) in Sternal Deposits of the Crustacea Porcellio Scaber., *Dalton Trans.*, **4**, 551-55 (2003).

[13]M. S. Rao, Kinetics and Mechanism of the Transformation of Vaterite to Calcite, *Bull. Chem. Soc. Jpn.*, **46**, 1414-17 (1973).

[14]A. Vecht and T. G. Ireland, The Role of Vaterite and Aragonite in the Formation of Pseudo-biogenic Carbonate Structures: Implications for Martian Exobiology, *Geochim. Cosmochim. Acta*, **64**, 2719-25 (2000).

[15]Y. S. Han, G. Hadiko, M. Fuji, and M. Takahashi, Effect of Flow Rate and CO$_2$ Content on the Phase and Morphology of CaCO$_3$ Prepared by Bubbling Method, *J. Cryst. Growth*, **276**, 541-48 (2005).

[16]D. Kralj, L. Brecevic, and A. E. Nielson, Vaterite Growth and Dissolution in Aqueous Solution I. Kinetics of Crystal Growth, *J. Cryst. Growth*, **104**, 793-800 (1990).

[17]N. Spanos and P. G. Koutsoukos, The Transformation of Vaterite to Calcite: Effect of the Conditions of the Solutions in Contact with the Mineral Phase, *J. Cryst. Growth*, **191**, 783-90 (1998).

[18]J. P. Andreassen and M. J. Hounslow, Growth and Aggregation of Vaterite in Seeded-batch Experiments, *AIChE J.*, **50**, 2772-82 (2004).

[19]J. Aizenberg, J. Hanson, T. F. Koetzle, S. Weiner, and L. Addadi, Control of Macromolecule Distribution within Synthetic and Biogenic Single Calcite Crystals, *J. Am. Chem. Soc.*, **119**, 881-6 (1997).

[20]S. Stocks-Fisher, J. K. Galinat, and S. S. Bang, Microbiological Precipitation of CaCO$_3$, *Soil Biol. Biochem.*, **31**, 1563-71 (1999).

[21]L. F. Wang, I. Sondi, and E. Matijevic, Preparation of Uniform Needle-like Aragonite Particles by Homogeneous Precipitation, *J. Colloid Interf. Sci.*, **218**, 545-53 (1999).

[22]I. Sondi and E. Matijevic, Homogeneous Precipitation of Calcium Carbonates by Enzyme Catalyzed Reaction, *J. Colloid Interf. Sci.*, **238**, 208-14 (2001).

[23]Q. Li, Y. Ding, F. Li, B. Xie, and Y. Qian, Solvothermal Growth of Vaterite in the Presence of Ethylene Glycol, 1,2-propanediol and Glycerin, *J. Cryst. Growth*, **236**, 357-62 (2002).

[24]F. G. Ferris, V. Phoenix, Y. Fujita, and R. W. Smith, Kinetics of Calcite Precipitation Induced by Ureolytic Bacteria at 10 to 20°C in Artificial Groundwater, *Geochim. Cosmochim. Acta*, **68**, 1701-10 (2004).

[25]I. Sondi and B. Salopek-Sondi, Influence of the Primary Structure of Enzymes on the Formation of CaCO$_3$ Polymorphs: A Comparison of Plant (*Canavalia ensiformis*) and Bacterial (*Bacillus pasteurii*) Ureases, *Langmuir*, **21**, 8876-82 (2005).

[26]Z. F. Mao and J. H. Huang, Habit Modification of Calcium Carbonate in the Presence of Malic Acid, *J. Solid State Chem.*, **180**, 453-60 (2007).

[27]H. Wakayama, S. R. Hall, and S. Mann, Fabrication of CaCO$_3$-Biopolymer Thin Films using Supercritical Carbon Dioxide, *J. Mater. Chem.*, **15**, 1134-6 (2005).

[28]E. Dalas and P. G. Koutsoukos, Crystallization of Calcite on Collagen Type-I, *Langmuir*, **4**, 907-10 (1988).

[29]G. Falini, M. Gazzano, and A. Ripamonti, Calcite Crystallization on Gelatin Films containing Polyelectrolytes, *Adv. Mater.*, **6**, 46-8 (1994).
[30]G. Falini, S. Fermani, M. Gazzano, and A. Ripamonti, Biomimetic Crystallization of Calcium Carbonate Polymorphs by means of Collagenous Matrices, *Chem-Eur. J.*, **3**, 1807-14 (1997).
[31]G. Falini, S. Fermani, M. Gazzano, and A. Ripamonti, Oriented Crystallization of Vaterite in Collagenous Matrices, *Chem-Eur. J.*, **4**, 1048-52 (1998).
[32]F. Manoli, J. Kanakis, P. Malkaj, and E. Dalas, The Effect of Amino Acids on the Crystal Growth of Calcium Carbonate, *J. Cryst. Growth*, **236**, 363-70 (2002).
[33]F. H. Shen, Q. L. Feng, and C. M. Wang, The Modulation of Collagen on Crystal Morphology of Calcium Carbonate, *J. Cryst. Growth*, **242**, 239-44 (2002).
[34]O. Grassmann, G. Muller, and P. Loebmann, Organic-Inorganic Hybrid Structure of Calcite Crystalline Assemblies Grown in a Gelatin Hydrogel Matrix: Relevance to Biomineralization, *Chem. Mater.*, **14**, 4530-5 (2002).
[35]O. Grassmann and P. Loebmann, Biomimetic Nucleation and Growth of CaCO₃ in Hydrogels Incorporating Carboxylate Groups, *Biomaterials*, **25**, 277-82 (2004).
[36]Y. Jiao, Q. L. Feng, and X. M. Li, The Co-effect of Collagen and Magnesium Ions on Calcium Carbonate Biomineralization, *Mater. Sci. Eng. C*, **26**, 648-52 (2006).
[37]C. Combes, B. Miao, R. Bareille, and C. Rey, Preparation, Physical-Chemical Characterisation and Cytocompatibility of Calcium Carbonate Cements, *Biomaterials*, **27**, 1945-54 (2006).
[38]A. C. Tas, Porous, Biphasic CaCO₃-Calcium Phosphate Biomedical Cement Scaffolds from Calcite (CaCO₃) Powder, *Int. J. Appl. Ceram. Tec.*, **4**, 152-63 (2007).
[39]J. M. Didymus, P. Oliver, S. Mann, A. L. DeVries, P. V. Hauschka, and P. Westbroek, Influence of Low-molecular-weight and Macromolecular Organic Additives on the Morphology of Calcium Carbonate, *J. Chem. Soc. Faraday Trans.*, **89**, 2891-2900 (1993).
[40]ICDD PDF: International Centre for Diffraction Data, Powder Diffraction File. Newtown Square, PA, USA.
[41]S. R. Kamhi, On the Structure of Vaterite, *Acta Crystallogr.*, **16**, 770-2 (1963).
[42]C. Gabrielli, R. Jaouhari, S. Joiret, and G. Maurin, In Situ Raman Spectroscopy Applied to Electrochemical Scaling. Determination of the Structure of Vaterite, *J. Raman Spectrosc.*, **31**, 497-501 (2000).
[43]H. Wei, Q. Shen, Y. Zhao, D. J. Wang, and D. F. Xu, Influence of Polyvinylpyrrolidone on the Precipitation of Calcium Carbonate and on the Transformation of Vaterite to Calcite, *J. Cryst. Growth*, **250**, 516-24 (2003).
[44]F. A. Andersen and D. Kralj, Determination of the Composition of Calcite-Vaterite Mixtures by Infrared Spectrophotometry, *Appl. Spectrosc.*, **45**, 1748-51 (1991).
[45]H. H. Willard and N. K. Tang, A Study of the Precipitation of Aluminum Basic Sulphate by Urea, *J. Am. Chem. Soc.*, **59**, 1190-2 (1937).
[46]B. Mavis and M. Akinc, Kinetics of Urea Decomposition in the Presence of Transition Metal Ions: Ni²⁺, *J. Am. Ceram. Soc.*, **89**, 471-7 (2006).
[47]J. A. Arnesen and A. Gildberg, Preparation and Characterisation of Gelatine from the Skin of Harp Seal (*phoca groendlandica*), *Bioresource. Technol.*, **82**, 191-4 (2002).
[48]L. Guo, R. H. Colby, C. P. Lusignan, and T. H. Whitesides, Kinetics of Triple Helix Formation in Semidilute Gelatin Solutions, *Macromolecules*, **36**, 9999-10008 (2003).
[49]H. Yoshioka, Y. Mori, S. Tsukikawa, and S. Kubota, Thermoreversible Gelation on Cooling and on Heating of an Aqueous Gelatin-Poly(*N*-isopropylacrylamide) Conjugate, *Polym. Advan. Technol.*, **9**, 155-8 (1998).
[50]M. Djabourov, Y. Grillon, and J. Leblond, The Sol-Gel Transition in Gelatin Viewed by Diffusing Colloidal Probes, *Polym. Gels. Netw.*, **3**, 407-28 (1995).

[51]C. Joly-Duhamel, D. Hellio, and M. Djabourov, All Gelatin Networks: 1. Biodiversity and Physical Chemistry, *Langmuir*, **18**, 7208-17 (2002).

[52]M. Djabourov, J. Leblond, and P. Papon, Gelation of Aqueous Gelatin Solutions. I. Structural Investigation, *J. Phys-Paris*, **49**, 319-32 (1988).

[53]W. F. Harrington and N. V. Rao, Collagen Structure in Solution. I. Kinetics of Helix Regeneration in Single-Chain Gelatins, *Biochemistry*, **9**, 3714-24 (1970).

[54]P. K. Jana and S. P. Moulik, Interaction of Amino Acids and Gelatin with Urea, *Indian J. Biochem. Biophys.*, **30**, 297-305 (1993).

[55]S. F. Chen, S. H. Yu, J. Jiang, F. Li, and Y. Liu, Polymorph Discrimination of $CaCO_3$ Mineral in an Ethanol/Water Solution: Formation of Complex Vaterite Superstructures and Aragonite Rods, *Chem. Mater.*, **18**, 115-122 (2006).

NOVEL DNA SENSOR BASED ON CARBON NANOTUBES ATTACHED TO A PIEZOELECTRIC QUARTZ CRYSTAL

Jessica Weber,[1,2] Deena Ashour,[3] Shreekumar Pillai,[3] Shree R. Singh,[3] Ashok Kumar [1,2]

[1]Department of Mechanical Engineering, University of South Florida, Tampa, Florida, USA

[2]Nanomaterials and Nanomanufacturing Research Center, University of South Florida, Tampa, Florida, USA

[3]Center for NanoBiotechnology Research, Alabama State University, Montgomery, Alabama, USA

ABSTRACT

A nanoscale molecular sensing device has been designed and developed for the detection of *Salmonella enterica* serovar Typhimurium. *Salmonella* Typhimurium is a common food borne pathogen that affects millions of people and a quick, responsive sensor is needed for its detection. Our device has been fabricated to detect very low levels of *Salmonella* by attaching single stranded DNA (ssDNA) specific to *Salmonella* to carbon nanotubes (CNTs), which are aligned vertically and attached to a gold electrode. The carbon nanotubes have been modified with different functional groups (i.e. carboxylic acid, thiol) to facilitate hybridization of the DNA. These nanotubes have been chemically assembled onto the surface of a quartz piezoelectric crystal with gold sensing electrodes. Atomic force microscopy (AFM) and scanning force microscopy (SEM) were utilized to verify CNT alignment on the gold surface. Several different spectroscopy techniques were used to characterize the CNT-based sensor with the attachment of functional groups, such as Fourier Transform Infrared (FTIR), UV-Vis, and x-ray diffraction.

INTRODUCTION

The development of a highly sensitive device for the detection of specific biological analytes (i.e. proteins and DNA) remains the foremost challenge in biosensor production. In particular, the advancement of miniaturized DNA sensors remains the subject of intense research today. The addition of one-dimensional nanostructures such as gold nanoparticles [1,2] and carbon nanotubes (CNTs) [3-5] on the electrode surface is one method that has been examined in order to best optimize the detection limit and sensitivity of a biosensor. Carbon nanotubes, in particular, are the quintessential electrode material due to their excellent mechanical strength, high electrical and thermal conductivity. Additionally, their large length to diameter aspect ratio along with a high surface area to weight ratio is an ideal combination for providing surface functionalization of biomolecules.

Due to the lack of solubility of CNTs in many solutions, integration of CNTs in biosensor design remains a difficult challenge. While covalent modification on the nanotube surface was originally used to solve solubility issues, this technique is now used extensively for biological functionalization such as enzyme attachment to the tip of the CNT [6,7]. For example, treatment of CNTs with sulfuric and nitric acid under sonication will shorten and open the ends of the nanotubes while introducing carboxylic acid functional groups [8]. This technique is not only beneficial for nanotube solubility, but for functional group surface attachment as well. It has been shown previously that proteins will covalently attach to the ends of shortened SWNTs, which behave as molecular wires to allow electrical communication between the underlying electrode and the redox protein [9].

The piezoelectric quartz crystal microbalance (QCM) system is a powerful sensing technique used to detect mass changes on an electrode surface with nano – sensitivity. An applied AC voltage causes the crystal to oscillate and the resonance frequency (f) of the crystal depends on the total oscillating mass, including water coupled to the oscillation. When a thin film (i.e. proteins, cells, DNA, etc.) is attached to the sensor crystal, the frequency decreases. In this way, the QCM operates as a very sensitive balance. Recently, it has been used in a wide array of biological sensor analysis, including DNA hybridization [10], bacterial growth [11], enzymatic activity [12] and cell adhesion [13]. Although there have been many publications on the QCM method for detection for biological analytes, few researchers are incorporating carbon nanotubes into the piezoelectric sensor to achieve an increased sensitivity. Recently, one group [13] has incorporating multi-walled carbon nanotubes (MWNTs) onto the surface of a QCM gold electrode to monitor the adhesion of mammalian cells in real time. However, the researchers used a simple drop – cast method of attaching the MWNTs to the gold surface. In this work, we discuss a surface condensation method for attaching SWNTs to a piezoelectric gold electrode in preparation for *Salmonella* DNA hybridization. It is believed that a carbon nanotube forest aligned perpendicular to the gold surface will provide a more robust electrode, with a larger active surface area which will increase the locations for hybridization to occur.

EXPERIMENTAL

Preparation and Characterization of the Film

A 5 MHz piezoelectric quartz crystal with a gold electrode (Maxtek, Inc.) was modified with functionalized single walled carbon nanotubes (SWNTs). In preparation for surface condensation of SWNTs on the gold surface, the bare electrode was first treated with a piranha etch (3:1 sulfuric acid to H_2O_2) for 30 min. at 60 °C and then rinsed with deionized water followed by drying in a N_2 stream. Single walled carbon nanotubes were chemically shortened and oxidized in a solution of 3:1 sulfuric and nitric acids under ultrasonication for 8 hrs. This treatment causes carboxylic acid groups to form on the ends of the CNTs. The functionalized nanotubes were then filtered under vacuum and rinsed to clear any excess acid. The obtained nanotubes are shown to be soluble in dimethylformamide (DMF), ethanol, acetone, and water [14]. The CNTs were collected from the filter paper and sonicated in DMF for an additional 4 hrs. Figure 1 shows a schematic of the surface condensation process utilized to modify the gold electrode surface. The cleaned electrode was immersed in a 2.0 mM $NH_2(CH_2)_{11}SH$ ethanol solution for 2.5 hrs. Afterwards the electrode was rinsed with ethanol and deionized water and dried in a N_2 stream. This procedure creates an amino-thiol self-assembled monolayer on the gold surface, which becomes a platform for the SWNTs to attach to in an orderly fashion to the gold. About 20 mg of dicyclohexylcarbodiimide (DCC) was added to 30 ml of the CNT/DMF suspension and sonicated for 30 minutes. The gold electrode was then immersed in this resulting solution for 2.5 hrs at 60 °C to allow for the surface condensation reaction to occur.

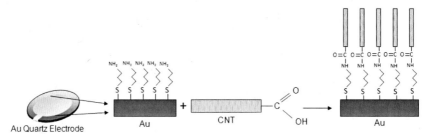

Figure 1. Schematic of surface condensation self-assembly of CNTs on gold electrode.

RESULTS AND DISCUSSION

Characterization of the Functionalized CNTs

Micro Raman spectroscopy was conducted on the gold electrode before and after modification with CNTs with a *Renishaw* 1000 Raman spectrometer using an argon laser (514.5 nm) at a laser power of 24.8 mW and a spot size of 1 μm. The laser was introduced to the sample through a 50X objective lens. The Raman scattering was collected with an exposure time of 10 s to the sample.

Raman scattering is one of the most useful and powerful techniques available to characterize carbon nanotube samples. For carbon nanotubes, the G mode corresponds to the stretching mode in the graphite plane and is centered around 1580 cm^{-1}. As seen in Figure 2, no observable Raman spectra is apparent for the untreated gold sample, while the modified electrode shows the G mode at about 1590 cm^{-1}. The D mode (the disorder band) is located between 1330 – 1360 cm^{-1}. While this band is expected to appear in multi-wall carbon nanotubes, it is usually attributed to defects in the tubes when observed in single-wall tube samples. Figure 2 shows that this band appears at 1351 cm^{-1} for the modified sample and the poor intensity of this peak can be interpreted as an indication of the high purity of the functionalized SWNTs. The radial breathing mode is unique to single walled carbon nanotubes and is directly related to the diameter of the tube. This mode usually appears around 200 cm^{-1}. The Raman peaks centered at 174 cm^{-1} and 1590 cm^{-1} on the treated sample indicates that surface condensation of carboxylated carbon nanotubes onto the gold film.

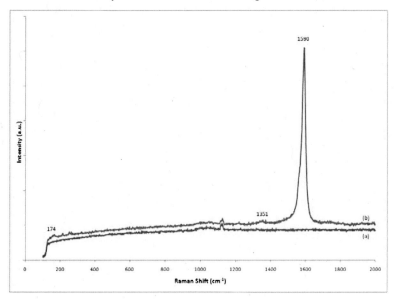

Figure 2. Raman spectra of (a) bare gold and (b) gold with functionalized CNTs immobilized onto the surface.

Atomic force microscopy was performed on the sample in order to obtain structural information about the electrode before and after surface condensation of the carbon nanotubes. Tapping mode AFM was performed over a 10 × 10 µm area. The functionalized carbon nanotubes were suspected to attach to the gold in bundles and Figure 3 clearly indicates that is the case. Due to the limitations of the AFM tip, we cannot clearly indicate the lateral dimensions of the nanotubes. Instead, we can observe an increase in roughness after modification of the sample.

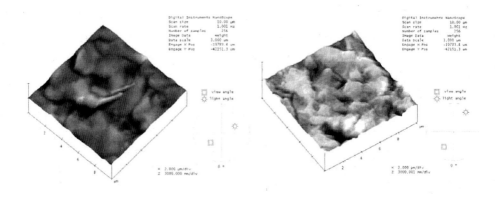

Figure 3. AFM images of (a) bare gold electrode and (b) immobilized CNTs on gold surface.

X-ray diffraction (XRD) was performed before and after electrode modification with functionalized CNTs with a Panalytical X'Pert Diffractometer in the range of 2θ from 20 to 100 degrees. Several peaks are observable from the bare gold surface. The gold (111) peak displays the highest intensity at 38°. The other gold peaks are (200) and (220) at about 44° and 64°, respectively. For the single-walled carbon nanotube bundles, the (002) peak appears at 26°. It has been reported previously that the intensity of this carbon peak in particular is a function of carbon nanotube alignment to the substrate [15]. The XRD spectra in Figure 4 confirm the morphology seen in the previous AFM images. Figure 3(b) shows that the CNT bundles seem to be leaning on one another. While the nanotubes are attached to the gold at their ends, they are probably not perfectly aligned to the substrate. The intensity of the CNT (002) peak confirms this – it is not the dominating peak, which would indicate complete disarray of CNTs; however, the peak is not completely diminished either.

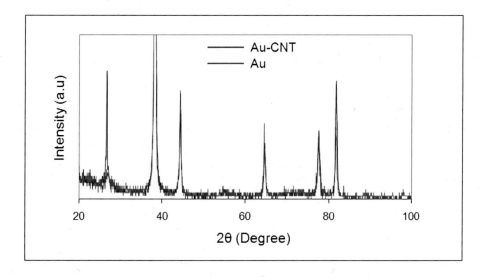

Figure 4. XRD plot of (a) bare gold electrode and (b) immobilized CNTs on the gold surface.

CONCLUSION

In conclusion, a single walled carbon nanotube – based electrode has been fabricated, characterized, and prepared for quartz crystal microbalance detection of DNA from *Salmonella enterica* serovar Typhimurium. Functionalization of carbon nanotubes was verified with different spectroscopy tools and the topology of the sensor was characterized with AFM. The surface condensation method was utilized to attach the SWNTs to a gold surface and some alignment of nanotubes perpendicular to the electrode has been observed. It can be noted that further studies will be performed with increasing the time of the SAM of cysteamine as well as the surface condensation reaction in order to optimize the alignment of the SWNTs on the gold surface. The authors believe that this method of CNT attachment to the gold piezo electrode will provide the most robust and reliable type of genosensor. This research represents a stepping stone towards the construction of an electrode with potentially low DNA concentration sensitivity using quartz crystal microbalance. A carbon nanotube based assay, with a larger surface area and excellent electrical properties will dramatically increase DNA attachment and complimentary single strand DNA detection sensitivity versus a simple gold piezo electrode. QCM experiments are in progress and will be presented in a future paper. Also, there is an interest to further investigate the electrochemical properties of the fabricated sensor. This CNT – based electrode has the possibility for identification of several types of biological analytes and molecules.

ACKNOWLEDGMENT

The authors would like to acknowledge the generous support of the National Science Foundation. This research was supported by the following NSF grants: NIRT #0404137, Crest #0734232, IGERT #0221681 and GK12 #0638709.

REFERENCES

[1] L. Authier, C. Grossiord, P. Brossier, and B. Limoges, Gold Nanoparticle-Based Quantitative Electrochemical Detection of Amplified Human Cytomegalovirus DNA Using Disposable Microband Electrodes, *Analytical Chemistry*, **73**, 4450-4456 (2001).

[2] W. Fritzsche, DNA-gold conjugates for the detection of specific molecular interactions, *Reviews in Molecular Biotechnology*, **82**, 37-46 (2001).

[3] X. J. Tara Elkin, Shelby Taylor, Yi Lin, Lingrong Gu, Hua Yang, Jessica Brown, Susan Collins, Ya-Ping Sun,, Immuno-Carbon Nanotubes and Recognition of Pathogens, *ChemBioChem*, **6**, 640-643 (2005).

[4] J. Wang, M. Li, Z. Shi, N. Li, and Z. Gu, Direct Electrochemistry of Cytochrome c at a Glassy Carbon Electrode Modified with Single-Wall Carbon Nanotubes, *Analytical Chemistry*, **74**, 1993-1997 (2002).

[5] X. Tang, S. Bansaruntip, N. Nakayama, E. Yenilmez, Y.-l. Chang, and Q. Wang, Carbon Nanotube DNA Sensor and Sensing Mechanism, *Nano Letters*, **6**, 1632-1636 (2006).

[6] S. B. Sinnott, Chemical Functionalization of Carbon Nanotubes, *Journal of Nanoscience and Nanotechnology*, **2**, 113-123 (2002).

[7] X. Yu, D. Chattopadhyay, I. Galeska, F. Papadimitrakopoulos, and J. F. Rusling, Peroxidase activity of enzymes bound to the ends of single-wall carbon nanotube forest electrodes, *Electrochemistry Communications*, **5**, 408-411 (2003).

[8] J. Liu, A. G. Rinzler, H. Dai, J. H. Hafner, R. K. Bradley, P. J. Boul, A. Lu, T. Iverson, K. Shelimov, C. B. Huffman, F. Rodriguez-Macias, Y.-S. Shon, T. R. Lee, D. T. Colbert, and R. E. Smalley, Fullerene Pipes, *Science*, **280**, 1253-1256 (1998).

[9] J. J. Gooding, R. Wibowo, Liu, W. Yang, D. Losic, S. Orbons, F. J. Mearns, J. G. Shapter, and D. B. Hibbert, Protein Electrochemistry Using Aligned Carbon Nanotube Arrays, *Journal of the American Chemical Society*, **125**, 9006-9007 (2003).

[10] F. Lucarelli, S. Tombelli, M. Minunni, G. Marrazza, and M. Mascini, Electrochemical and piezoelectric DNA biosensors for hybridisation detection, *Analytica Chimica Acta*, **609**, 139-159 (2008).

[11] Y. Wu, Q. Xie, A. Zhou, Y. Zhang, L. Nie, S. Yao, and X. Mo, Detection and Analysis of Bacillus subtilis Growth with Piezoelectric Quartz Crystal Impedance Based on Starch Hydrolysis, *Analytical Biochemistry*, **285**, 50-57 (2000).

[12] X. Turon, O. J. Rojas, and R. S. Deinhammer, Enzymatic Kinetics of Cellulose Hydrolysis: A QCM-D Study, *Langmuir*, **24**, 3880-3887 (2008).

[13] X. Jia, L. Tan, Q. Xie, Y. Zhang, and S. Yao, Quartz crystal microbalance and electrochemical cytosensing on a chitosan/multiwalled carbon nanotubes/Au electrode, *Sensors and Actuators B: Chemical*, **134**, 273-280 (2008).

[14] Z. Liu, Z. Shen, T. Zhu, S. Hou, L. Ying, Z. Shi, and Z. Gu, Organizing Single-Walled Carbon Nanotubes on Gold Using a Wet Chemical Self-Assembling Technique, *Langmuir*, **16**, 3569-3573 (2000).

[15] A. Cao, C. Xu, J. Liang, D. Wu, and B. Wei, X-ray diffraction characterization on the alignment degree of carbon nanotubes, *Chemical Physics Letters*, **344**, 13-17 (2001).

THERMAL CONDUCTIVITY OF LIGHT-CURED DENTAL COMPOSITES: IMPORTANCE OF FILLER PARTICLE SIZE

Michael B. Jakubinek,[1,2] Richard Price,[2,3] and Mary Anne White[1,2,4*]
[1] Dept. Physics, Dalhousie University, Halifax, Nova Scotia B3H 3J5, Canada
[2] Institute for Research in Materials, Dalhousie University, Halifax, Nova Scotia B3H 1W5, Canada
[3] Dept. Dental Clinical Sciences, Dalhousie University, Halifax, Nova Scotia B3H 1W2, Canada
[4] Dept. Chemistry, Dalhousie University, Halifax, Nova Scotia B3H 4J3, Canada

ABSTRACT
Light-cured dental composite materials are composed of resins filled with micro- to nano-sized ceramic particles. We have previously shown that the curing light intensity and exothermicity of the polymerization reaction during the setting of the composites are the main factors determining the magnitude of the resulting temperature excursion at the pulp-dentin junction (M.B. Jakubinek, C. O'Neill, C. Felix, R.B. Price, and M.A. White, Temperature excursions at the pulp-dentin junction during the curing of light-activated dental restorations, *Dent. Mater.*, **24**, 1468-76 (2008)), which could be detrimental to the tooth. Here we explore the role of filler particle size effect on the thermal conductivity of dental composite materials, both experimentally (commercial dental composites) and from the perspective of idealized spheres in a matrix. We found that, in principle, thermal conductivity could be further reduced by employing smaller ($r \sim 1 - 10$ nm) filler particles. However, in practice, the materials studied here are already good thermal insulators, and further reductions in thermal conductivity would have only minor effects on the maximum temperature experienced by the tooth pulp during the light-curing process.

INTRODUCTION
        Most commonly used, tooth-colored, dental restorative materials are composites of acrylic resins filled with micro to nano-sized glass or ceramic particles. These composites come in flowable (or more viscous, paste-like) forms and also contain photoinitiators (e.g., camphorquinone) to allow curing by irradiation with light in the 400 – 500 nm range after insertion into the tooth. The restoration process has two energy/heat sources: the energy of irradiation, and the exothermic polymerization. Excessive heat applied to a tooth can cause pulpitis and even pulp death.[1] The heat produced during the restoration process is more of a risk than hot or cold liquids in the mouth because the source, light absorption and exothermicity of the restoration material, can be very close to the sensitive pulp tissue (especially in deep restorations). While the range of safe pulp temperatures is not accurately known, it is often assumed, based on the work of Zach and Cohen,[1] that temperatures higher than 42.5 °C can cause irreversible damage to the pulp. Temperature increases during light-curing of dental composites have been studied by in vitro experiments[2,3,4,5,6,7,8,9,10] and, under certain circumstances, the temperatures observed are sufficient to pose a hazard to the tooth.
        Mechanical properties (e.g., strength, stiffness, bond strength, wear)[11], biocompatibility, and aesthetic factors (e.g., color, opacity, stain resistance)[11] are most commonly optimized in dental composites; however, thermal properties such as thermal conductivity, κ, and heat capacity, C_p, also are important to the performance of dental materials.[11,12,13] Given that the typical thermal conductivities of their constituent materials (polymers, glasses, ceramics) are low in comparison to amalgam or gold alloy fillings, a resin-based composite restoration offers better insulation against temperature changes due to hot or cold liquids in the mouth. However, based on the preceding discussion, one might expect that further reduction in thermal conductivity would be another important design goal for composites in order to best protect the tooth pulp from temperature changes during the curing process.

Typical values of thermal conductivity for dental composites range from somewhat less than that of dentin ($\kappa_{dentin} \sim 0.6$ W m^{-1} K^{-1}) to $2 - 3$ times its thermal conductivity.[12,13] Watts et al.[14] measured the thermal diffusivity for 21 commercial dental composites, the majority of which were light-cured. They found that most composites investigated did not have thermal diffusivities greatly in excess of dentin, and observed only a gradual trend of increasing thermal diffusivity with volume fraction of filler.[14] Both observations are somewhat counter-intuitive based on the rule of mixtures, given that the thermal conductivity of the filler can safely be assumed to be greater than 1 W m^{-1} K^{-1} (e.g., $\kappa_{a-SiO_2} = 1.38$ W m^{-1} K^{-1})[15] and the composites are highly filled (> 50% by volume). However, reduced thermal conductivity can result from nanoscale structure. This is the case with ivory, a tooth-like biomaterial composed of nanosized hydroxyapatite crystals in a collagen matrix hierarchically organized from nano- to macro-scales,[16] which has quite low thermal conductivity for a strong material, lower than either of its components.[17] (Dentin and enamel are similar materials and their thermal conductivities also fall below rule of mixture predictions based on their compositions.[17])

Here we report the thermal conductivity of several commercial dental composites and apply a literature model for composites with spherical inclusions[18] to explore the hypothesis that the low thermal conductivity is due to a filler particle size effect. We then apply our previously reported finite element method model for temperature increases during the curing of a composite filling[19] to explore whether maximizing the size effect-induced thermal resistivity could produce significant reductions in the maximum temperature increase at the tooth pulp during the light-curing process.

PARTICLE SIZE EFFECTS IN COMPOSITES

Particle/matrix composites with small filler particles are known to exhibit properties that are different than composites containing larger fillers. In the case of mechanical properties, nano-sized filler particles have been shown to result in different properties than composites with micron-sized fillers.[20,21] The large surface-to-volume ratio in nanoparticle-filled composites is a key factor responsible for their different properties. The polymer chains have a strong affinity for the particle surface and, therefore, the polymer chains surrounding a particle lose some of their mobility creating a low-mobility polymer shell around each particle.[21,22] The affected volume of polymer is much larger for nanofillers than for micron-sized fillers. The current investigation focuses exclusively on thermal conductivity but size effects also could prove useful in the optimization of mechanical properties of dental composites.

Increased surface-to-volume ratio also increases the contribution of the thermal boundary resistance, resulting in a lower effective thermal conductivity. Therefore, when the size of the filler particles in a filler/matrix composite is sufficiently small, the effective thermal conductivity of the composite depends not only on the thermal conductivity, geometric distribution and the volume fraction of each component but also on the particle size.[18,23,24,25,26,27,28]

For a composite with well-dispersed, spherical filler particles of radius r in a matrix, effective medium theory consideration of the thermal conductivity, κ, gives[18]

$$\frac{\kappa_{composite}}{\kappa_{matrix}} = \frac{\dfrac{\kappa_{filler}}{\kappa_{matrix}}(\gamma+2)+2\gamma+2f\left[\dfrac{\kappa_{filler}}{\kappa_{matrix}}(\gamma-1)-\gamma\right]}{\dfrac{\kappa_{filler}}{\kappa_{matrix}}(\gamma+2)+2\gamma-f\left[\dfrac{\kappa_{filler}}{\kappa_{matrix}}(\gamma-1)-\gamma\right]}, \quad (1)$$

where $\gamma = \dfrac{r}{R_{bd}\kappa_{matrix}}$, f is the volume fraction of filler, and R_{bd} is the thermal boundary resistance per unit area of filler/matrix interface. As one would expect, equation (1) shows that the addition of any filler with $\kappa_{filler} < \kappa_{matrix}$ will reduce the thermal conductivity of the composite. Less intuitively, the addition of higher thermal conductivity filler (i.e., $\kappa_{filler}/\kappa_{matrix} > 1$, which is the situation in dental composites) can either increase or decrease the thermal conductivity of the composite. Further, in the small γ limit, $\kappa_{composite}$ is independent of κ_{filler}. Figure 1 shows the results of equation (1) in the small particle/high R_b limit (i.e., $\gamma \ll 1$) and large particle/low R_b limit (i.e., $\gamma \gg 1$). Although the model predicts that κ will change monotonically with filler volume, one should consider that this might not be valid for very large volume fractions. This is typically the case for mechanical properties, where an optimum property is achieved for a particular volume fraction of filler, and is attributed to the tendency of the filler particles to agglomerate when the filler content exceeds some threshold value.[20] Agglomeration decreases the surface-to-volume ratio and also could create voids, which are preferential sites for cracking.[21] From the thermal perspective, a decrease in surface area reduces the contribution of the thermal boundary resistance.

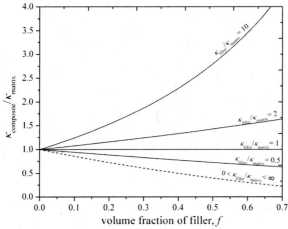

Figure 1. Influence of filler content on $\kappa_{composite}/\kappa_{matrix}$ for filler/matrix composites with well-dispersed, spherical filler particles in the big particle/low R_{bd} limit (———) for $\kappa_{filler}/\kappa_{matrix} = 10, 2, 1,$ and 0.5. The small particle/high R_{bd} limit (– – – –) is a universal curve, independent of the $\kappa_{filler}/\kappa_{matrix}$ ratio.

It follows that, for any given filler and matrix combination where $\kappa_{filler} > \kappa_{matrix}$, there exists a critical filler particle radius (r_c), defined as the radius below which the thermal conductivity of the resulting composite will be decreased relative to the matrix and above which the thermal conductivity of the composite will be increased.[29] The critical radius,

$$r_c \equiv \frac{1}{\left(\dfrac{\kappa_{filler}}{\kappa_{matrix}} - 1\right)} \kappa_{filler} R_{bd} \qquad (2)$$

is obtained by setting equation (1) equal to 1 and solving for r. Putnam et al.[29] determined the thermal boundary conductance, $G_{bd} = 1/R_{bd}$, for a ceramic/polymer interface (alumina/polymethylmethacrylate) to be 30 ± 10 MW m^{-2} K^{-1} near room temperature. This corresponds to a critical particle radius of 7.5 ± 2 nm,[29] below which the effective thermal conductivity of the composite would be decreased by the addition of the higher thermal conductivity, alumina filler. The value of G_{bd} for alumina/PMMA provides a good indication of what would be expected for the interfaces in dental composites, which also are ceramic/acrylic polymer interfaces, although different types of composites will have different conductance. For example, G_{bd} is higher for metal/ceramic interfaces: ~700 MW m^{-2} K^{-1} for SiC/Al metal-matrix composites.[27]

This theory suggests that, under appropriate circumstances, filler/polymer matrix composites could be used either to increase thermal conductivity, such as for thermal interface materials (e.g., adhesives and pads, and pastes) or to reduce thermal conductivity. The latter could prove useful in minimizing the thermal conductivity of dental composites.

MATERIALS AND METHODS

Commercial dental composites

Thermal conductivities of four commercial dental composites, Filtek[TM] Supreme (3M/ESPE, shade: A2), PermaFlo® (Ultradent, shade: A2), Vit-l-esence[TM] (Ultradent, shade: A2), and Heliomolar® Flow (Ivoclar Vivadent, shade: A2), were investigated. Compositional information, as specified by the manufacturers, is listed in Table I.

For comparison, the thermal conductivity of an unfilled, cured resin also was measured. The resin, which is used in the Filtek[TM] Supreme composite, is denoted here as "Filtek[TM] Supreme-Unfilled resin". Note that equation (1) requires the volume fraction but the manufacturers specify the filler content as mass%, and the corresponding volume fraction is 10 – 20% lower. This estimate is based on the density the Filtek[TM] Supreme-Unfilled resin (1.2 g/cm^3, determined from the mass and dimensions of a cured sample) being approximately half that of typical fillers.

Samples were prepared by dispensing the uncured, flowable forms into the center of a rubber washer, which was used as a mold. Transparent plastic was used as the top and bottom of the mold and the materials were cured by the application of light, from the top and the bottom, using an Allegro LED curing light (1250 mW cm^{-2} intensity; repeated light exposures were used to ensure full curing). This mold produced pill-shaped samples approximately 4.8 mm in diameter and 2 mm in height.

Table I. Summary of compositional information for the measured, commercial dental composites (shade: A2), as specified by their manufacturers.[30, 31, 32, 33]

Composite	Resin	Filler
Filtek[TM] Supreme (3M ESPE)	bis-GMA,[a] bis-EMA,[b] UDMA,[c] and TEGDMA[d]	78.5 mass % filler; non-agglomerated/non-aggregated 20 nm silica filler; loosely bound agglomerate silica nanoclusters (ca. 1 μm).
PermaFlo® (Ultradent)	methacrylate-based	68 mass % filler; 1 μm average particle size.
Vit-l-esence[TM] (Ultradent)	bis-GMA-based	75 mass % filler; 700 nm average particle size.
Heliomolar® Flow (Ivoclar Vivadent)	bis-GMA, UDMA, and DDDMA[e]	59 mass % filler; highly dispersed SiO$_2$, ytterbium trifluoride, and copolymer; particle size: 40 – 200 nm.

[a]bis-GMA = bisphenol A diglycidyl ether dimethacrylate; [b]bis-EMA = bisphenol A polyethylene glycol diether dimethacrylate; [c]UDMA = urethane dimethacrylate; [d]TEGDMA = triethylene glycol dimethacrylate; [e]DDDMA = decandiol dimethacrylate.

Thermal conductivity measurements

Thermal conductivity was determined using the thermal transport option on a Physical Property Measurement System (PPMS; Quantum Design, San Diego, CA). The PPMS provides for measurements under vacuum ($< 10^{-4}$ Torr), and employs a radiation shield and corrective terms, to minimize and account for non-conductive (i.e., convective and radiative) heat transfer. The PPMS is designed for both steady-state and transient thermal conductivity measurements. The samples in this study were measured using a transient method, in which curve fitting is used to determine thermal conductivity from measurements of the temperature difference across a sample as a function of time resulting from the application of a known heating power.[34]

RESULTS AND DISCUSSION

The room-temperature thermal conductivities for four commercial dental composites, along with a comparison to representative thermal conductivities for the unfilled, cured resin (Filtek[TM] Supreme-Unfilled resin) and the filler particles (amorphous silica, a-SiO$_2$), are shown in Figure 2. We previously reported the thermal conductivities of the composites from 275 to 350 K and found that they did not change dramatically within this temperature range.[19] The thermal conductivity of cured Filtek Supreme[TM]-Unfilled resin is somewhat higher than that of many common, chain polymers (e.g., $\kappa_{PMMA} = 0.2$ W m^{-1} K^{-1})[29] but reasonable as it is crosslinked, which is important for the mechanical properties required for dental materials. Types of filler particles added to dental resins include silica, Ti-, Zr- and Al-oxides, LiAlSiO$_2$, ytterbium trifluoride, and Ba and Sr glasses.[11,35] Given the composition of the fillers, it is assumed that their thermal conductivity is approximately constant at orally relevant temperatures and that 1.4 W m^{-1} K^{-1} (i.e., κ for amorphous SiO$_2$ at 300 K)[15] is an accurate estimate for the silica filler used in Filtek[TM] Supreme and representative of low-κ filler particles in general.

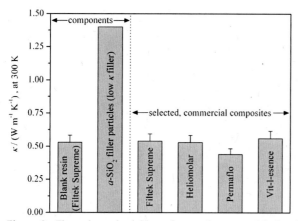

Figure 2. Thermal conductivities of several commercial dental composites (Filtek[TM] Supreme, Heliomolar® Flow, PermaFlo® and Vit-l-esence[TM]) at 300 K in comparison to an unfilled acrylic resin used in dental composites (Filtek[TM] Supreme-Unfilled resin) and amorphous SiO$_2$ (a-SiO$_2$), which is thought to be representative of κ_{filler} for Filtek[TM] Supreme and other, typical fillers.

As is the case for dentin and other natural, tooth-like biomaterials,[11] the thermal conductivity of Filtek[TM] Supreme falls below the rule of mixture prediction based on the volume fractions and thermal conductivities of its unfilled resin and a silica filler. The thermal conductivities of the other commercial composites are comparably low, which suggests that similar conclusions will apply to many commercial materials. However, more information about their constituent materials is required to evaluate this hypothesis.

Using the thermal conductivity values for the Filtek[TM] Supreme-Unfilled resin and amorphous SiO_2 as κ_{matrix} and κ_{filler}, respectively, in equation (2), and assuming $G_{bd} = 1/R_{bd} \approx 30$ MW m^{-2} K^{-1} from the boundary conductance for alumina/PMMA,[29] we find a critical particle radius of $r_c \approx 30$ nm. Figure 3 shows the thermal conductivity for such a composite, calculated from equation (1), as a function of filler volume fraction for several different sized fillers. Considering the assumptions about the filler and R_{bd}, Figure 3 appears reasonable in comparison to the observation that $\kappa_{composite}/\kappa_{matrix} \sim 1$ for Filtek[TM] Supreme, which contains 20 nm ($r = 10$ nm) and agglomerated ~ 1 μm ($r = 500$ nm) particles. This figure illustrates the advantage of nano-sized fillers in comparison to micro-sized from the perspective of producing low-κ dental composites. The influence of filler particle size and loading on thermal conductivity could be explored more quantitatively by preparing composites in the lab and better controlling/quantifying the filler. If done with a range of volume fractions, such experiments also would allow for determination of R_{bd} for resin/filler combinations used in dental composites. Note that Filtek[TM] Supreme and many composites are hybrids consisting of larger and smaller filler particles. With information on the particle size distribution, equation (1) can be adapted to include more than one filler particle size. A "double effective medium theory" was employed by Davis and Artz[36] for a bimodal distribution by treating the larger particles as embedded in an "effective matrix", which consisted of a composite of the blank matrix filled with the smaller particles.

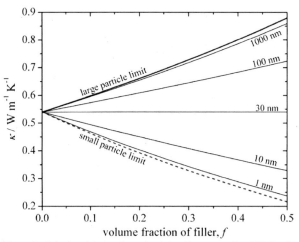

Figure 3. Calculated thermal conductivity (from equation (1)) for the filler/matrix composite described in the text. The results for $r = 1$ nm, 10 nm, 30 nm, 100 nm, and 1000 nm filler particles are shown along with the limiting results for "small" (– – – – –) and "large" (————) filler particles.

Thermal conductivity measurements and application of literature models suggest that, as a result of a size effect (increased contribution from thermal boundary resistance), the thermal conductivity of many commercial dental composites will be less than expected based on their composition and the rule of mixtures approximation. Further reductions in κ should be possible by reducing the filler particle size to the 1 – 10 nm range, or using hybrids with sufficient content of similarly small particles. However, this may have unintended effects on other properties and it becomes more difficult to achieve high loadings as particle size is decreased. Fortunately, in practice, further lowering κ is not required for thermal insulation in a post-restored tooth. For the composites measured here, $\kappa_{composite} < \kappa_{dentin}$ and we all know from experience that natural tooth materials form an effective barrier against typical hot and cold temperatures in the mouth.

The value of further reducing κ in order to protect against temperature increases during the light curing of a composite restoration is less intuitive. We recently reported finite element method (FEM) simulations on a model tooth that explored the effects of many different factors on the temperature increases during the curing process.[19] That model, shown in Figure 4, was based on a 2D, axisymmetric model of an extracted tooth that was used in comparable in vitro experiments and is described in detail elsewhere.[19] The model was shown to provide a good approximation of the experimentally measured temperature increases and, through a series of simulations, demonstrated that the curing light intensity, curing time, and the composite's enthalpy of polymerization are the most important factors in the temperature increase near the pulp.[19] The greatest risks occur when using the light to cure a thin layer of bonding resin, or in deep restorations without a thermal barrier layer.[19] An interest in mitigating the thermal hazard by tailoring the thermal properties of the materials was part of the original motivation for this work; however, the simulations show that viable changes in thermal conductivity and heat capacity would produce only minor reductions in the maximum temperature experienced by the tooth pulp. Even a highly optimistic reduction in thermal conductivity from 0.5 to 0.1 W m^{-1} K^{-1}, combined with densities and heat capacities at the high end of those measured for commercial composites, reduces the maximum temperature increase at the pulp by less than 2 °C. Such a decrease could be useful but would be difficult to achieve through changes in κ (and heat capacity). As demonstrated in the simulations,[19] other approaches, such as lower light intensity over a longer irradiation time, would be more effective in minimizing the thermal risk during light curing.

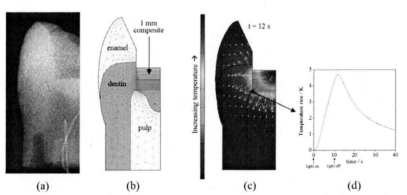

(a)   (b)   (c)   (d)

Figure 4. Temperature increases during light curing were studied using in vitro experiments on an extracted tooth (a) and in finite element method simulations (b). In (c) and (d), sample simulation results are shown for curing a 1 mm layer of composite in a tooth with 0.8 mm of dentin remaining between the base of the restoration and the closest point on the pulp-dentin junction.

CONCLUSIONS

Given their composition and filler size, the thermal conductivity of light-cured dental composites seems to be controlled by size effects that can be understood using literature models for spheres in a matrix that include the effect of thermal boundary resistance. In principle, thermal conductivity is a consideration in developing/selecting dental composites and it could be further reduced by employing smaller ($r \sim 1 - 10$ nm) filler particles. However, in practice, the composites measured here already provide good insulation in post-restored teeth (i.e., $\kappa_{composite} \leq \kappa_{dentin}$) and potential, further reductions in thermal conductivity would have only minor effects even on the maximum temperature experienced by the tooth pulp during the light-curing process.

ACKNOWLEDGEMENTS

We gratefully acknowledge Christopher Felix and Catherine O'Neill for their contributions to related work. This work was supported by NSERC of Canada, the Killam Trusts and the Sumner Foundation, along with the Canada Foundation for Innovation, Atlantic Innovation Fund and other partners that fund the Facilities for Materials Characterization managed by the Institute for Research in Materials at Dalhousie University.

FOOTNOTES
*Corresponding author. Electronic mail: Mary.Anne.White@Dal.Ca

REFERENCES
[1]L. Zach, and G. Cohen, Pulp response to externally applied heat, *Oral Surg. Med. Oral Pathol.*, **19**, 515-30 (1965).
[2]D.L. Hussey, P.A. Biagioni, and P.-J. Lamey, Thermographic measurement of temperature change during resin composite polymerization in vivo, *J. Dent.*, **23**, 267-71 (1995).
[3]A. C. Shortall, and E. Harrington, Temperature rise during polymerization of light-activated resin composites, *J. Oral Rehab.*, **25**, 908-13 (1998).
[4]M. Hannig, and B. Bott, In-vitro pulp chamber temperature rise during composite resin polymerization with various light-curing sources, *Dent. Mater.*, **15**, 275-81 (1999).
[5]A. Uhl, R.W. Mills, K.D. Jandt, Polymerization and light-induced heat of dental compostes cured with LED and halogen technology, *Biomaterials,* **24**, 1809-20 (2003).
[6]S. Bouillaguet, G. Caillot, J. Forchelet, M. Cattani-Lorente, J.C. Wataha, and I. Krejci, Thermal risks from LED- and high intensity QTH-curing units during polymerization of dental resins, *J. Biomed. Mater. Res. B*, **72**, 260-7 (2005).
[7]A.E. Peutzfeldt, Temperature rise induced by some light emitting diode and quartz-tungsten-halogen curing units, *Eur. J. Oral. Sci.*, **113**, 96-8 (2005).
[8]A.A. Al-Qudah, C.A. Mitchell, P.A. Biagioni, and D.L. Hussey, Thermographic investigation of contemporary resin-containing dental materials, *J. Dent.*, **33**, 593-602 (2005).
[9]A. Uhl, A. Völpel, and B.W. Sigusch, Influence of heat from light curing units and dental composite polymerization on cells in vitro, *J. Dent.*, **34**, 298-306 (2006).
[10]A.A. Al-Qudah, C.A. Mitchell, P.I. Biagioni, and D.L. Hussey, Effect of composite shade, increment thickness and curing light on temperature rise during photocuring, *J. Dent.*, **35**, 238-45 (2007).
[11]L.-Å. Lindén, *Photocuring of polymeric dental materials and plastic composite resins, in Radiation curing in polymer science and technology, Vol. IV: Practical aspects and applications*, edited by J.P. Fouassier (Kluwer Academic Publishers, 1993).
[12]R.G. Craig, Dental Materials, in *Handbook of Biomaterials Evaluation*, 2nd Ed., edited by A.F. von Recum (Plenum, New York, NY, 1970).

[13] J.L. Ferracane, *Materials in Dentistry*, 2nd Ed. (Lippincott Williams and Wilkings, Hagerstown, MD, 2001).

[14] D.C. Watts, R. McAndrew, and C.H. Lloyd, Thermal diffusivity of composite restorative materials, *J. Dent. Res.*, **66**, 1576-8 (1987).

[15] Y.S. Touloukian, R.W. Powell, C.Y. Ho, and P.G. Klemens, Eds. *Thermal Conductivity: Nonmetallic Solids* (Plenum, New York, NY, 1970).

[16] X.W. Su, and F.Z. Cui, Hierarchical structure of ivory: from nanometer to centimeter, *Mater. Sci. Eng. C*, **7**, 19-29 (1999).

[17] M.B. Jakubinek, C.J. Samarasekèra, and M.A. White, Elephant ivory: a low thermal conductivity, high strength nanocomposite, *J. Mater. Res.*, **21**, 287-92 (2006).

[18] C.-W. Nan, R. Birringer, D.R. Clarke, and H. Gleiter, Effective thermal conductivity of particulate composites with interfacial thermal resistance, *J. Appl. Phys.*, **81**, 6692-9 (1997).

[19] M.B. Jakubinek, C. O'Neill, R.B. Price, and M.A. White, Temperature excursions at the pulp-dentin junction during the curing of light-activated dental restorations, *Dent. Mater.*, **24**, 1468-76 (2008).

[20] Y. Ou, F. Yang, and Z.-Z. Yu, A new conception on the toughness of nylon 6/silica nanocomposite prepared via *in situ* polymerization, *J. Polym. Sci. B.: Polymer Physics*, **36**, 789-795 (1998).

[21] B.J. Ash, D.F. Rogers, C.J. Wiegand, L.S. Schadler, R.W. Siegel, B.C. Benicewicz, and T. Apple, Mechanical properties of Al2O3/polymethylmethacrylate nanocomposites, *Polymer Composites*, **23**, 1014-25 (2002).

[22] W. Hergeth, U. Steinau, H. Bittrich, G. Simon, and K. Schmutzler, Polymerization in the presence of seeds. IV: Emulsion polymers containing inorganic filler particles, *Polymer*, **30**, 254-58 (1989).

[23] D.P.H. Hasselman, and L.F. Johnson, Effective thermal conductivity of composites with interfacial thermal barrier resistance, *J. Compos. Mater.*, **21**, 508-15 (1987).

[24] Y. Beveniste, Effective thermal conductivity of composites with a thermal contact resistance between the constituents: nondilute case, *J. Appl. Phys.*, **61**, 2840 (1987).

[25] A.G. Every, Y. Tzou, D.P.H. Hasselman, and R. Raj, The effect of particle size on the thermal conductivity of ZnS/diamond composites, *Acta Metall. Mater.*, **40**, 123-9 (1992).

[26] D.P.H. Hasselman, K.Y. Donaldson, and A.L. Greiger, Effect of reinforcement size on the thermal conductivity of a particulate-silicon carbide reinforced aluminum matrix composites, *J. Am. Ceram. Soc.*, **75**, 3137-40 (1992).

[27] A.L. Greiger, D.P.H. Hasselman, and K.Y. Donaldson, Effect of reinforcement particle size on the thermal conductivity of a particulate silicon carbide-reinforced aluminum-matrix composite, *J. Mater. Sci. Lett.*, **12**, 420 (1993).

[28] D.P.H. Hasselman, K.Y. Donaldson, J. Liu, L.J. Gauckler, and P.D. Ownby, Thermal conductivity of a particulate-diamond reinforced cordierite matrix composite, *J. Am. Ceram. Soc.*, **77**, 1757-60 (1994).

[29] S.A. Putnam, D.G. Cahill, B.J. Ash, and L.S. Schadler, High-precision thermal conductivity measurements as a probe of polymer/nanoparticle interfaces, *J. Appl. Phys.*, **94**, 6785-88 (2003).

[30] *Filtek™ Supreme universal restorative system: Technical product profile*, 3M ESPE (2002).

[31] *PermaFlo flowable composite*, Ultradent Products Inc., (2003).

[32] *Vit-l-esence™ esthetic restorative material*, Ultradent Products, Inc., (2003).

[33] *Heliomolar® Flow: Instructions for Use*, Ivoclar Vivadent (2001).

[34] N.R. Dilley, R.C. Black, L. Montes, A. Wilson, and M.B. Simmonds, Commercial apparatus for measuring thermal transport properties from 1.9 to 390 Kelvin, *Mat. Res. Soc. Symp. Proc.*, **691**, 85-90 (2002).

[35] N. Moszner, and U. Salz, Composites for dental restoratives, in Polymers for dental and orthopedic applications, edited by S.W. Shalaby and U. Salz (CRC Press, Boca Raton, FL, 2006).

[36] L.C. Davis, and B.E. Artz, Thermal conductivity of metal-matrix composites, *J. Appl. Phys.*, **77**, 4954-60 (1995).

Porous Bioceramics

MANUFACTURING OF POROUS PPLA-HA COMPOSITE SCAFFOLDS BY SINTERING FOR BONE TISSUE ENGINEERING

Ana Paula M. Casadei, Fabrício Dingee, Tatiana E. da Silva, André L.G. Prette, Carlos R. Rambo and Marcio C. Fredel
Departament of Mechanical Engineering, Federal University of Santa Catarina, UFSC, Florianópolis, Brazil

Eliana A.R. Duek
Centre of Medical and Biological Sciences, PUC, Sorocaba, Brazil

ABSTRACT

This work reports on the production of porous ceramic-polymer scaffolds based on poli(L- lactic acid) (PLLA) and hydroxyapatite (HA) with porosity controlled by adequate sintering conditions. The composites were prepare with three different mass fraction of PLLA/HA (90/10, 80/20, and 70/30) and two different polymer particle size range (106-212 μm and 212-300 μm). The mixture was sintered at 185ºC, since low densification degree was observed at this temperature. Different sintering times were established, based on the time to achieve partial sintering and neck formation and taking into account the reproducibility of the process. The influence of sintering time, temperature, PLLA particle size and HA content on the morphology of the scaffolds was evaluated. The scaffolds were characterized by scanning electron microscopy (SEM) and compressive tests. The results revealed that the scaffolds exhibited interconnectivity between pores, pore sizes between 100 to 300 μm and suitable compressive strength to be used in bone tissue engineering.

INTRODUCTION

The recent advances in the basic science of molecular cell biology combined with advances in materials science and engineering is pushing the field of biomaterials into new applications, such as bone tissue engineering[1]. There are multiple clinical reasons to develop bone tissue-engineering alternatives, including the need for better filler materials that can be used in the reconstruction of large orthopedic defects and the need for orthopedic implants that are mechanically suitable to their biological environment.

To induce new bone formation some elements are necessary: a morphogenetic signal, such as bone morphogenetic proteins (BMP), osteogenic cells derived from host bone that will respond to the signal, a suitable carrier of this signal that can deliver it to specific sites that serves as scaffold for the growth of the responsive host cells and a viable, well vascularized host bed[2,3,4,5]. The carrier (scaffold) must have the appropriate three-dimensional (3-D) structure to serve as an osteoconductive matrix for bone-forming cells; it should be biocompatible to minimize interference with bone induction from an inflammatory reaction; the chemistry of carrier materials should be tailored to support cell adhesion and proliferation; it must be biodegradable by enzymes and/or circulating biological fluid; it must persist *in vivo* long enough to maintain bioactive elements at the site of implantation and optimize their release profile. The particular challenge for the design of bone scaffolds include maintaining appropriate mechanical stability that should be comparable to that of adjacent bone when implanted, while possessing high pore density to allow cellular infiltration and ingrowth[6,7]. The adequate microarchitecture of a porous scaffold to be used as bone substitute must take into account its porosity, as well as pore size and interconnectivity[8].

A variety of materials, including ceramics, collagen, noncollagenous proteins, bioactive glasses, and biodegradable polymers, have been widely used to develop porous 3-D scaffolds using various fabrication techniques. Ceramics comprised of hydroxyapatite (HA), SiO_2, Al_2O_3, ZrO_2 and TiO_2 are nowadays used for various medical applications due to their positive interactions with human tissues[9]. Among the possible candidate materials for bone engineering scaffolds, calcium phosphates present several advantages, derived from their similarity and high compatibility with natural bone. Porous HA materials have been used as bone scaffolds to promote an improved bone ingrowth and osteointegration. Recent *in vivo* study indicated that the dissolution rate of sintered carbonated HA ceramics implanted subcutaneously was intermediate between tricalcium phosphate (b-TCP) and pure HA[10]. However, to be used effectively in load bearing compartments, the mechanical properties of the HA scaffolds should be improved[11].

Biodegradable polymers, mainly polyesters such as poly(lactic acid) (PLA), poly(glycolic acid) (PGA) and their copolymers (PLGA), have been also widely used to develop porous 3-D scaffolds using various fabrication techniques. The hydrophobicity of PLLA and its acid degradation products can difficult its resorption and result in aseptic inflammation of tissue.

Therefore, only bioceramic or biopolymer alone is not fully adequate to fulfill well the requirements for bone repair application. To combine the osteoconductivity of calcium phosphates and good biodegradability of polyesters, polymer/ceramic composite scaffolds have been developed for bone tissue engineering[12]. PLLA combined with non-sintered HA particles exhibit high mechanical strength and a degradation rate similar to the rate of bone regeneration. This composite is totally resorbable, biocompatible and bioactive[13]. Shikinami et al observed that plates and screws made of PPLA and HA particles exhibited higher mechanical strength than those produced of pure PLLA[14]. Hydroxyapatite particles in polymer appeared to provide an anchor for the attachment of bone cells (osteoblasts). Nanometric HA particles can be dispersed on the pore walls of the scaffolds and bound to the polymer very well. Apatite crystals, furthermore, keep the pH of the environment within the physiological range. Acid reaction around the implantation site with PLA and PGA implants can be prevented when polymers are used together with apatites[15,16].

Conventional methods for several porous scaffolds manufacturing techniques have been proposed, such as combination on porogen leaching and freeze-drying techniques[17], gas foaming[18], thermally induced phase separation[19], foaming and hydrolysis[20], rapid prototyping[21], among others. In most of these methods organic agents as pore formers are used, which may lead to residues of solvents or from thermal decomposition of the polymeric templates. Moreover, high temperature processes are also commonly involved, which makes the methods not cost-effective. To transpose these drawbacks a simple method is proposed in this work, which makes use only of the precursor materials. It consists on the production of porous ceramic-polymer scaffolds based on poli(L- lactic acid) (PLLA) and hydroxyapatite (HA) with porosity controlled by adequate sintering conditions.

MATERIALS AND METHODS

The polymer used as matrix was pellets of poly(L-lactic acid) (PLLA) (CCMB/PUC Sorocaba) with average particle size of 2.4 ± 0.7 mm, molecular weight of 100.000 g/mol and density of 1.15 g/cm^3. The bioceramic used was nano-hydroxyapatite-carbonated calcined at 800°C (Biomaterials group/UFSCar, Brazil), with density of 2.24 g/cm^3.

The PLLA was previously dry milled, using a processor, and then sieved at 300 μm, 212 μm and 106 μm and stored to avoid degradation due to humidity. The composites were prepared with three different mass fraction of PLLA/HA (90/10, 80/20, and 70/30) and two different polymer particle size range (106-212 μm and 212-300 μm).

The average weight of each sample subjected to sintering was 50 mg. After weighing of raw materials, the mixtures were homogenized in ultrasound in the form of suspension in isopropyl alcohol to distribute the particles of HA. After drying the mixture was placed in an aluminium mould and sintered in a laboratory made furnace, pdesigned to reduce thermal gradients along the sample. The sintering temperature was defined from differential thermal analysis of PLLA (DTA, Shimadzu, TGA 50). The glass transition temperature (Tg) of PLLA was 177 °C and the melting temperature (Tm) 187 °C. The temperature range where the sintering was more efficient for all the samples was 184-187 °C with a heating rate of 20 °C/min and air-cooled. After sintering, cylindrical samples with average dimensions about 3 mm diameter and 9 mm height were obtained.

The morphology and microstructure of the fractured surfaces of the scaffolds were evaluated by scanning electron microscopy (SEM, Philips, XL30) coupled with energy dispersive X-ray spectroscopy (EDS, Oxford, EDAX. The samples were cooled in liquid nitrogen, fractured and coated with gold.

The density of the composite was determined according to Eq. 1:

$$\rho = \frac{1}{\frac{m_{HA}}{\rho_{HA}} + \frac{m_{PLLA}}{\rho_{PLLA}}} \qquad (1)$$

Where ρ is density of the composite, m_{HA} and m_{PLLA} are the masses of hydroxyapatite and PLLA, respectively; ρ_{HA} and ρ_{PLLA} are the densities of HA and PLLA, respectively. The porosity was calculated from the relation between the solid (composite) density and the geometrical density, according to the expression: $P = (1 - \rho_G/\rho)$, where ρ_G is the bulk density of the PLLA/HA scaffolds.

The mechanical strength was determined by uniaxial compression tests in a universal mechanical test device EMIC DL 2000, with load cell of 50 kgf, according to an adaptation of the standard ASTM695[22]. The speed of the crosshead was set to 1.3 mm/min.

RESULTS AND DISCUSSION

The PLLA melting temperature, obtained from DTA was 183°C. Although temperatures between 180-187°C were tested to sinter the composites, to obtain porous scaffolds by sintering complete densification must be avoided. Therefore, 185°C was chosen as the sintering temperature, since low densification degree was observed at this temperature. Sintering times were determined experimentally taking into account the reproducibility of the process. Different sintering times were established, based on the time to achieve partial sintering and neck formation, according to previous work[22]. The table I shows the isotherm time for each studied sample. The processing time is longer for PLLA with particles size 212-300 µm than for the 106-212 µm particles size range. Two factors are very important to the sintering process, which are the atomic mobility and surface energy. The surface energy is assessed by the surface area, so small particles that have higher surface area and more energy per unit volume, will sinter faster than large particles[23].

Table I – Sintering time for each particle size range of the porous PLLA/HA scaffolds.

PLLA/HA wt (%)	Sintering time (min)	
	PLLA-106-212 µm	PLLA-212-300 µm
90/10	4	13
80/20	16	34
70/30	46	85

Sintering lowers the surface energy by reducing surface area with concomitant formation of interparticle bonds. This fact can promote in many cases an increase in density due to dimensional shrinkage. The particles actually attract each other and self-compress to eliminate pores. However, sometimes a minimal-dimensional change is desired in many components, it is the case of porous scaffolds for tissue engineering. In these cases, it is necessary to find the adequate adjustment between neck formation and sintering time.

The sintered samples exhibited porosity between 48 and 61% (table II). The porosity increases with increasing the ceramic content. Moreover, the porosity in the PLLA with 106-212 µm particle size is higher than in the scaffold prepared with PLLA of 212-300 µm particle size.

Table II - Mean porosity of the studied samples.

PLLA/HA wt (%)	Mean porosity (%)	
	PLLA-106-212 µm	PLLA-212-300 µm
90/10	55.5 ± 1.6	48.5 ± 1.6
80/20	57.8 ± 2.1	51.7 ± 3.4
70/30	58.6 ± 2.9	61.8 ± 2.0

The statistical analysis using the ANOVA/MANOVA method showed that both variables, HA concentration and polymer size influence significantly the porosity (Fig. 1). Analysis revealed that porosity varies with HA content ($P=10^{-6}$) and particle size of PLLA (2×10^{-5}). Fig. 1 shows the dependence of HA content and PLLA particle size on the porosity of the composites. Scaffolds with higher HA concentration exhibited higher porosity. Larger PLLA particles have lower superficial reactivity, which extends the sintering time for the composites with particle sizes in the range of 212-300 µm. Addition of HA hinders sintering for each PLLA particle size range. Although for both particle size range the porosity increased with HÁ content for the 106-212 µm range this increasing tendency is less pronounced with addition of HA. On the other hand, for the particle size range of 212-300 µm the porosity of the scaffolds exhibits a more pronounced increasing tendency with addition of HA. Additionally, there is an equivalence tendency of the two particle size range, suggesting that after a particular sintering time the HA content is more influent on porosity than particle size.

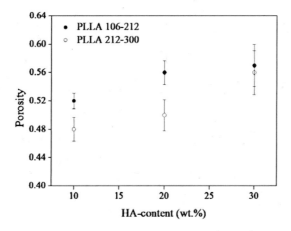

Fig. 1: Porosity of the PLLA/HA scaffolds in function of HA content for both particle size range.

The cell proliferation and growth in the scaffolds interior depends of many others factors beyond porosity. The pore size and interconnectivity are very important to allow cells and blood vessels migration. The pore size must vary between 100 and 250 μm[24], 100 and 400 μm[25] and 150 and 500 μm, while the interconnectivity shall not be smaller than 40 μm[26]. The scaffold with lower ceramic concentration (90/10) and smaller PLLA particle size (106-212 μm) presents adequate pore morphology with suitable pore size and connectivity, as shown in the SEM micrographs of Fig. 2. The presence of necks can be observed between the particles (Fig. 2a) and the HA particles are homogeneously distributed over the PLLA surface (Fig. 2b).

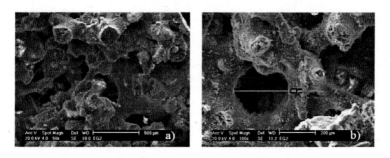

Fig. 2: SEM micrographs of a scaffold with 90/10 (PLLA/HA) concentration and

PLLA 106-212 μm particle size.

More important than only a high porosity in scaffolds for bone tissue repairing, is the combination between porosity and mechanical properties. As references, compressive strengths in the range of of 2-12 MPa for cancellous bone and 100-230 MPa for compact bone are reported[27]. Table III shows the mean compressive strength of the porous PLLA/HA scaffolds.

Table III: Mean compressive strength of the porous PLLA/HA scaffolds.

HA [%]	PLLA [μm]	Compressive strength [MPa]
10	106-212	9.50 ± 5.06
	212-300	1.24 ± 4.42
20	106-212	2.08 ± 1.53
	212-300	5.84 ± 3.96
30	106-212	4.26 ± 3.20
	212-300	3.74 ± 0.27

The mean compressive strength off the PLLA/HA scaffolds varied between 2 and 12 MPa. The highest value was reached for the sample with smaller ceramic concentration (90/10) and PLLA 212-300 μm particles size. The scaffolds exhibit compressive strength in the range of the cancellous bone, which makes them suitable for bone regeneration.

CONCLUSIONS

Porous PLLA/HA composite scaffolds were successfully produced by partial sintering. Porosity was controlled only by adjusting the processing parameters, without any additives. The sintering time increased with PLLA particle size and HA concentration. The porosity increased with HA concentration and decreased with increasing of PLLA particle size. The scaffolds exhibited adequate pore morphology, size and interconnectivity to be used in bone tissue engineering.

ACKNOWLEDGEMENTS
The authors thank CNPq-Brazil for financial support.

REFERENCES
[1]L. G. Griffith, Polymeric Biomaterials, Acta mater., 48, 263±277 (2000).
[2]K. J. L. Burg, S. Porter, Biomaterial Developments for Bone Tissue Engineering, Biomaterials 21, 2347-2359 (2000).
[3]A. J. Burdick, M. N. Mason, A. D. Hinman, K. Thorne, K. S. Anseth, Delivery of Osteoinductive Growth Factors from Degradable PEG Hydrogels Influences Osteoblast Differentiation and Mineralization, Research Article, Journal of Controlled Release, 83, 53–63 (2002).
[4]M. C. Kruyt, W. J. A. Dhert, H. Yuan, C. E. Wilson, C. A. Blitterswijk, A. J. Verbout, J. D. Bruijn, Bone Tissue Engineering in a Critical Size Defect Compared to Ectopic Implantations in the Goat, Journal of Orthopaedic Research, 22, 544–551 (2004).
[5]N. Saito, N. Murakami, J. Takahashi, H. Horiuchi, H. Ota, H. Kato, T. Okada, K. Nozaki, K. Takaoka, Synthetic Biodegradable Polymers as Drug Delivery Systems for Bone Morphogenetic Proteins, Advanced Drug Delivery Reviews, 57, 1037– 1048 (2005).
[6]A. R.Vaccaro, The Role of the Osteoconductive Scaffold in Synthetic Bone Graft, Orthopedics, 25, no 5/supplement (2002).

[7]S. H. Lee, H. Shin, Matrices and Scaffolds for Delivery of Bioactive Molecules in Bone and Cartilage Tissue Engineering, Advanced Drug Delivery Reviews, 59, 339–359 (2007).

[8]D. Flautre, M. Descamps, C. Delecourt, M. C. Blary, Porous HA Ceramic for Bone Replacement: Role of the Porous and Interconnections – experimental study in rabbit, Journal of Material Science: Materials in Medicine, 12, 679-682 (2001).

[9]W. J. E. M. Habraken, J. G. C. Wolke, J. A. Jansen, Ceramic Composites as Matrices and Scaffolds for Drug Delivery in Tissue Engineering, Advanced Drug Delivery Reviews, 59, 234–248 (2007).

[10]J. Barralet, M. Akao, H. Aoki, Dissolution of Dense Carbonate Apatite Subcutaneously Implanted in Wistar rats, J. Biomed. Mater. Res., 49, 176–182 (2000).

[11]H. W. Kima, J. C. Knowlesa, H. E. Kimb, Hydroxyapatite/poly(e-caprolactone) Composite Coatings on Hydroxyapatite Porous Bone Scaffold for Drug Delivery, Biomaterials, 25, 1279–1287 (2004).

[12]Y. Kang, X. Xu, G. Yin, A. Chen, L. Liao, Y. Yao, Z. Huang, X. Liao, A Comparative Study of the in vitro Degradation of Poly(L-lactic acid)/b-tricalcium phosphate Scaffold in Static and Dynamic Simulated Body Fluid, European Polymer Journal, 43, 1768–1778 (2007).

[13]Y. Shikinami, M. Okuno, Bioresorbable devices madeof forged composites of hydroxyapatite (HA) particles and poly L – lactide (PLLA). Part I: basic characteristics. Biomaterials, 20, 859-877 (1999).

[14]Y. Shikinami, M. Okuno, Bioresorbable devices made of forged composites of hydroxyapatite (HA) particles and poly L – lactide (PLLA). Part II: practical properties of miniscrews and miniplates. Biomaterials, 22, 3197-3211 (2001).

[15]M. Yaszemski, D. J. Trantolo, K. U. Lewandrowski, V. Hasirci, D. E. Altobelli, D. L. Wise, Biomaterials in Orthopedics, Marcel Dekker, Inc, (2004).

[16]G. Wei, P. X. Ma, Structure and Properties of nano-Hydroxyapatite/polymer Composite Scaffolds for Bone Tissue Engineering, Biomaterials, 25, 4749-4757 (2004).

[17]G. Chen, T. Ushida, T. Tateishi, Development of Biodegradable Porous Scaffolds for Tissue Engineering, Materials Science and Engineering C, 17, 63–69 (2001).

[18]D. J. Mooney, F. Daniel, D. F. Baldwin, N. P. Suht, J. P. Vacanti, R. Langer, Novel Approach to Fabricate Porous Sponges of Poly(D,L-lactic-co-glycolic acid) without the use of Organic Solvents. Biomoterials, 17, 1417-1422 (1996).

[19]Y. S. Nam, T. G. Park, Biodegradable Polymeric Microcellular Foams by Modified Thermally Induced Phase Separation Method, Biomaterials, 20, 1783-1790 (1999).

[20]A. Almiralla, G. Larrecqa, J. A. Delgadoa, S. Martinezb, J. A. Planella, M. P. Ginebraa, Fabrication of Low Temperature Macroporous Hydroxyapatite Scaffolds by Foaming and Hydrolysis of an a-TCP paste, Biomaterials, 25, 3671–3680 (2004).

[21]W. Y. Yeong, C. K. Chual, K. F. Leong, M, Chandrasekaran, Rapid Prototyping in Tissue Engineering: Challenges and Potential, Trends in Biotechnology, 22, 12 (2004).

[22]F. Dingee, Sintering of PLLA and PLLA-HA composites for fabrication of three-dimensional porous matrixes. M.Sc. dissertation, Federal University of Santa Catarina, Brazil, 2007.

[23]R. M. German, Sintering Theory and Practice, John Wiley,7 Sons inc.

[24]K. Whang, C. H. Thomas, K. E. Healy, G. Nuber, A Novel Method to Fabricate Bioabsorbable Scaffolds, Polymer, 36, 4, 837-842 (1995).

[25]M. Yoneda, Repair of an Intercalated Long Bone Defect with a Synthetic Biodegradable Bone-inducing Implant, Biomaterials, 26, 5145-5152 (2005).

[26]V. Oliver, N. Faucheux, P. Hardouin, Biomaterial Challenges and Approaches to Stem Cell Use in Bone Reconstructive Surgery, DDT, 9, 18 (2004).

[27]T. Kokubo, H. Kim, M. Kawashita, Novel Bioactive Materials with Different Mechanical Properties. Biomaterials, 24, 2161-2175 (2003).

EFFECT OF ZINC ON BIOACTIVITY OF NANO-MACROPOROUS SODA-LIME
PHOSPHOFLUOROSILICATE GLASS-CERAMIC

H.M. Moawad, S. Wang, H. Jain
Department of Materials Science & Engineering
Lehigh University, Bethlehem, PA 18015, USA

M. M. Falk
Department of Biological Sciences
Lehigh University, Bethlehem PA 18015, USA

ABSTRACT
Multi-scale porosity is desirable for the use of biocompatible glass and glass-ceramics as a bioscaffold material because porosity promotes cell attachment and thus better acceptance of the scaffolds. Recently, we demonstrated the fabrication of soda-lime phosphofluorosilicate glass-ceramics with porosity ranging from several nanometers to >100 micrometers by the melt-quench-heat-etch method, and optimized the phase distribution for rapid growth of hydroxyapatite (HA). In this work we have extended the usefulness of this class of biomaterial by adding ZnO that is believed to stimulate bone growth. Starting with $48SiO_2-2.7P_2O_5-4CaF_2-xZnO-yCaO-zNa_2O$ glasses with x=0.0, 0.5, 1, 3, 5, 8, 10, y+z =45.3-x (mol %), nano-macro porous glass-ceramics were fabricated. We observed the formation of many crystalline phases, but mainly sodium calcium silicate, calcium phosphate, and fluorapatite. In general, the addition of ZnO improves glass durability. For $x \leq$ 3 mol% HA is formed in simulated body fluid, but for higher concentration the surface layer deviates from HA composition significantly.

INTRODUCTION
        Tissue engineering has been rapidly emerging as a viable option for the repair of skeletal tissues. One approach of tissue engineering considers the implantation of cells onto bioactive and degradable scaffolds that serve as temporary physical support.[1] Therefore, the design and construction of scaffold are attracting increasing attention for tissue engineering. An ideal bioscaffold should have the following characteristics: a) highly porous three-dimensional, interconnected pore network for cell growth and for the transport of nutrients and metabolic waste, b) biocompatible at all stages, c) bioresorbable with controlled degradation and resorption rate to match cell or tissue growth in vitro or in vivo, d) suitable surface chemistry for cell attachment, proliferation, and differentiation, and e) mechanical properties to match those of the tissues at the site of implantation.[2]
        The pore fraction and structure are key parameters that determine the properties and applicability of scaffolds in a specific application. For glass as a candidate bone scaffold material, porosity is desired for cellular growth and attachment to the implant material's surface. It has been reported that bioactive glass forms a better bond with bone if it is porous.[2] There are advantages for nano-macroscale porosity, and the ideal bioactive glass should have this biomodal, even multi-modal porosity.[3,4] Irrespective of the size, it is important that the pores are interconnected for cell attachment, proliferation, differentiation, etc. Several methods have been developed in recent years for the fabrication of glass samples with these desired characteristics.[5-8] Here we focus on the melt-quench-heat-etch method developed in our laboratory recently, which is expected to yield samples of superior mechanical strength than those prepared by most other methods.[9,10] It is based on multi-scale spinodal phase separation, and therefore can be highly sensitive to the composition of glass.
        Fluoride ions are usually added to drinking water and toothpastes, indicating their importance in bone repair or prevention of bone damage. In addition, the crystal structure of fluorapatite

$(Ca_5(PO_4)_3F)$ is very similar to the crystal structure of hydroxyapatite (HA), $(Ca_5(PO_4)_3(OH))$ that is the main inorganic mineral component of all bones. Therefore, recently we developed new soda lime phosphosilicate bioactive glass series doped with CaF_2 as a source of fluoride ions, and successfully introduced multi-modal interconnected porosity by the melt-quench-heat-etch method.[11] Our results indicated that there are four important parameters, which may affect the formation of HA layer on the surface of nano-macro porous soda-lime phosphofluorosilicate: initial glass composition, temperature of crystallization, type of crystalline phases and leached amount of useful $Na_2Ca_2Si_3O_9$ and $Ca_5(PO_4)_3F$ phases. As an example, Figure 1 shows remarkable effect of crystallization of the base glass of the present study on the compatibility of human MG63 osteoblast cells.[12] Note that the cells become rounded on the surface of untreated glass indicating not as good compatibility as with the partially crystallized glass-ceramic of the same overall composition where they stretch out and proliferate rapidly. The best results were obtained for our starting composition based on 48S glass series that contained 4-8 mol % CaF_2 and crystal growth temperature of 750°C.

It has been reported that incorporation of zinc into an implant material could promote bone formation around the implant and accelerate recovery of a patient.[13] Zinc is an essential trace element having stimulatory effects on bone formation in vitro and in vivo.[14] Therefore, in this work we have extended the usefulness of bioactive soda-lime phosphofluorosilicate based nano-macro porous scaffolds by doping the starting glass with ZnO as a source of zinc for the repair and reconstruction of hard tissues. Toward this goal, we have utilized X-ray diffraction, scanning electron microscopy and energy dispersive X-ray spectroscopy to identify the composition, phases and microstructures of our samples. In addition, mercury porosimeter and inductive coupling plasma methods were used to quantitatively characterize the pore size distribution and the effect of Zn content on the concentration of P, Ca and Si in simulated body fluid after the immersion of our porous glass-ceramic samples.

II. EXPERIMENTAL PROCEDURE

The glasses of composition $48SiO_2-2.7P_2O_5-4CaF_2-xZnO-yCaO-zNa_2O$; x=0.0, 0.5, 1, 3, 5, 8, 10, y+z =45.3-x (mol %) were prepared with SiO_2 (99.99%), $CaCO_3$ (99%), Na_2CO_3 (99%), $Ca_5(OH)(PO_4)_3$ (99%), CaF_2 (99%) and ZnO (99%) as starting materials. The calculated batch of powders was mixed and ground using an alumina mortar and pestle. It was melted in a platinum crucible at 1300°C for 2 hours. The homogenized melt was poured into a stainless steel mold and then the so formed glass was annealed at 500°C to relax residual stresses. The result was a glass phase-separated on nm scale with interconnected spinodal texture. To induce additional larger scale phase separation, the samples were subjected further to a devitrification heat treatment consisting of nucleation at 670°C for 1h, followed by crystal growth at 750°C for 6h. To create nano-macro porosity, the heat treated glasses were leached for 1 hour in 0.3N HCl at 85 °C. The glasses are identified such that the numbers preceding S, F and Z refer to the mol% of silica, calcium fluoride and zinc oxide, respectively. To indicate the transformation of a glass to glass-ceramic, letter G is added at the end. Thus, for example, 48S4F10ZG, represents a glass-ceramic made from glass containing 48 mol% SiO_2, 4 mol% CaF_2, 10 mol% ZnO.

To identify the phases and observe microstructure the samples were analyzed by X-ray diffraction (XRD) (Rigaku X-ray diffractometer) and scanning electron microscopy (SEM). Hitachi 4300 Field Emission SEM was used to examine sectioned and polished samples of each glass to elucidate the phase separation and microstructure. The elemental distribution in different phases was determined by energy dispersive X-ray (EDX) spectroscopy, using Cu K and Cu L as reference for peak position. The parameters for data acquisition (time, full scale for intensity, pulse processing time) were kept the same for all the samples. The pore size distribution of the samples was determined by mercury porosimeter (Micromeritics Auto pore IV).

The in vitro formation of apatite layer was observed in conventional simulated body fluid (SBF), where the fluid contained inorganic ions in concentration corresponding to human blood plasma. For the preparation of 1 liter of SBF, the following reagents were dissolved in distilled water in indicated amounts: 7.996g NaCl, 0.350g NaHCO$_3$, 0.224g KCl, 0.228g K$_2$HPO$_4$.3H$_2$O, 0.305g MgCl$_2$.6H$_2$O, 40ml 1N-HCl, 0.278g CaCl$_2$, 0.071g Na$_2$SO$_4$, 6.05g NH$_2$C(CH$_2$OH)$_3$.[4,15-17] The fluid was buffered at physiological pH of 7.4 at 37°C. Each glass or glass-ceramic specimen (2 mg) was immersed in 1 ml of SBF in a polyethylene bottle covered with a tight lid. The HA layer formed on the surface of the solid and porous glass-ceramic samples after soaking in SBF for 7 days was characterized by SEM, XRD, EDX and the solution was analyzed by the inductive coupling plasma (ICP) method (Perkin Elmer model 7300V ICP-OES).

III. RESULTS AND DISCUSSION

Figure 2 shows the microstructure of 48S4FG, 48S4F1ZG, 48S4F3ZG, 48S4F10ZG samples after the heat treatment. The micrographs of heat treated 48S4FxZG samples indicate that there are many crystalline phases. The XRD patterns of the 48S4F1ZG and 48S4F10ZG compositions are shown in Figure 3 as examples of our glass-ceramics. The location of most diffraction peaks matches the standard ICDD (International Center for Diffraction Data) powder diffraction file card numbers 15-177, 1-1078, 12-671 and 4-7-5856.[16-21] Accordingly, four distinct crystal phases are identified: Ca$_4$P$_6$O$_{19}$, Na$_2$Ca$_2$Si$_3$O$_9$, Na$_2$CaSi$_3$O$_8$, and Ca$_5$(PO$_4$)$_3$F. At the same time, there remain a few unidentified peaks indicating the presence of at least one new phase yet to be determined. None of the identified phases in 48S4FxZG appeared to be zinc compounds.

Figure 4 shows the development of porous structure with increasing zinc content when the heat treated samples of the 48S4FxZG (x= 1, 3, 8, 10 mol%) glass-ceramic series are subjected to chemical etching in 0.3N HCl at 85°C for 1h. The inset in Fig. 3(c) shows the microstructure of 48S4F3ZG sample at a much higher magnification, where nanoscale pores can be seen readily. It is clear from the micrographs in Fig. 4 (c), obtained at relatively low and high magnifications, that the present melt-quench-heat-etch method has produced a structured network of interconnected nano-macro porosity. The nano-macro interconnected porosity in our glass-ceramic samples has been confirmed by mercury porosimeter. In Fig. 5 we show an example of multi-modal nano-macro porosity in the present glass-ceramic series. The mechanism for the creation of multi-modal nano-macro interconnected porosity in these samples appears to be similar to that observed in zinc-free composition investigated previously.[9] It is observed from a comparison of micrographs of 48S4FxZG glass-ceramic samples in Fig. 3 that the macropore density decreases with increasing ZnO content from 1 to 10 mol%.

Figure 6 shows SEM micrographs of the 48S4FxZG series of glass-ceramics (for x=0, 0.5, 3, 8 and 10) after soaking in SBF for 7 days. Note that a layer is formed on the surface of all the samples. Presumably, the so formed layer is HA enriched with Ca and P. It covers the whole surface of our porous samples. Note that the deposited large particles aggregate on the surface of 48S4FxZG samples with x =0-3 mol% ZnO. By comparison, there are smaller particles aggregated on the surface of samples containing x = 8 and 10 mol% ZnO.

The composition of the layer formed on the surface of glass-ceramic samples is determined by EDX. Figure 7 shows selected examples of EDX spectra for the surface layer on various nano-macro porous glass-ceramic samples with x = 0, 3, 8 mol% ZnO. We note that the intensity of two key elements, Ca and P, decreases with increasing ZnO. It is known that Na$_2$Ca$_2$Si$_3$O$_9$ and Na$_2$CaSi$_3$O$_8$ strongly enhance the bioactivity of glass-ceramics.[8,19] At the same time, the crystal structure of crystalline Ca$_5$(PO$_4$)$_3$F (fluorapatite) is very similar to the structure of HA.[22] Thus, fluorapatite crystalline phase acts as a seed or nucleation center for the formation of HA phase. Following our previous work,[11] we propose that it is predominantly the existence of Na$_2$Ca$_2$Si$_3$O$_9$ and Ca$_5$(PO$_4$)$_3$F that causes the enhancement of HA formation on the surface of 48S4FxZG porous glass-ceramic samples.

On the other hand, it seems that the addition of more than 3 mol% ZnO decreases apatite forming ability on the surface of the samples.

Figure 8 shows changes in the elemental concentration of SBF after soaking of the porous glass-ceramic 48S4FxZG series for 1 to 5 days. Here we note the effect of soaking time and the concentration of ZnO in starting glass on the leaching of ions into SBF. Specifically, the phosphorus concentration decreases with increasing soaking time for porous glass-ceramics containing 0 to 1 mol% ZnO. On the contrary, it increases with increasing time in the range of composition with x=5 to 10 mol% ZnO. For the sample made with x=3 mol% ZnO glass, P concentration decreases with time, but its magnitude is significantly higher than that for x=0-1.

A similar leaching trend is observed for calcium whose concentration decreases with increasing time for x=0-3 mol% ZnO. Again as for phosphorous, its concentration in solution increases with time for x=5-10 mol% ZnO. We believe that the decrease in phosphorous and calcium concentration is due to the consumption of phosphate and calcium ions needed for the formation of apatite on the surface of samples with x=0-3 mol% ZnO. In contrast to P and Ca, the concentration of silicon in solution increases with the soaking time, but decreases with increasing x for the whole 0-10 mol% ZnO composition range. The increase in silicon with soaking time in the SBF solution indicates the release of silicate ions from the dissolution of samples. Evidently, the release of phosphorous, calcium and silicon from porous glass-ceramic samples is significantly affected by the concentration of ZnO in the starting glass.

To understand the effect of zinc oxide addition on apatite formation on the surface of these novel porous glass-ceramics, one should consider the reaction between the samples and SBF. The reaction of Si-Na-K-Ca-Mg-P-O glasses with SBF has been described by Hench.[23] Briefly, it takes place in following steps: (a) The exchange of Na^+ and K^+ from the glass with H^+ or H_3O^+ in SBF. This ion exchange is accompanied by the loss of silica into the solution and the formation of silanol on the glass surface. This step is followed by the condensation and polymerization of a SiO_2-rich layer on the surface. (b) Migration of Ca^{2+} and PO^{3-}_4 through the silica-rich layer forming a CaO-P_2O_5-rich film that also incorporates calcium and phosphate from SBF. (c) The final step is the crystallization of the amorphous calcium phosphate film to form an apatite layer. Preliminary results show that the glass-ceramic dissolution rate decreases with increasing ZnO, especially for x >3 mol%. Any of these steps can impact the dissolution process, and it is difficult to establish the rate determining step from the present data. The improvement in durability and apatite forming ability shows opposite dependence on ZnO content. Thus, it is very important to balance the two factors in optimizing the composition of present nano-macro porous glass-ceramics.

IV. CONCLUSION

Nano-macro porous glass-ceramics containing $Ca_4P_6O_{19}$, $Na_2Ca_2Si_3O_9$, $Na_2CaSi_3O_8$ and $Ca_5(PO_4)_3F$ crystalline phases were fabricated starting from soda-lime phosphofluorosilicate glasses doped with 0-10 mol% ZnO. The HA forming ability on the surface of porous glass-ceramic samples and hence bioactivity is significantly influenced by the mol % of ZnO in the initial glass and consequently the crystalline phases remaining in the samples after the chemical treatment. Specifically, the formation of HA remains unaffected by the addition of up to 3 mol% ZnO, but deteriorates for higher ZnO content. On the other hand, the chemical durability exhibits an inverse trend such that the glass-ceramic becomes significantly more durable for more than 3 mol% ZnO. Our processing parameters have been optimized with respect to the creation of multi-modal porosity and the formation of HA layer on 48S4FxZG (x=0-3 mol% ZnO) porous glass-ceramic.

ACKNOWLEDGEMENT
This work was initiated and continued as an international collaboration with support from National Science Foundation (International Materials Institute for New Functionality in Glass (DMR-0409588) and Materials World Network (DMR-0602975) programs). MMF is funded by the National Institutes of Health (NIH, NIGMS, grant GM55725). We thank Drs. Arup K. Sengupta and Sudipta Sarkar for help with the ICP measurements.

REFERENCES

[1] E.A. Abou Neel, I. Ahmed, J.J. Blaker, A. Bismarck, A.R. Boccaccini, M.P. Lewis, S.N. Nazhat and J.C. Knowles, *Acta Biomat.*, 1 553-563 (2005).

[2] A.R. Boccaccini and V. Maquet, *Compos. Sci. and Tech.*, 63 2417-2429 (2003).

[3] F. Balas, D. Arcos, J. Perez-Pariente and M. Vallet-Regi, *J. Mater. Res.*, 16 [5] 1345-8 (2001).

[4] P. Sepulveda, J.R. Jones and L.L. Hench, *J. Biomed. Mater. Res.*, 59 [2] 340-348 (2002).

[5] A.C. Marques, H. Jain and R.M. Almeida, *Eur. J. Glass. Sci. Tech.* 48 65-68 (2007).

[6] N. Li, Q. Jie, S. Zhu and R. Wang, *Ceram. Internat.*, 31 [5] 641-646 (2005).

[7] M.N. Rahaman, W. Liang and E. Day, *Ceram. Eng. Sci. Proceeding*, 26 [6] 3-10 (2005).

[8] T. Peltola, M. Jokinen, H. Rahiala, E. Levanen, J.B. Rosenholm, I. Kangasniemi and A. Yli-Urpo, *J. Biomed Mat. Res.*, 44 12-21 (1999).

[9] H.M.M. Moawad and H. Jain, *J. Am. Ceram. Soc.,* 90 [6] 1934-6 (2007).

[10] H.M.M. Moawad and H. Jain, *Ceram. Eng. Sci. Proc.,* Development in Porous, Biological and Geopolymer Ceramics, 28 [9] 183-195 (2008).

[11] H.M.M. Moawad and H. Jain, *J. Mater. Sci. Mater. Med.*, in press.

[12] A. Billiau, V.G. Edy, H. Heremans, J. Van Damme, J. Desmyter, J.A. Georgiades and P. De Somer, *Human interferon*, 12 11-25 (1977).

[13] A. Ito, K. Ojima, H. Naito, N. Ichinose and T. Tateishi, J. Biomed. Mat. Res., 50 178-183 (2000).

[14] A. Ito, H. Kawamura, M. Otsuka, H. Ikeuchi, H. Ohgushi, K. Ishikawa, K. Onuma, N. Kanzaki, Y. Sogo and N. Noboru, *Mat. Sci. Eng.,* C22 21-25 (2002).

[15] H.M. Kim, T. Miyazaki, T. Kokubo and T. Nakamura, *Key Eng. Mater.* 192-195 47-50 (2001).

[16] H.A. El-Batal, M. A. Azooz, E.M.A. Khalil, A.S. Monem and Y.M. Hamdy, *Mater. Chem. Phys.,* 80 599-609 (2003).

[17] S. Jalota, S.B. Bhaduri, A.C. Tas, *J. Mater. Sci. Mater. Med.,* 17 697-707 (2006).

[18] T.H. Elemer, M.E. Nordberg, G.B. Carrier and E.J. Korda, *J. Amer. Ceram. Soc.,* 53 171-175 (1970).

[19] O.P. Filho, G.P. LaTorre and L.L. Hench, *J. Biomed. Mater. Res.,* 30 509-514 (1996).

[20] O. Peitl, E.D. Zanotto and L.L. Hench, *J. Non-Cryst. Solids,* 292 115-126 (2001).

[21] X. Chen, L.L. Hench, D. Greespan, J. Zhong and X. Zhang, *Ceramic Internat.,* 24 401-410 (1998).

[22] M. Mathew and S. Takagi, *J. Res. Nat. Inst. Stand. Technol.* 106 1035-1044 (2001).

[23] L.L. Hench, *J. Am. Ceram. Soc.,* 74(4) 1487-1510 (1991).

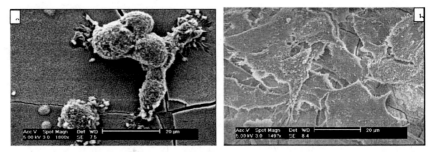

Figure 1. SEM picture of human MG63 osteoblast cells on the surface of $48SiO_2$-$2.7P_2O_5$-$4CaF_2$-$24.2CaO$-$21.1Na_2O$ (a) glass and (b) glass-ceramic of the same composition.

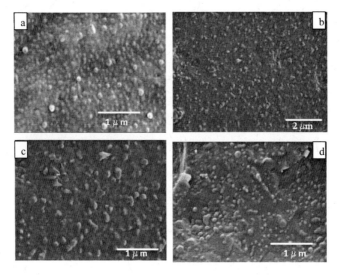

Figure 2: Low magnification SEM micrographs of 48S4FxZG specimens containing x mol% ZnO after the heat treatment: a) x=0, b) x=1, c) x=3, d) x=10.

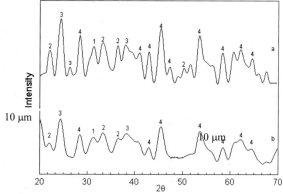

Figure 3: X-ray diffraction patterns of glass-ceramic samples after the nucleation and growth heat treatments: a) 48S4F1ZG, b) 48S4F10ZG, the source of diffraction peaks: 1) $Ca_4P_6O_{19}$, 2) $Na_2Ca_2Si_3O_9$, 3) $Na_2CaSi_3O_8$, 4) $Ca_5(PO_4)_3F$.

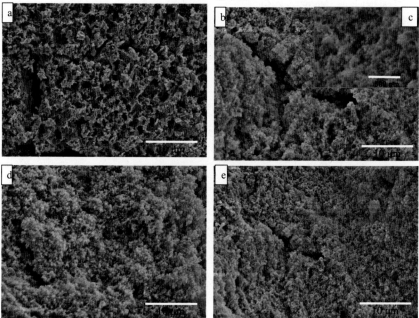

Figure 4: Low magnification SEM micrographs of 48S4FxZG specimens after heat treatment + chemical leaching: a) x=1 mol%, b, c) x=3 mol%, d) x=8 mol%, e) x=10 mol%. The inset (c) is a high magnification image of the sample (b).

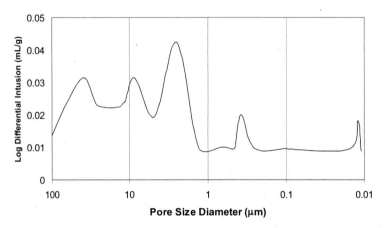

Figure 5: Mercury porosimeter data showing the distribution of pore size in 48S4F3ZG specimens after heat treatment + chemical leaching.

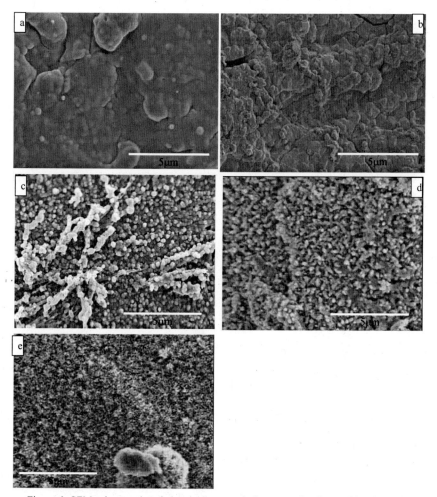

Figure 6: SEM micrographs of chemically treated glass-ceramic after soaking for 7 days in SBF: a) 8S4FG, b) 48S4F0.5ZG, c) 48S4F3ZG, d) 48S4F8ZG, e) 48S4F10ZG.

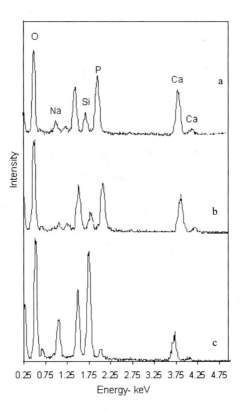

Figure 7: EDX spectra of chemically treated glass-ceramic after soaking for 7 days in SBF: a) 48S4FG, b) 48S4F3ZG, c) 48S4F8ZG. The peak at ~1.4 eV is from aluminum stub holding the sample.

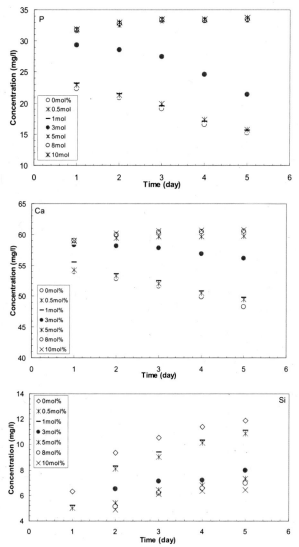

Figure 8: The change of P, Ca and Si concentration in SBF with immersion time of porous glass-ceramic 48S4FxZG with various ZnO content.

POROUS SCAFFOLDS USING NANOCRYSTALLINE TITANIA FOR BONE GRAFT
APPLICATIONS

Arun Kumar Menon and Samar Jyoti Kalita
Department of Mechanical, Materials and Aerospace Engineering
University of Central Florida
P.O. Box 162450
Orlando, FL 32816-2450

ABSTRACT
The need for a better scaffold to treat large bone-defects resulting from skeletal conditions still prevails, and it has been a challenge to meet all the biomedical requirements. It is established that an ideal bone scaffold should: (a) be osteogenic and resorbable, (b) have porous structure with interconnectivity, (c) possess suitable surface chemistry for cell attachment, differentiation and proliferation, and (d) match with the mechanical properties of the tissue being replaced. It is hypothesized that these requirements could be fulfilled by creating scaffolds using nanocrystalline titanium dioxide (TiO_2) via proper manufacturing technique/s. Nano-TiO_2 has shown improved mechanical strength and osteoblast functions, and proved to be a promising orthopedic biomaterial.
Formerly, we developed a simple sol-gel technique to synthesize 5-12 nm anatase powder. In this work, we used the same technique to produce TiO_2 nanopowder and fabricated porous scaffolds, using Poly (ethylene glycol) (PEG) as the pore-forming agent. The calcined nanopowder was homogenously mixed with 5, 10 and 15 wt. % of PEG and then cold-die compacted in a steel mold at 19.4 MPa. Controlled porosity scaffolds were fabricated through the indirect Fused Deposition Modeling process. The green samples/ molds were sintered in air, at 1500°C to obtain the porous scaffolds with interconnectivity.
The porous scaffolds were characterized for microstructure, phase purity, porosity, surface pore size, compression and biaxial flexural strength. XRD technique was used to analyze the phase/s. SEM was used to study the microstructure. Porosity was measured using immersion technique. Bulk density decreased with the increase in porosity of the structures. Density decreased from 3.79 g/cc to 3.03 g/cc as porosity increased from 9% to 22%. Biaxial flexural strength test was performed as per ASTM F-394. Flexural strength of 128(±7) MPa and 85(±6) MPa were achieved in structures with 9% and 18% porosity, respectively.

INTRODUCTION
Porous ceramics have gained significant attention in various industries due to many high temperature applications which include filters, catalyst supports, separation membranes, thermal barrier coatings, insulating layers [1, 2], fuel cell electrodes [3], bioceramics [4] and biomedical scaffolds [5] where high chemical stability and good corrosion resistance is required. The major advantage of using these materials is that they are mechanically strong, inert and can be cleaned and reused. Therefore, there is increased research going on in designing and developing these materials.
The most common method of producing these materials involves pressing a powder containing a mixture of ceramic particles together with the pore-formers and then sintering this compacted mixture at very high temperatures, wherein the pore-forming agent subsequently gets

eliminated during heating, as these are generally organic materials like synthetic polymers or substances of biological origin, which decompose on heating at relatively low temperatures. These pore-formers leave behind voids (pores) in the material hence producing a porous ceramic material. The advantages of using the pore formers are they are of relatively low cost, environment friendly, non-toxic and above all in most cases they act as binding agents, thus binding the ceramic particles together during compaction hence providing improved densification and strength to the green structures. For all the applications, a good porosity (> 40%) and a well controlled pore size distribution is needed.

During the recent years, significant research has been done to develop porous ceramic structures of alumina, SiN [6], YSZ [7] and other bioceramics to mimic internal bone structure. However, not much work has been done to produce porous TiO_2 structures, particularly using nanocrystalline powder. The present work deals with this aspect. Researchers have shown that sintered structures prepared from nanophasic anatase exhibits improved osteoblast adhesion and functions [8]. It is known that TiO_2 exists in three major polymorphs $viz.$, anatase, rutile and brookite [9]. Both anatase and rutile have tetragonal symmetry (anatase has a body centered tetragonal structure whereas rutile is simple tetragonal). Rutile belongs to the D_{4h}^{14} -P4$_2$/mnm space group, while anatase belongs to the D_{4h}^{19} -I4$_1$/amd space group [10]. Anatase transforms into rutile phase at higher temperature which depends on crystallite size of the staring powder. For example, retulization takes place around 700°C for micon-size powder and is seen about 500°C for nano TiO_2 and 1000°C for very coarse powder. The primary objective of the work was to investigate the effect of differing amounts of polyethylene glycol (PEG) on the porosity and mechanical properties of TiO_2 ceramic scaffolds sintered at various temperatures. Also, the bioactivity and biodegradation behavior these sintered scaffolds were studied in simulated body fluid maintained at 32°C.

MATERIALS AND METHODS

Powder Synthesis and Processing of Green Structures

Nanocrystalline titanium dioxide powder with an average crystallite size of 5-15 nm was synthesized using sol-gel technique [11]. Titanium tetraisopropoxide ($Ti[OCH(CH_3)_2]_4$) was used as the precursor. The flowchart showing the steps involved in the synthesis of nanocrystalline TiO_2 power is as shown in Fig.1. The hydrolysis reaction involved in the formation of TiO_2 can be represented by the following reaction:

$$Ti[OCH(CH_3)_2]_4 + 2H_2O \rightarrow TiO_2 + 4C_3H_7OH \qquad (1)$$

Measured quantity of polyethylene glycol (PEG, Fisher Scientific, NJ), was introduced into pure TiO_2 ceramics, in differing amounts, through homogeneous mixing, using a mortar and pestle. The doped TiO_2 powders were then compacted into pellets using cold die compaction technique at a pressure of 31 MPa, using a uniaxial manual hydraulic press from Carver, inc. (Webash, IN).

The green structures of pure TiO_2, 10% PEG and 20% PEG doped TiO_2 were used for this study. These green structures were sintered in a high temperature programmable muffle

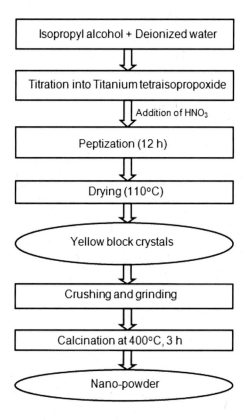

Figure 1. Sol-Gel Synthesis of nano titanium dioxide powder.

furnace (Model 46100, Barnstead International Co., Dubuque, IA), in an atmosphere of air. For the first set of samples, the sintering cycle employed consisted of several steps: first holding at 150°C to stabilize the furnace; second holding at 400°C to remove residual stresses; finally holding at 1500°C for 3 h for attaining good densification. A heating rate of 3°C min^{-1} and a cooling rate of 10°C min^{-1} were used in order to minimize thermal stress-induced cracking. For the second set of samples, the program remained the same; however the final holding temperature was reduced to 1400°C.

Microstructural Analysis
The microstructure of the sintered ceramics was observed using a Scanning Electron Microscope (SEM). The sintered structures were coated with gold-palladium for 1 min using a magnetron sputter coater from Emitech Inc. JOEL SEM (Model 6400F, JEOL, Tokyo, Japan) was used to observe the microstructures.

The percentage of porosity in the samples was evaluated by a simple immersion technique [12] , wherein the samples were immersed in distilled water. The following equation was used to calculate the percentage of porosity (ξ_a) in the samples:

$$\xi_a = [(m_s - m_d)/(m_s - m_i)] \qquad (2)$$

Where m_s is saturated mass, m_w is the mass in water and m_d is the dry mass.

Phase Analysis
Phase characterization after sintering at 1500°C was performed on composition containing 10 wt% of PEG in the initial powder. XRD patterns were recorded in the 2θ range of 20-60 degrees with an automated X-ray diffractometer (Model D/MAX-B, Rigaku Co., Tokyo, Japan) using Cu Kα radiation (λ=1.5418 Å) in the step scanning mode, with tube voltage of 35 kV and tube current of 30 mA. The 2θ step size was 0.04° and a scanning rate of 1.5 deg/min was used.

Mechanical Characterization
Biaxial Flexural testing
Flexural strength properties of porous TiO_2 structures were studied for the structures sintered at 1500°C and 1400°C. The tests were performed according to the ASTM F-394 standard with some alteration in fixture dimension in order to accommodate the specimen size. The tests were carried out in a tensile tester (Model 3369, Instron Co., USA) at a constant crosshead speed of 0.05 mm/min. The test consists of supporting TiO_2 disc pellet by three ball bearings near its periphery positioned 120° apart on a circle, 7.5 mm in diameter. Uniform loading is attained by placing a layer of non-ridged material such as polythene between the piston and test specimens.
The fracture load was recorded at the end of the test and this was used to calculate the biaxial fracture strength which is determined using the following equations [13]:

$$S = -0.2387\, P\, (X-Y)/d^2 \qquad (3)$$
$$X = (1+v)\ln (B/C)^2 + [(1-v)/2]\,(B/C)^2$$
$$Y = (1+v)\,[1+ \ln (A/C)^2] + [(1-v)]\,(A/C)^2$$

Where **S** is maximum center tensile stress in MPa and **P** is the total load causing fracture in N. **v** is Poisson's ratio (taken as 0.27), **A** is radius of support circle in mm, **B** is the radius of loaded area or ram tip in mm, **C** is the radius of specimen in mm and, **d** is specimen thickness at fracture origin in mm.

Vickers hardness testing
The hardness values of porous TiO_2 structures sintered at 1500°C and 1400°C were evaluated by Vickers hardness tester (Model LV-700, LECO Co., MI). During the hardness test, a load of 9.8 N with a loading time of 5 s was applied. The pellets used for the testing had average dimensions of 9.9 mm in diameter and 1.6 mm in thickness. One sample of each composition was tested for the hardness at five different locations, the average of these five values were evaluated in determining the final value of hardness.

Biodegradation analysis

The rate of biodegradation of porous TiO_2 structures sintered at 1500°C was determined on the basis of decrease in the mechanical properties as a function of time, in a protein-free dynamic simulated body fluid (SBF). Acellular SBF has been widely used by researchers to test bioactivity of materials by examining the formation of apatite layer on the surface of testing materials. The *in vivo* formation of an apatite layer on the surface of a bone grafting material can be reproduced in acellular SBF; which is prepared to have an ion concentration nearly equal to that of human blood plasma (Na^+ 142.0, K^+ 5.0, Ca^{2+} 2.5, Mg^{2+} 1.5, Cl^- 147.8, HCO_3^- 4.2, HPO_4^{2-} 1.0, and SO_4^{2-} 0.5 mM, and a pH of 7.3) [14, 15].

Porous TiO_2 structures sintered at 1500°C for 3 h were prepared for the biodegradation study. These structures were immersed in SBF solution, maintained inside an incubator at a constant temperature of 36.5°C. A set of eight samples of each composition were placed in perforated plastic trays filled with freshly prepared SBF, which were then placed inside the incubator. The structures were left in the plastic trays. The SBF was replaced every three days with a freshly prepared one, to maintain its ionic concentration and the pH. At the end of day 21 and day 42, the samples were taken our and dried at 100°C in a furnace and measured for their weight to calculate weight change as a result of time. The change in geometric density was calculated based on mass-loss in the samples at different time points.

RESULTS AND DISCUSSION

Sintering and densification

The green as well as the sintered ceramic structures were measured for their density values. All the green structures had an average bulk density of 1.79 g cm^{-3}. About five to eight structures of each of the different compositions were sintered at 1500°C for 3 h and 1400°C for 3 h respectively. After sintering, the density of all the samples was calculated separately and the effect of sintering temperature and quantity of pore-former added on the densification of the structures were studied. The average sintered densities were plotted as a function of composition as shown in Fig.2. The pure TiO_2 structures possessed the highest sintered density of 3.8 g cm^{-3} and 3.77 g cm^{-3} when sintered at 1500°C and 1400°C respectively, as the pore forming agents were absent. The density of the porous structures decreased as the amount of pore former increased. The structures sintered at 1400°C possessed lower density than the structures sintered at 1500°C. This is as illustrated in Fig.2. When the amount of pore-former was more, there were more voids on those structures as the pore-former being a volatile material gets eliminated during the firing process in the initial stages itself.

Porosity and Microstructural analysis

Porosity in the structures sintered at 1400°C and 1500°C were evaluated using equation 2. The percentage porosity in the structures as a function of composition is illustrated in Fig.3. The highest percentage porosity was found out to be 38% for 20% PEG doped structures sintered at 1400°C, but when the sintering temperature was increased to 1500°C, the porosity decreased to 37%. The percentage of porosity was however, 19% and 17% for 10% PEG doped structures sintered at 1400°C and 1500°C respectively. The percentage porosity is affected by two different factors namely: sintering temperature and the amount of pore former. It was observed that as the sintering temperature increased and amount of pore former decreased the porosity decreased.

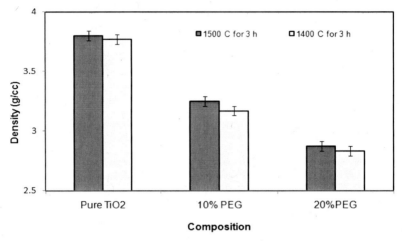

Figure 2. Variation in density for the pure and porous TiO$_2$ structures

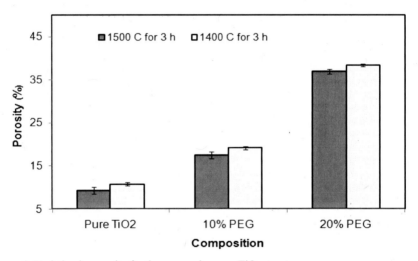

Figure 3. Variation in porosity for the pure and porous TiO$_2$ structures

The SEM micrographs for pure as well as porous TiO$_2$ structures are shown in Fig.4 and 5. The SEM micrographs also give out a clear indication of the pores present on the surface of all the sintered structures. The micrographs also indicate that the grain size increases as the sintering temperature increases. The grain size was determined by linear intercept method. From all the micrographs, it is evident that the addition of pore formers inhibited the grain growth. It was found that at 1500°C, grain size of 25 μm was obtained for the pure TiO_2 structures. Grain size of 16 μm was obtained for 10% PEG doped structures sintered at 1500°C. It can be seen in all the micrographs that the addition of pore-formers have suppressed the grain growth.

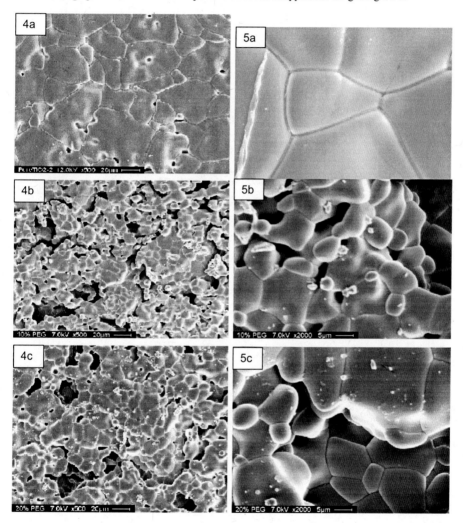

Figures 4 and 5. SEM images (a) Pure TiO_2 (b) 10% PEG doped TiO_2 (c) 20% PEG doped TiO_2.

Phase Analysis

Fig. 6 shows XRD pattern of sintered structure at 1500°C. The original green structure contained 10 wt% of PEG. Phase identification and analysis was performed using PDF cards #21-1272 for anatase and #21-1276 for rutile. It was found that all peaks were of rutile phase.

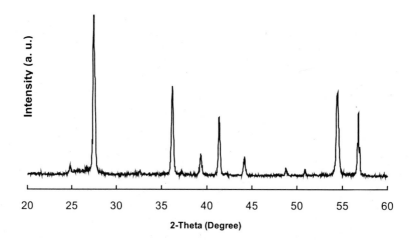

Figure 6: XRD pattern after sintering at 1500°C (Initial composition had 10 wt% PEG). All peaks belong to rutile phase.

Mechanical Characterization

Biaxial flexural testing

All the sintered structures were subjected to biaxial flexural testing separately. The results of the biaxial testing are shown in Fig. 7. Where biaxial strength (in MPa) is plotted against composition. From the plots it is seen that the biaxial flexural strength improves as the sintering temperature increases, whereas it decreases as the amount of pore-former added increases. For the porous TiO_2 structures, the maximum strength of 110 (±10 MPa) was obtained for 10% PEG doped structures sintered at 1500°C. Whereas the minimum strength of 78 (±2.6 MPa) was obtained for 20% PEG doped structures sintered at 1400°C. The biaxial flexural strength decreases as the amount of pore former increased.

Hardness testing

TiO_2 structures sintered at 1400°C and 1500°C were evaluated for hardness using Vickers hardness tester. One structure of each composition was measured for hardness at five different points and the average of these values was reported and plotted in Fig. 8. It is clear that increase in sintering temperature and decrease in the amount of pore former helped improving the hardness. A maximum of 223 (±2 HV) was obtained for the 10% PEG doped structures sintered at 1500°C and a minimum hardness of 163 (±1.5 HV) was obtained for 20% PEG doped structures sintered at 1400°C. Higher sintering temperature helped achieve higher hardness.

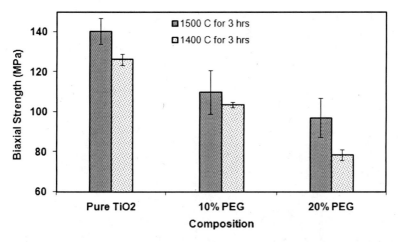

Figure 7. Variation of biaxial flexural strength for the pure and porous TiO_2 structures

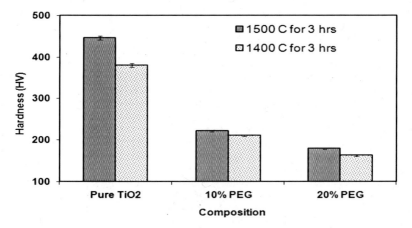

Figure 8. Variation of hardness of the pure and porous TiO_2 structures

Biodegradation Studies

The results of biodegradation analysis of all the structures in SBF maintained at a constant temperature of 36.5°C were evaluated. It was seen that for the porous TiO_2 structures there was an increase in weight of the structures at the end of day 21 and day 42. The increase in the weight of the structures is due to the formation of apatite layer on the surface and sides of the structures. The presence of the apatite layer was also later confirmed by X-ray diffraction. In case of pure TiO_2 structures, there was a decrease in the weight of the structures at the end of 42

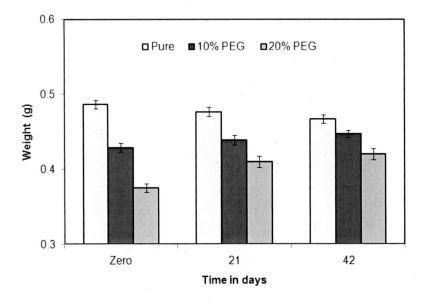

Figure 9. Variation in mass for pure and porous TiO$_2$ structures placed in SBF

days. This is possibly due to the degradation of the structures in SBF with time. The results clearly indicate that the rates of biodegradation are different for pure and porous TiO$_2$ structures. The change in weight of all the structures with time is shown in Fig. 9. The variation in mechanical properties such as biaxial flexural strength and hardness of all the structures used for biodegradation study with time is also being investigated.

CONCLUSIONS

Nanocrystalline TiO$_2$ was successfully synthesized via a simple sol-gel process. The porous TiO$_2$ structures were made with polyethylene glycol (PEG) as the pore-former. The densification results indicate that increase in the amount of PEG leads to decrease in the density as the porosity would be more in these structures. Sintering at 1500°C gave better densification results. The SEM micrographs indicate the addition of PEG suppressed the grain growth for all the TiO$_2$ structures. The grain size of all the porous structures was reduced to about one-half times by the addition of PEG. Even though, the porous structures possessed lower hardness and biaxial flexural strength values than the pure TiO$_2$ structures, the porous structures degraded at a slower rate in simulated body fluid when compared to the pure TiO$_2$ structures. The XRD results confirmed the presence of apatite on the surface of all the structures at the end of the bioactivity study using simulated body fluid.

*Credit for this work is equally shared by the authors.

REFERENCES

1. Scheffler, M., Colombo, P . *"Cellular Ceramics - Structure, Manufacturing, Properties and Applications"*. 2005: p. 342-360, 401-620.

2. W., R.R., *"Porosity of Ceramics"*. 1998(Marcel Dekker, Newyork,): p. 315-532.

3. Minh , N.Q., *"Ceramic Fuel Cells"*. Journal of American Ceramic Society, 1993. **76**: p. 563-588.

4. Yoon, B.H., et al., *"Aligned porous alumina ceramics with high compressive strengths for bone tissue engineering"*. Scripta Materialia, 2008. **58**(7): p. 537-540.

5. Zhuang, H., Y. Han, and A. Feng, *"Preparation, mechanical properties and in vitro biodegradation of porous magnesium scaffolds"*. Materials Science and Engineering: C, 2008. **28**(8): p. 1462-1466.

6. Ohji, T., *"Microstructural design and mechanical properties of porous silicon nitride ceramics"*. Materials Science and Engineering: A, 2008. **498**(1-2): p. 5-11.

7. Albano, M.P., et al.," *Processing of porous yttria-stabilized zirconia by tape-casting"*. Ceramics International, 2008. **34**(8): p. 1983-1988.

8. T.J. Webster, R.W. Siegel, and R. Bizios,*"Osteoblast Adhesion on Nanophase Ceramics"*, Biomaterials, 1999. **20**: p. 1221-1227.

9. L.S. Dubrovinsky, N.A. Dubrovinskaia, V. Swamy, J. Muscat, N.M. Harrison, R. Ahuja, B. Holm, and B. Johansson, *"The hardest known oxide"*, Nature, 2001. **410**: p. 653-654.

10. U. Diebold, *"The surface* science *of titanium dioxide"*, Surf. Sci. Rep, 2003. **48**: p. 53-229.

11. Qiu, S. and Kalita, S.J., *"Synthesis, processing and characterization of nanocrystalline titanium dioxide"*. Materials Science and Engineering: A, 2006. **435-436**: p. 327-332.

12. Akiyoshi, M.M., Da Silva, A. P., Da Silva, M. G., Am Ceram. Soc. Bull., 2002. **81**: p. 39.

13. Kalita, S.J., Qiu, S. and S. Verma, *"A quantitative study of the calcination and sintering of nanocrystalline titanium dioxide and its flexural strength properties"*. Materials Chemistry and Physics, 2008. **109**(2-3): p. 392-398.

14. Kokubo, T., *"Surface chemistry of bioactive glass-ceramics"*. Journal of Non-Crystalline Solids, 1990. **120**(1-3): p. 138-151.

15. Kokubo, T., *"Ca, P-rich layer formed on high-strength bioactive glass-ceramic"*. J. Biomed. Mater. Res., 1990. **24**: p. 331.

POROUS BIOMORPHIC SiC FOR MEDICAL IMPLANTS PROCESSED FROM NATURAL AND ARTIFICIAL PRECURSORS

J. Ramirez-Rico, C. Torres-Raya, D. Hernandez-Maldonado, C. Garcia-Gañan, J. Martinez-Fernandez, A.R. de Arellano-López
Dpto. Física de la Materia Condensada-ICMSE
Universidad de Sevilla-CSIC
Sevilla, Spain

ABSTRACT
Porous biomorphic SiC (bioSiC) is a monolithic, porous ceramic obtained by melt infiltration of carbon precursors obtained from wood and subsequent removal of residual Si by acid etching. The microstructure of the resulting material resembles that of the original wood, with anisotropic porosity and a high strength to density ratio. The properties of the porous bioSiC depend on those of the original wood and can be tailored by adequate selection of the precursor. Its biocompatibility and potential as a host for bone growth have been demonstrated in previous works.

We have successfully fabricated porous bioSiC from Sipo and Medium Density Fiberboard woods, and studied the kinetics of Si removal by acid reaction. Our results suggest that the etching reaction is diffusion limited and does not depend on the morphology of the porosity, but only on the total fraction of pores. Etching kinetics follows a $t^{-0.5}$ law and is found to be anisotropic, which can be explained considering the anisotropic nature of the pore distribution. Compressive strength was studied as a function of etching time for porous bioSiC obtained from the two precursors, and was found to compare favorably with that of dry human bone in the range of attainable densities.

INTRODUCTION
Biomorphic silicon carbide (bioSiC) is Si/SiC composite obtained by reactive melt infiltration of a porous carbon precursor obtained from wood pyrolysis[1-8]. The microstructure of the resulting material closely resembles that of the original wood precursor, and is usually anisotropic due to the anisotropic character of the wood channels, which are often filled with residual Si after infiltration[9]. BioSiC materials have good thermomechanical properties [10, 11], making them good candidates for high temperature structural applications, but also as a cost-effective ceramic material in applications where ceramics are not normally used due to the associated manufacturing costs. Properties like density, elastic modulus, anisotropy or strength of bioSiC depend on the microstructure and density of the wood precursor [12]. Therefore, it is possible to fabricate different bioSiC materials with tailored properties by adequate selection of the carbon template. It has been demonstrated that the volume change of the template before and after the infiltration step is very small [13]. This allows the carbon precursor to be cut and machined in complex shapes with relatively low-cost tools and processes, and then infiltrated to obtain bioSiC pieces in near net-shape, significantly lowering manufacturing costs.

Because the microstructure of bioSiC consists of a hierarchical distribution of pores and channels that closely resemble that of bone and allows the internal growth and vascularisation of tissue inside the ceramic pores, bioSiC has been explored as a material for use in bone implants [14, 15]. Although the optimal degree of porosity is still uncertain, it is known that when pore sizes exceed 100 μm, bone will grow within the interconnecting pore channels near the surface and

203

maintain its vascularity and long-term viability, and the porous ceramic serves as a structural scaffold for bone formation[16]. The light, high strength porous bioSiC can be coated with bioactive layers of hydroxyapatite, substituted apatites or silica-based glasses to further improve the fixation and osteointegration performance of the material [17]. *In vitro* biocompatibility studies demonstrate that the biological response of this ceramic product is similar to titanium controls [18], and *in vivo* implantation experiments show how it gets colonized by the hosting bone tissue due to its unique interconnected hierarchic porosity [19].

From the mechanical point of view, porous bioSiC compares reasonably well with human bone. For the range of densities that can be obtained from wood, compressive stiffness of bioSiC is in the range of 10-100 GPa[11], which is similar to that of dry human and bovine cortical bone[20]. The main drawback of these materials comes from their low fracture toughness, which at around 1 MPa m$^{1/2}$ is significantly lower than the range of 4-7 MPa m$^{1/2}$ usually found in human cortical bone[11, 21, 22]. Since the high toughness of bone is mainly due to the role of collagen lamellae as crack-bridging agents and is less dependent on mineral content [23, 24], it is expected that the toughness of bioSiC will rise after being colonized by bone tissue.

Since bioSiC is fabricated by liquid infiltration techniques, some residual Si is often present, filling some of the original plant's vessels. If the application requires a material with low density and open porosity, residual Si needs to be removed. Several approaches exist, such as the use of carbon wick at high temperatures, however the most common is Si etching using a mixture of acids. Although both techniques are sometimes not completely successful in eliminating all residual Si, it has been shown that the release of Si into simulated body fluid after one week of immersion was in the ppm range for bioSiC, suggesting that the presence of some residual Si should have no negative effects[15].

One drawback of using natural wood as a starting material for bioSiC fabrication is the inherent variability in its properties associated to different environmental factors that affect tree growth. For that reason, we have explored the use of artificial, industrial woods such as fiberboards and particleboards. These wood precursors are usually manufactured by binding wood fibers or particles with different resins and/or wax and are pressed at high temperatures into flat panels, therefore introducing anisotropy in the material's properties.

In this paper, we study the etching kinetics of bioSiC fabricated from Sipo and MDF woods as a function of time and orientation, and the compressive strength along the channels of the porous material, as a function of density and etching time.

EXPERIMENTAL

BioSiC samples were fabricated using Sipo (Entandrophragma Utile) and Medium Density Fiberboard (MDF) as starting precursors, using a method previously described elsewhere. In brief, samples of wood were pyrolyzed in a flowing Ar atmosphere at 1000 °C (0.5 °C/min heating rate). Subsequently, carbon samples of different dimensions were cut and infiltrated with liquid Si (Silgrain HQ-99.7% purity; Elkem Silicon, Oslo, Norway) at 1450°C in a tube furnace in vacuum. Original carbon precursors were all cut from the same piece of wood. Samples were made in batches and each carbon precursor was covered with powdered Si using a Si/C weight ratio of 3.2. This ratio is higher than stoichiometric and was chosen to ensure an almost complete conversion of C into SiC. A solid Si layer was sometimes found covering parts of the external sample surfaces, and was removed by mechanical grinding. No bulk volume increase was observed.

Wet chemical etching was performed using a mixture of HF and HNO_3 at a molar ratio of 1.66 in water. This ratio corresponds to stoichiometry of the reactions:

$$3Si+4HNO_3 \rightarrow SiO_2+4NO+4H_2O$$
$$3SiO_2+12HF \rightarrow 3SiF_4+6H_2O$$

(1)

Resulting bioSiC samples were etched for different times without stirring, and their weight loss was measured after each etching step. After etching, all samples were submerged in an ultrasonic bath until neutral pH was reached, and then carefully rinsed in distilled water. Their density was measured by the Archimedes' method.

For the study of etching kinetics, cubic bioSiC samples measuring 1 cm^3 in volume were produced and changes in weight and density were monitored as a function of etching time, both for samples obtained from Sipo and MDF wood. For determination of the etching rate as a function of time, samples were cut after etching along relevant directions as described in Figure 1 and the reaction interface was tracked as function of immersion time by means of scanning electron microscopy (SEM) observation. The rate of advance of the reaction front could therefore be determined for the different relevant directions in both cases, as a function of etching time. Microstructure observations were carried out using SEM in both the carbon precursor and bioSiC material. Samples were prepared using conventional metallographic techniques.

Compressive strength was measured at room temperature using constant cross-head compression experiments at a velocity that resulted in an initial strain rate of $2 \cdot 10^{-4}$ s^{-1}, using a screw-driven universal testing machine (Microtest EM1/50/FR, Madrid, Spain). The compressive load was applied using Al_2O_3 rods, which were protected from direct contact with the samples by sintered SiC pads. Samples from both Sipo and MDF woods were tested after etching for different periods of time. Sample dimensions were (3x3x6) mm^3 and the compressive load was applied along the longest direction. Sample orientation was chosen so that the applied load is parallel to the axial direction in Sipo wood and the longitudinal direction in MDF (see Figure 1). The faces where load was applied were ground flat-parallel. Sample dimensions were chosen to follow a length/diameter < 2.0 to minimize barrelling and buckling artifacts due to friction. Samples were always tested to failure. Wherever reported, error bars in figures throughout correspond to one standard deviation.

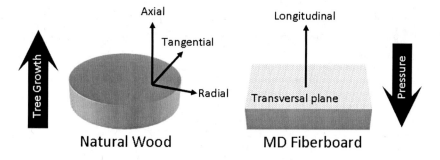

Figure 1. Definition of relevant directions in both natural wood (left) and MD Fiberboard (right)

RESULTS AND DISCUSSION

Microstructure of the carbon precursors

Figure 2 shows relevant micrographs for both Sipo and MDF carbon precursors after pyrolisis. The differences in their porosity distributions are evident, as is also the anisotropy of the microstructure. In the case of Sipo wood, it can be seen that the microstructure consists of a bimodal distribution of channels parallel to the direction defined as axial in Figure 1. These channels are connected by transversal pores in the radial-tangential plane. The anisotropy of the porosity is relevant in determining the final properties of the material, and also will determine the infiltration and etching kinetics. In the case of MDF, the original wood consisted of wood fibers bonded together by resin and other binding agents that also transform to C after high-temperature exposition to an inert atmosphere. The anisotropy is again evident in this case, as most of the carbon fibers are oriented so their axis is contained in the transversal plane. From the longitudinal sections it can be seen that the fibers themselves are also porous.

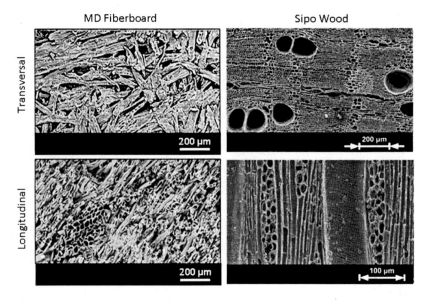

Figure 2. Microstructure of carbon precursors used in this study. The longitudinal direction is defined as the direction of tree growth for Sipo wood and as the pressing direction during manufacturing of MDF. The transversal plane is perpendicular to longitudinal direction.

Microstructure of porous bioSiC

Figure 3 shows the resulting microstructure after Si removal by means of the chemical reaction described in Eq. (1). By comparison with Figure 2 it can be seen that the porous bioSiC microstructure closely resembles that of the original carbon precursor, although the volume expansion that occurs during SiC formation tends to close some of the pores. Average pore diameter reduction was reported as less than 10%[25]. It is important to note that this volume changes produced a change in the total porosity, but no macroscopic volume or shape changes in the samples were observed after reaction.

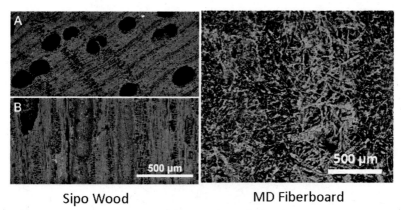

Sipo Wood   MD Fiberboard

Figure 3. Porous bioSiC obtained after chemical etching of melt infiltrated bioSiC.

Chemical etching of bioSiC and production of porous bioSiC

Figure 4 shows the weight loss as function of etching time for bioSiC samples originating from both Sipo and MDF wood. In this case, samples were $(10mm)^3$ cubes. This weight loss is attributed to removal of residual Si and not to etching of the SiC phase. In Figure 5 the evolution of sample density with etching time is presented. From both figures it can be seen that, although the weight loss kinetics is similar for both types of wood, the actual density is different due to the differences in density of the precursors. Additionally, the data scatter for Sipo samples is higher than the scatter for MDF samples, which can be attributed to the lower reproducibility of the natural wood. Even though all Sipo samples were cut from the same piece of wood to eliminate variability due to growing conditions and/or tree age, there are local variations in the wood microstructure that account for the slightly larger dispersion in density.

Wet chemical etching is a sequential process that involves advective transport of the reactants through liquid infiltration, followed by reaction at the liquid-solid interface. The liquid is therefore depleted of reactants and a concentration gradient is established that produces diffusion according to Fick's law. Each of these processes (advection, reaction and diffusion) can be the limiting step in the reaction and thus determine the overall kinetics of the process. For the simplest case where reaction at the interface is of first order and limiting, the etching rate should be time-independent if the temperature of the reaction vessel is kept constant.

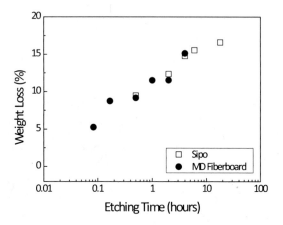

Figure 4. Weight loss as a function of etching time for $(10mm)^3$ bioSiC samples obtained from both Sipo and MDF wood precursors. It can be seen that the weight loss kinetics does not depend on the actual sample microstructure, suggesting that the etching process is diffusion limited.

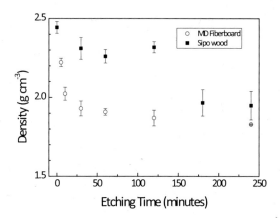

Figure 5. Evolution of density for bioSiC obtained from Sipo and MDF woods, as a function of etching time.

Previous studies on both planar[26-28] and powdered[29] Si have shown that diffusion is usually the limiting step in the reaction. If this is the case, then the Si weight should change according to:

$$\frac{dm}{dt} = -DA\frac{dc}{dx} \qquad (2)$$

Where m is the Si mass, D is the diffusion coefficient of the limiting species in the liquid, A is the area of the surface being etched and dc/dx is the concentration gradient near the liquid-solid interface. Mass conservation imposes that:

$$\frac{dm}{dt} = RA\rho \qquad (3)$$

Where ρ is the density of the solid and R is the etching rate. For simplicity, we approximate the concentration profile as linear so we can write:

$$\frac{dc}{dx} = \frac{c_l - c_s}{w} \qquad (4)$$

In the previous equation, c_l is the concentration in the liquid far from the reaction front and c_s is the concentration at the interface, while w is the boundary layer thickness. The previous reasoning is valid for diffusion-limited reactions in the cuasi steady-state. The solution of the associated, one dimensional diffusion problem allows us to estimate $w = \sqrt{4Dt}$ and combining equations (2) to (4) we can then write:

$$R = \frac{D}{\rho\sqrt{4Dt}}(c_l - c_s) \qquad (5)$$

Equation (5) therefore establishes that for diffusion limited reactions, the etching rate will depend on time as $t^{-0.5}$. To test this hypothesis samples etched for different times were cut in the directions specified in Figure 1 and the reaction front was tracked by means of SEM observation. As an example, sections perpendicular to the axial direction in bioSiC obtained from Sipo are depicted in Figure 6. There it can be seen that the reaction front is relatively flat for low etching times. As the reaction proceeds, however, the samples are also etched from the inside as the larger channels are free of Si and allow for liquid flow into the material.

The reaction front was tracked using image analysis in all relevant directions for bioSiC made from both Sipo and MDF woods, and the results of this study are summarized in Figure 7. Although in all cases a similar $t^{-0.5}$ law is followed, there are obvious differences between the samples obtained from Sipo and from MDF, and in the case of Sipo, even for different directions in the same type of samples. These differences can be understood in the light of figures Figure 2Figure 3. The channel diameter distribution is different in each direction, with larger channels in the axial, then the radial and finally the tangential direction. These differences will lead to an anisotropic permeability that can be accounted for in equation (5) by means of an effective diffusion coefficient following Archie's law:

$$D_{eff} = D\phi^n \qquad (6)$$

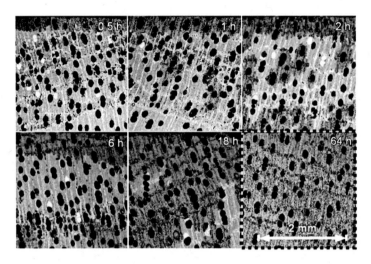

Figure 6. Scanning electron micrographs showing the evolution of microstructure during etching for bioSiC obtained from Sipo wood, as a function of etching time. From left to right and top to bottom, etching times are 0.5, 1, 2, 6, 18 and 64 h.

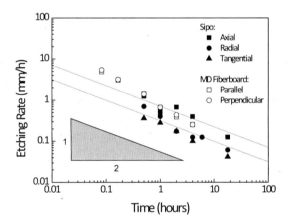

Figure 7. Reaction front etching rates obtained from scanning electron micrographs of sections following the directions depicted in Figure 1, as a function of etching time. The etching rate approximately follows a $t^{-0.5}$ law that confirms the diffusion-limited nature of the reaction at the conditions used in this study. Straight lines correspond to fits assuming a $t^{-0.5}$ for the directions showing higher and lowe etching rates, and are included as a visual aid.

In equation (6) ϕ is the effective porosity in a particular direction, expressed as the ratio of pore area to total area for each section parallel to the directions depicted in Figure 1, and n is an exponent in the 1.8-2 range. Fluid transport through the pores probably plays a role at the early stages of the process, however once the reaction layer is depleted reactants have to diffuse through the pores and the reaction becomes diffusion-limited. In this scenario it is expected that total porosity will play a bigger role than pore diameter distribution in determining etching kinetics.

Compressive strength of porous bioSiC as a function of etching time

Figure 8 shows the compressive strength of porous bioSiC from Sipo and MDF as a function of etching time, when the load is applied parallel to the axial direction in Sipo and the longitudinal direction in MDF. It can be seen that the scatter in the case of Sipo is higher, which is again a consequence of the higher variability on the properties of natural wood. The strength values are also significantly higher, which can be explained considering the microstructural differences between the two. In the case of Sipo, the resulting microstructure is connected in three dimensions and the SiC struts carry most of the load themselves. For MDF, which is composed of fibers, the etching decreases the connectivity in the microstructure and significantly lowers the properties of the material.

It has been found that the compressive strength of both cortical[30] and trabecular[20] bone exhibits a quadratic dependence on the relative dry density when the load is applied parallel to the main vascular channels. The compressive strength of samples from both Sipo and MDF is plotted as a function of density in Fig. 9, where the compressive strength of human bone, taken from Ref. 20, is included for comparison. The figure is plotted in log-log scale for ease of view. Remarkably, bioSiC from Sipo performs similarly or better than dry human bone in the range of densities studied, and the compressive strength is found to follow a quadratic dependence on density as well.

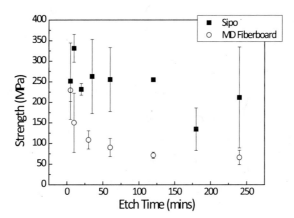

Figure 8. Compressive strength as a function of etching time for bioSiC materials obtained from Sipo and MDF woods.

Although bioSiC obtained from Sipo follows the same trend as human bone, probably due to the similarities in their microstructures as interconnected scaffolds, the compressive strength of bioSiC from MDF shows a different trend and, interestingly, it decreases at a much higher rate with density than bioSiC from Sipo. This can be explained in the same fashion as before, considering that etching in MDF affects not only density but also the connectivity of the microstructure.

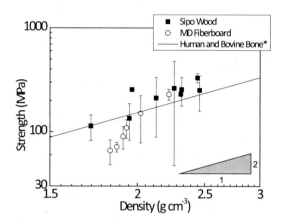

Figure 9. Compressive strength as a function of density for samples etched from 5 to 420 minutes, for bioSiC samples made from Sipo and MDF woods. Scale is logarithmic on both axes. The straight line represents the behavior of human and bovine bone, taken from ref. 20.

CONCLUSION

We have successfully fabricated porous bioSiC scaffolds by pyrolysis of both natural and artificial woods, followed by reactive infiltration of molten Si. The Si removal kinetics and compressive strength as a function of etching time were studied. The etching kinetics was found to be diffusion limited in all cases, as confirmed by the etching rate dependence with time. The etching rate was found to be anisotropic and faster in those directions that present higher apparent porosity.

The compressive strength in porous bioSiC is found to vary with etching time, and this dependence is stronger in the material obtained from artificial rather than natural wood precursors. The strength bioSiC obtained from Sipo was found to compare favorably with human bone in the attainable range of densities, and to follow a similar trend as bone. The strength of MDF-bioSiC degraded rapidly with etching time due to loss of connectivity in the microstructure. The dispersion in strength was higher for the bioSiC obtained from Sipo, which is attributed to microstructural variability.

ACKNOWLEDGEMENTS

The authors gratefully acknowledge the European Commission for financial support under the FP6 Contract No. NMP4-CT-2006-033277 (TEM-PLANT). The authors are grateful to the CITIUS at the University of Sevilla for the use of their electron microscopy facilities. J.R.-R. is grateful to the Junta de Andalucía for a PhD grant.

REFERENCES

1. M. Singh, J. Martinez-Fernandez and A.R. de Arellano-Lopez: Environmentally conscious ceramics (ecoceramics) from natural wood precursors. *Curr. Op. Solid State Mat. Sci.* **7**, 247-254 (2003).
2. A.R. de Arellano-Lopez, J. Martinez-Fernandez, P. Gonzalez, C. Dominguez, V. Fernandez-Quero and M. Singh: Biomorphic SiC: A new engineering ceramic material. *Int. J. Appl. Ceram. Tech.* **1**, 56-67 (2004).
3. P. Greil: Biomorphous ceramics from lignocellulosics. *J. Eur. Ceram. Soc.* **21**, 105-118 (2001).
4. T. Ota, M. Takahashi, T. Hibi, M. Ozawa, S. Suzuki, Y. Hikichi and H. Suzuki: Biomimetic process for producing SiC 'wood'. *J. Am. Ceram. Soc.* **78**, 3409-3411 (1995).
5. H. Sieber, C. Hoffmann, A. Kaindl and P. Greil: Biomorphic Cellular Ceramics. *Adv. Eng. Mater.* **2**, 105-109 (2000).
6. C.E. Byrne and D.C. Nagle: Cellulose derived composites - A new method for materials processing. *Materials Research Innovations* **1**, 137-144 (1997).
7. C.E. Byrne and D.C. Nagle: Carbonization of wood for advanced materials applications. *Carbon* **35**, 259-266 (1997).
8. C.E. Byrne and D.C. Nagle: Carbonized wood monoliths - Characterization. *Carbon* **35**, 267-273 (1997).
9. C. Zollfrank and H. Sieber: Microstructure evolution and reaction mechanism of biomorphous SiSiC ceramics. *J. Am. Ceram. Soc.* **88**, 51-58 (2005).
10. J. Martinez-Fernandez, F.M. Valera-Feria and M. Singh: High temperature compressive mechanical behavior of biomorphic silicon carbide ceramics. *Scripta Mater.* **43**, 813-818 (2000).
11. V.S. Kaul, K.T. Faber, R. Sepulveda, A.R. de Arellano-Lopez and J. Martinez-Fernandez: Precursor selection and its role in the mechanical properties of porous SiC derived from wood. *Mater. Sci. Eng. A* **428**, 225-232 (2006).
12. F.M. Varela-Feria, M.J. Lopez-Robledo, J. Martinez-Fernandez, A.R. de Arellano-Lopez and M. Singh: Precursor selection for property optimization in biomorphic SiC ceramics. *Ceram. Eng. Sci. Proc.* **23**, 681-685 (2002).
13. F.M. Varela-Feria, J. Ramirez-Rico, A.R. de Arellano-Lopez, J. Martinez-Fernandez and M. Singh: Reaction-formation mechanisms and microstructure evolution of biomorphic SiC. *J. Mater. Sci.* **43**, 933-941 (2008).
14. J.P. Borrajo, P. Gonzalez, J. Serra, S. Liste, S. Chiussi, B. Leon, A. De Carlos, F.M. Varela-Feria, J. Martinez-Fernandez and A.R. de Arellano-Lopez: Cytotoxicity study of biomorphic SiC ceramics coated with bioactive glass. *Bol. Soc. Esp. Ceram. Vid.* **45**, 109-114 (2006).
15. P. Gonzalez, J.P. Borrajo, J. Serra, S. Liste, S. Chiussi, B. Leon, K. Semmelmann, A. De Carlos, F.M. Varela-Feria, J. Martinez-Fernandez and A.R. de Arellano-Lopez: Extensive Studies on Biomorphic SiC Ceramics Properties for Medical Applications. *Key Eng. Mater.* **254-256**, 1029-1032 (2004).

16.   O. Gauthier, J.M. Bouler, E. Aguado, P. Pilet and G. Daculsi: Macroporous biphasic
       calcium phosphate ceramics: Influence of macropore diameter and macroporosity
       percentage on bone ingrowth. *Biomaterials* **19**, 133-139 (1998).
17.   P. Gonzalez, J. Serra, S. Liste, S. Chiussi, B. Leon, M. Perez-Amor, J. Martinez-
       Fernandez, A.R. de Arellano-Lopez and F.M. Varela-Feria: New biomorphic SiC
       ceramics coated with bioactive glass for biomedical applications. *Biomaterials* **24**, 4827-
       4832 (2003).
18.   A. De Carlos, J.P. Borrajo, J. Serra, P. Gonzalez and B. Leon: Behaviour of MG-63
       osteoblast-like cells on wood-based biomorphic SiC ceramics coated with bioactive glass.
       J. Mater. Sci. **17**, 523-529 (2006).
19.   P. Gonzalez, J.P. Borrajo, J. Serra, S. Chiussi, B. Leon, J. Martinez Fernandez, F.M.
       Varela Feria, A.R. de Arellano-Lopez, A. De Carlos, A. Munoz, M. Lopez and M. Singh:
       A new generation of bio-derived ceramic materials for medical applications. *J. Biom.
       Mater. Res. A* (2008).
20.   T.S. Keller: Predicting the compressive mechanical behavior of bone. *Journal of
       Biomechanics* **27**, 1159-1168 (1994).
21.   R.K. Nalla, J.J. Kruzic, J.H. Kinney and R.O. Ritchie: Mechanistic aspects of fracture
       and R-curve behavior in human cortical bone. *Biomaterials* **26**, 217-231 (2005).
22.   W. Bonfield: Advances in the fracture mechanics of cortical bone. *Journal of
       Biomechanics* **20**, 1071-1081 (1987).
23.   P. Zioupos and J.D. Currey: Changes in the stiffness, strength, and toughness of human
       cortical bone with age. *Bone* **22**, 57-66 (1998).
24.   D. Vashishth, J.C. Behiri and W. Bonfield: Crack growth resistance in cortical bone:
       Concept of microcrack toughening. *Journal of Biomechanics* **30**, 763-769 (1997).
25.   F.M. Valera-Feria, *Fabricación, Caracterización y Propiedades Mecánicas del Carburo
       de Silicio Biomórfico*. 2004, Universidad de Sevilla.
26.   D.J. Monk, D.S. Soane and R.T. Howe: Hydrofluoric acid etching of silicon dioxide
       sacrificial layers. II. Modeling. *J. Electrochem. Soc.* **141**, 270-274 (1994).
27.   M.S. Kulkarni and H.F. Erk: Acid-based etching of silicon wafers: Mass-transfer and
       kinetic effects. *J. Electrochem. Soc.* **147**, 176-188 (2000).
28.   M. Elwenspoek, U. Lindberg, H. Kok and L. Smith: Wet chemical etching mechanism of
       silicon. *Proceedings of the IEEE Micro Electro Mechanical Systems* 223-228 (1994).
29.   M. Steinert, J. Acker, A. Henßge and K. Wetzig: Experimental studies on the mechanism
       of wet chemical etching of silicon in HF/HNO$_3$ mixtures. *J. Electrochem. Soc.* **152**,
       C843-C850 (2005).
30.   D.R. Carter and W.C. Hayes: Bone compressive strength: The influence of density and
       strain rate. *Science* **194**, 1174-1175 (1976).

Porous Ceramics

STRENGTH AND PERMEABILITY OF OPEN-CELL MACRO-POROUS SILICON CARBIDE AS A FUNCTION OF STRUCTURAL MORPHOLOGIES

Joseph R. Fellows, Hyrum S. Anderson, James N. Cutts, Charles A. Lewinsohn, and Merrill A. Wilson
Ceramatec, Inc.
Salt Lake City, Utah, U.S.A.

ABSTRACT

Open-cell, macro-porous silicon carbide foams are ideal candidate materials for heat exchange (HX) structures where a high surface area to mass ratio is required. Silicon carbide (SiC) possesses excellent thermal conductivity, strength, and corrosion resistant properties and therefore would also be ideal in many heat exchange systems that operate in high temperature corrosive environments. Strength and flow characteristics are important parameters in the design of any HX system; therefore we have studied these characteristics as they relate to the foam's structural morphology which differs as a function of total porosity, using foams produced by the replica technique and synthetic templates. The objective of this work is to present results of replica formed SiC foams so that a choice of foam strengths and attendant permeability can be selected better for a HX system. Foam strengths are dependent more on porosity characteristics rather than pore spacing. Compressive strengths are low (0.8-1.6 MPa) for foams with porosities ranging between 77 and 93% porosity. Permeability is affected equally by porosity and pore distribution, but results show an even greater correlation between window area and permeability due to the localized Reynolds number through the window. Friction factors range between 5 and 45. Sintered foams contain structural irregularities, due to the dip-coat method utilized in the replicate approach, but data remain consistent with established models for foam compression and permeability, thereby validating our results. Selection of SiC foam for a specific application, requiring specific permeability characteristics, is made easier as the result of this work.

INTRODUCTION

Open-cell macro-porous silicon carbide (SiC) foams, which provide high surface area to mass ratio, are well suited for heat exchange (HX) designs requiring good corrosion resistance and strengths, at elevated temperatures,[1,2] while also providing excellent thermal conductivity. Strength and permeability, however, are among several considerations that are required in the proper selection of SiC foams used in any particular HX design. The characterization of strength or permeability is often a function of the morphology of a given foam structure, with differing porosities, strut configurations and associated pore distributions. To make proper choices in the selection of a foam used within an HX device, it is important to understand the correlation between the structural morphology and the attendant effects on strength and flow characteristics. Several SiC foams, processed using a replica technique, and designated initially by pore spacing, pore per millimeter (p/mm), have been processed and prepared to study how the morphology of the sintered foam impacts the strength of the foam body under compressive loads, as well as how the flow rates vary as a function of pore distribution and porosity values.

EXPERIMENTAL PROCEDURE

Processing of ceramic foams, as described by Studart *et al.*[3] was accomplished by utilizing the replica technique, with synthetic templates. Commercial polyurethane foam preforms, identified by the manufacturer with a pore per inch (ppi) designation of 20, 30, 60, 80, and 100, were obtained from Merryweather Foam, Inc. (Anthony, New Mexico). The traditional metric for differentiating foams uses the ppi designation, however, we have chosen to modify this metric and redefine the same foams using pore spacing (or distribution) by converting from ppi to pore per millimeter (p/mm). For example, a 20 ppi sample is now referred to as 0.8 p/mm foam in our work.

A proprietary silicon carbide slip was made using UF-15 sub-micron α-SiC powder, produced by H. C. Stark, and organic solvents and additives. Slip viscosity of 2700 cP was used for dip coating preforms with pore spacing of 0.8- 1.2 p/mm (20-30 ppi), and 1500 cP to coat preforms with pore spacing of 2.4 – 3.9 p/mm (60-100 ppi). Upon dip-coating the synthetic templates, samples underwent binder and template burnout, using a Lindberg box furnace fitted with an Inconel alloy retort that enables control of gas composition. Subsequent sintering to 2150°C was done using an AVS hot press, Model VHP-12-2200-N2.

Ten column-shaped specimens of sintered SiC foam, with approximate diameters of 19 mm and lengths of 57 mm were fabricated for each p/mm group, for a total of fifty specimens. Figure 1 shows optical photographs of typical sintered foam from each of the groups. Each pore size designation was characterized for porosity, average window areas, and average strut dimensions. Porosity was obtained using the Archimedes method.

The window size, or opening that allows access into the cells, was measured using the optical microscope. As seen in Figure 1, window sizes vary in geometry and were averaged assuming, for simplicity, that the geometries consisted of both rectangular and elliptical windows, with variations. Therefore widths and heights of random windows (random sampling of 10 to 20 windows for the different sintered p/mm groups) were calculated as both rectangles and ellipses, with an average value taken from these calculations. Average strut dimensions were also determined by obtaining optical microscope measurements of these features throughout each foam structure and taking an average value. The synthetic foams have thin random membranes covering the windows into pores of the foam. Thus, as noted in Figure 1, there are many pores that appear to be blocked due to the SiC slip that coated and replicated these membranes. Further, the sintered foams tested varied in average strut length and strut diameter. These physical characteristics, however, are similar within each family or group pore distribution.

For any given foam, derived by the replica technique used here, in conjunction with the synthetic foam templates, the strut diameter will vary along the length of the strut. It is not a simple task to assign a strut diameter that is characteristic of replica-derived, dip-coated foam, nor is it straight forward with respect to defining the hollow inner dimensions of each strut, as the template itself varies in thickness. However, when considering foam placed under compressive forces, the thinner struts are the most likely to fail, assuming all other flaws and failure mechanisms are temporarily neglected. Due to the nature of the synthetic template, the region where these struts are the thinnest is typically mid-span between adjoining junctions. Therefore, an average strut cross-sectional area was calculated based on the width of these smaller struts. These and other morphological characteristics for each group of sintered foam are averaged and tabulated in Table I. Upon crushing the foams under a compressive load, data was tabulated to determine the stress (σ^*) at which the cylindrical foam structure failed. Figure 2 shows an optical photograph of a typical 2.4 p/mm (60 ppi) foam and how the width, and thus the area, of each strut varies as a function of length, with the thicker features found at the strut junctions.

Table I. Tabulation of morphologies, associated average strengths, and constant C6 for SiC foams produced through the replica technique using synthetic preforms.

Group p/mm (ppi)	Average Sintered p/mm	Porosity (%)	Average Window Area (mm²)	Average Strut Cross-sectional Area (mm²)	Stress σ* (MPa)	Average Constant C6
0.8 (20)	0.8	93.5	2.45	1.25×10^{-2}	0.85	0.13
1.2 (30)	1.1	93.0	1.15	2.23×10^{-2}	0.91	0.12
2.4 (60)	2.6	77.1	0.11	1.31×10^{-3}	1.58	0.05
3.1 (80)	3.3	89.6	0.09	1.12×10^{-3}	1.34	0.10
3.9 (100)	3.8	88.0	0.09	1.22×10^{-3}	1.17	0.07

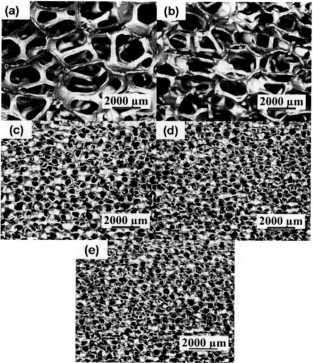

Figure1. Optical photographs of various sintered SiC foam groups used in this study: (a) 0.8, (b) 1.2, (c) 2.4, (d) 3.1, and (e) 3.9 p/mm.

Figure 2. Optical photograph of a 2.4 p/mm (60 ppi) sintered SiC foam. Strut width varies along the length with thicker portions found at cell wall junctions.

Within each group of ten, five specimens were tested in compression for strength measurements, while the other five where subjected to flow tests to determine permeability characteristics. Compression testing was accomplished using an Instron electromechanical test frame, Model 5566. All specimens were loaded at a rate such that failure of the foam column would occur within two minutes of initial loading (between 250 and 300 N/min, for the 0.8 – 1.2 p/mm (20 and 30 ppi) specimens, and at 200 N/min for the 2.4 – 3.9 p/mm (60 – 100 ppi) specimens, with each specimen having a 1:3 diameter to length ratio. A compliant foam layer was added to the top and bottom of each column to distribute the applied load uniformly, and to minimize uneven loads that may occur at the strut level, due to strut irregularities. A typical compression test set-up is seen in Figure 3, with 2.4 p/mm SiC foam being tested. Flow tests, using air at room temperature, were conducted by sealing individual specimens within a heat shrink wrap and flowing air at various flow rates from 5 to 20 standard liters per minute (slpm) through these columns as seen in Figure 4.

Figure 3. Compression test set-up.

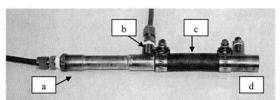

Figure 4. Flow test device: (a) gas inlet, (b) inlet pressure port, (c) shrink wrapped SiC foam, and (d) gas outlet.

RESULTS AND DISCUSSION

Compression Results

According to models provided by Gibson and Ashby,[4] compression of SiC foam has three stages: cell strut bending, cell strut collapse, and finally opposing cells collapsing on other cells with a rapid stress increase. The results of the collapse and failure of the foams tested in this study showed two basic types of overall failure through the main body of the column. The first type of failure is characterized by failure and collapse of the column starting at a cross-sectional plane of the column, normal to the loading axis: an initial plane is pulverized and subsequent lower planes pulverize one on top of the other in a pancake fashion. The plane that fails first can be located anywhere in the length of the column, but our results show that this failure plane was typically found on the top 1/2 of the cylinder. We noted variations of this type of failure whereby the cross-sectional plane was irregular, but typically the entire column would be pulverized with fragments that are not recognizable as coming from a column if the loading was continued beyond initial failure. This type of failure is shown in Figure 5(a), whereby the top half of the column is simply missing (pulverized), with an outline showing where the top part of the column used to be. The second type of failure we encountered, also showing variations, exhibited a fracture plane inclined at approximately 45°. Thus, the column simply broke into two or more parts, with the fragments very large compared to that seen in the first type of failure. This second type of foam failure is seen in Figure 5(b).

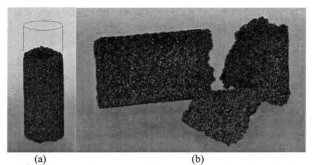

(a)   (b)

Figure 5. (a) 'Pancake' collapse of the SiC foam under a compressive load. The region where the foam was pulverized is outlined. (b) Main foam fragments have fracture planes that vary, but have some surfaces typically inclined around 45°, with multiple fragments noted in this type of failure.

Data for all specimens tested in compression was tabulated, with the relationship of strength (σ^*) to pore distribution (p/mm) shown in Figure 7. Here it can be seen that strength is not linearly dependent on pore distribution. Figure 8 shows the strength (σ^*) of the foams as a function of total porosity, with strength values being low, as would be predicted,[5,6] for this type of replicated foam having hollow struts. Here it is shown that strength increases as porosity decreases, in accordance with previous work,[2,4,5,7] with a nonlinear fit as would be expected.[5] Comparing Figures 7 and 8, it is clear that porosity is a more predictive parameter in determining strength of the foam and not pore distribution.

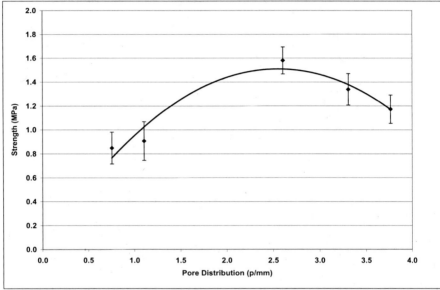

Figure 7. Strength as a function of pore distribution.

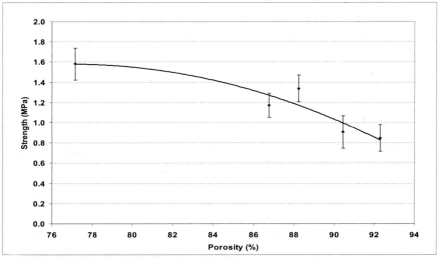

Figure 8. Strength as a function of porosity (%).

Gibson and Ashby have shown that for brittle foams placed in compression, a constant C_6 can be obtained which would include other factors of the foam morphology. A correlation between maximum strengths required of a brittle foam to its relative porosity, is given by[4]

$$\frac{\sigma^*}{\sigma_{fs}} \approx C_6 \left(\frac{\rho^*}{\rho_s} \right)^{3/2} \tag{1}$$

where σ^* is the maximum stress (MPa) at which the foam failed, σ_{fs} is the flexural stress of the dense SiC (measured to be 390 ± 58 MPa), C6 is the constant of proportionality, ρ^* is density of the SiC foam, and ρ_s is the theoretical density of SiC (3.21 g/cc). Values for C_6 have been reported as typically on the order of 0.2.[4]

To manufacture an array of SiC foams, with porosities tailored to give desired strengths, it would be helpful to have this constant to relate the maximum strength required of the foam to its relative porosity. In terms of porosity, Equation (1) becomes

$$\frac{\sigma^*}{\sigma_{fs}} \approx C_6 (1 - P)^{3/2} \tag{2}$$

with P being the volume fraction of porosity in the foam. The foams tested have values of C_6 near those reported by Gibson and Ashby.[4] The values of the constant C_6, for foams in this study, are shown in Figure 9, and average C_6 values are reported in Table I.

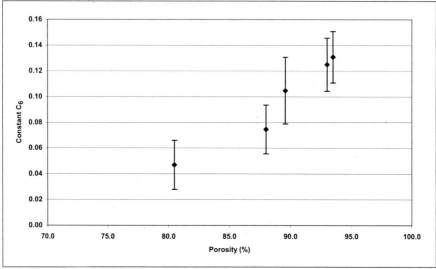

Figure 9. C_6 as a function of porosity (%).

Brezny and Green[7] have shown a correlation between increasing strut strength and decreasing cell size, and noted that their results may be explained due to the fact that as the cell size decreases, there is also a decrease in strut surface area and attendant decrease in flaw distribution. Typically, as the cell size decreases, the window size also decreases. Therefore porosity affects not only the strength, but also the permeability, because flow is constrained by access through the windows. Therefore, if any correlation is to be made between strength and permeability, the window size of the cells must be accounted for. In commercially available synthetic foams, average window sizes associated with the cells would be relatively constant and proportional, meaning that any cell would have a limited range of window sizes that would repeat throughout the structure. Assuming that the average window area, calculated here in terms of the window diameter, is directly proportional to the average strut length, since a strut forms the window and cannot be any longer than a side of that window, a correlation should exist between strength and window size. Figure 10 shows the correlation between strength $\sigma*$ and mean window diameter, for a range of pore spacing. As the window sizes become smaller, there is an increase in load capacity, associated with decreased strut lengths; however, as noted in this distribution, these foam samples exhibit a point where the maximum load varies a small amount as the average window size remains relatively constant (i.e.<0.5mm). This could be explained due to the fact that, as discussed previously, there are irregularities such as windows being covered by SiC membranes with increased surface areas, with corresponding changes of localized strengths and flaw distributions. It is important to note that the maximum strength $\sigma*$ attained, correlating to average window size, is associated with the 2.4 p/mm (60 ppi) foam (compare with Figure 7).

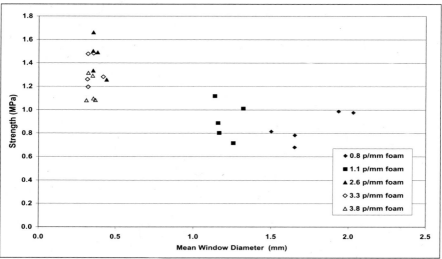

Figure 10. Strength as a function of window diameter.

Permeability Results

Figure 11 shows the general trend of permeability for all samples tested, with pressure drop (normalized) increasing as porosity decreases, in agreement with prior studies.[6] The change in pressure, or pressure drop ΔP, is a function of several variables given by[8]

$$\Delta P = f\left(\frac{l}{D}\right)\rho\frac{V^2}{2} \tag{3}$$

where f is the friction factor, l is the tube length, D is the tube diameter, ρ is density of the gas that flows through the tube, and V is the free stream velocity of the gas. Data collected from the flow tests allow one to note a very good correlation between the friction factor (f) and the Reynolds number (Re), shown in Figure 12. Such correlation shows that there is a physical relation, based on morphology, between f and Re[9]. Further, the correlation between both pore distribution and porosity to the friction factor are shown in Figure 13. Here is shown that both pore distribution and porosity have significant influence on the friction factor.

Seeing that the friction factor is significantly influenced by pore spacing and porosity, the ability to produce SiC foam tailored to achieve a specified pressure drop requires an additional characteristic of the foam based on its morphology. As discussed previously, the window area between cells limits and determines the flow characteristics of the foam. In the laminar flow regime, the friction factor is typically inversely proportional to the local Reynolds number; in this case the friction factor can be related to the Reynolds number based on the window diameter as shown in Figure 14. This last correlation shows that to best tailor foam to meet specified pressure drop performance, the characteristic of primary importance is the window diameter. It should be noted that the dimension of these windows will vary with pore spacing and porosity. In other words for a preferred window area, there is range of porosities that can be attained for a given pore spacing of the

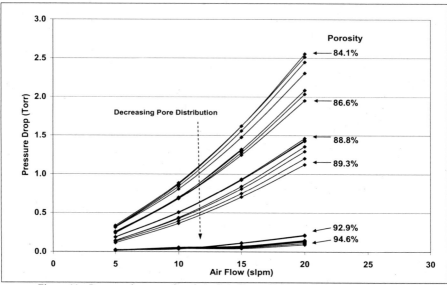

Figure 11. Pressure drop as a function of air flow. Pressure drop as related to total porosity and pore spacing is also noted.

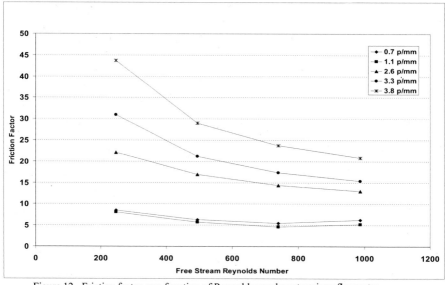

Figure 12. Friction factor as a function of Reynolds number at various flow rates.

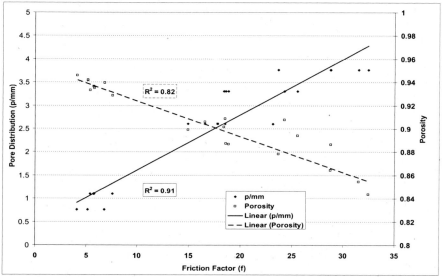

Figure 13. Friction factor as a function of pore spacing and total porosity.

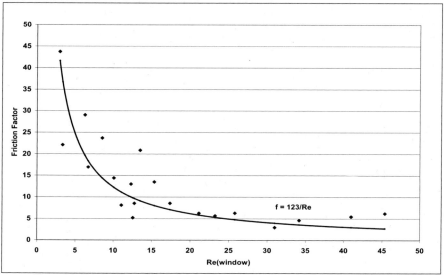

Figure 14. Friction factor as a function of Reynolds number through cell window.

synthetic template, and vice versa. Therefore, once the target flow and pressure drop have been determined, the necessary window diameter for the foam can be engineered and constructed using a specified range of pore openings per unit length and/or porosity characteristics.

CONCLUSION

SiC foam is an ideal candidate material for HX systems due to its thermal properties and strength. This work showed how to quantify how SiC foams, produced using the replica method and synthetic templates, behave so a selection of these foams for strength and permeability, necessary in a HX system, can be made. Given a specified strength and/or permeability, the choice of morphological parameters (pore per unit length, porosity and window diameter) may of necessity require that one characteristic is enhanced at the expense of another characteristic,[6] all predicated upon it's use within, and operational conditions of, the HX device.

The strength of the SiC foams tested is strongly dependent on porosity rather than pore spacing, with these foams behaving according to established models. As porosity decreases there is an increase in strength under compressive loading, with a maximum failure stress of 1.6 MPa, correlating to a porosity of 77%.

Further, flow tests of these foams have shown that pressure drop and friction factors are a function of both porosity and pore distribution. This data and correlations have shown that the relationship between the friction factor and the Reynolds number of the window between cells are of primary importance in determining flow characteristics and subsequent tailoring of foams.

REFERENCES

[1]T. Sato, K. Kubota, and M. Shimada, "Corrosion Kinetics and Strength Degradation of Sintered α-Silicon Carbide in Potassium Sulfate Melts", *J. Am. Ceram. Soc.*, 74 [9] 2152-55 (1991).

[2]P.Colombo, J. R. Hellmann, D. L. Shelleman, "Mechanical Properties of Silicon Oxycarbide Ceramic Foams", *J. Am. Ceram. Soc.,* 84 [10] 2245-51 (2001).

[3]A. R. Studart, U. T. Gonzenbach, E. Tervoort, and L. J. Gauckler, "Processing Routes to Macroporous Ceramics: A Review," *J. Am. Ceram. Soc.*, 89 [6] 1771-89 (2006).

[4]L. J. Gibson, and M. F. Ashby, "*Cellular Solids: Structure and Properties*", 2nd Ed., (1997), Cambridge University Press.

[5]R. W. Rice, "*Porosity of Ceramics*" (1998), CRC Press.

[6]M. D. M. Innocentini, P. Sepulveda, V. R. Salvini, and V. C. Pandolfelli, "Permeability and Structure of Cellular Ceramics: A Comparison between Two preparation Techniques", *J. Am. Ceram. Soc.*, 81 [12] 3349-52 (1998).

[7]R. Brezny, and D. J. Green, "The Effects of Cell Size on the Mechanical Behavior of Cellular Materials", *Acta Metall. Mater.*, 38[12] 2517-26 (1990).

[8]J. P. Holman, "*Heat Transfer*", 5th Ed., (1981), McGraw Hill Book Co.

[9]W. M. Kays, and A. L. London, "*Compact Heat Exchangers*", 3rd Ed – reprint (1998), Krieger Publishing Co.

DESIGN OF SILICA NETWORKS USING ORGANIC-INORGANIC HYBRID ALKOXIDES FOR HIGHLY PERMEABLE HYDROGEN SEPARATION MEMBRANES

Masakoto Kanezashi, Kazuya Yada, Tomohisa Yoshioka and Toshinori Tsuru
Department of Chemical Engineering, Hiroshima University
1-4-1 Kagamiyama, Higashi-Hiroshima, 739-8527, Japan

ABSTRACT
    Sol-gel method was applied for the preparation of organic-inorganic hybrid silica membranes. 1,3,5,7-tetramethylcyclotetrasiloxane (TMCTS) and bis (triethoxysilyl) ethane (BTESE) were used as silica precursors instead of tetraethoxysilane (TEOS) for the preparation of organic-inorganic hybrid silica membranes. Molecular weight of hybrid silica polymer was successfully controlled by reaction time and reaction temperature. Single gas permeation characteristics for hybrid silica membranes were examined to discuss the effect of silica precursor on amorphous silica networks. Pore size distribution suggested hybrid silica membranes derived by BTESE and TMCTS had loose amorphous silica structures compared to silica membranes derived by TEOS. Hybrid silica membranes derived by BTESE and TMCTS showed high hydrogen permeance (0.3-1 x 10^{-5} mol·m^{-2}·s^{-1}·Pa^{-1}) with high selectivity of H_2 to SF_6 (H_2/SF_6 permselectivity: ~1,000) and low H_2 to N_2 permselectivity (~10). It was found that hybrid silica membrane derived by TMCTS had high hydrothermal stability due to the presence of CH_3 groups in amorphous silica networks.

INTRODUCTION
    Silica is one of the most attractive materials for inorganic membranes because of their thermally stable amorphous structures in a wide temperature range up to 1000°C. Sol-gel and CVD methods have been applied in the preparation of silica membranes on porous substrate such as alumina and vycor glass [1,2]. In general, CVD derived silica membranes show high hydrogen to nitrogen permselectivity (>1,000) with low hydrogen permeance (<4 x 10^{-8} mol·m^{-2}·s^{-1}·Pa^{-1}) because of molecular sieving effect between permeating gas molecules and dense silica layer on porous substrate [3-6], i.e., hydrogen molecules can easily permeate through silica networks compared with nitrogen molecules because of much smaller molecular size (kinetic diameter of hydrogen: 0.26 nm, kinetic diameter of nitrogen: 0.364 nm). Compared to CVD derived silica membranes, sol-gel derived silica membranes show approximately one order magnitude high hydrogen permeance with moderate hydrogen selectivity (<1,000) [2,7-9]. This is because of rather wide pore size distributions comprised of spaces within amorphous silica networks and grain boundaries. Since the presence of grain boundaries in a membrane has a strong influence on hydrogen separation performance, the polymeric route for control of pore sizes is much preferable in the preparation of hydrogen separation membranes, because pore sizes by the polymeric route are considered to be made up of spaces within amorphous silica networks [10].
    Tetraethoxysilane (TEOS) is a commonly used precursor for the preparation of sol-gel derived silica membranes. In the process of the hydrolysis and polymerization reaction of TEOS, the Si-O-Si unit is a minimum unit for amorphous silica networks. There is a possibility that spaces within amorphous silica networks depend on the silica precursors, i.e., changing the silica precursors can create the different amorphous silica networks. Several research groups have reported the utilization of organic-inorganic hybrid alkoxides for the design of amorphous silica networks [11-17]. The inorganic networks prepared by the amorphous silica with pyrolysis of methyl groups gave high CO_2

permeance in the order of 10^{-7} mol·m^{-2}·s^{-1}·Pa^{-1} with CO_2/CH_4 selectivity higher than 70 [11].   Kusakabe et al. [13] investigated the effect of the length of alkyl groups on pore structure as well as the permeation properties for silica membranes, which were prepared by co-polymerization of TEOS and octyl-, dodecyl- or octadecyltriethoxysilane.   The results suggest pore size of silica membrane was successfully controlled by changing the length of alkyl groups.   However, there still has a problem for instability of the inorganic networks derived by organic-template method, because amorphous silica structures can be densified in the presence of water vapor even at room temperature [3,9].   To improve hydrothermal stability, we have reported metal doped silica membranes such as nickel and cobalt [18-22].   Castricum et al. [15,16] fabricated hydrothermally stable hybrid silica membranes for separation of water from by pervaporation.   Recently, Kanezashi et al. [17] proposed the strategy for the design of amorphous silica networks using the organic-inorganic hybrid alkoxides for the development of hydrogen separation membrane with hydrothermal stability.

In this work, 1,3,5,7-tetramethylcyclotetrasiloxane (TMCTS), which contains 4-membered Si ring with CH_3 groups and bis (triethoxysilyl) ethane (BTESE) with ethylene groups between Si atoms, as shown in Fig. 1, were used as silica precursors for the preparation of organic-inorganic hybrid amorphous silica networks.   Single gas permeation characteristics for hybrid silica membranes were examined to discuss the space size of amorphous silica networks.   Hydrothermal stability of hybrid silica membrane was also examined by measuring the gas permeation behavior before and after exposure to humid atmosphere.

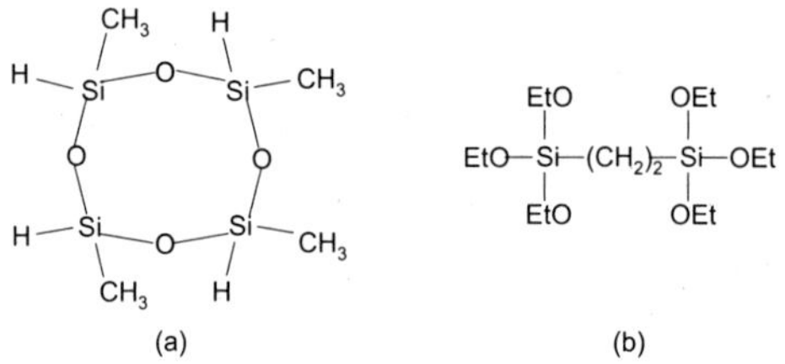

(a)   (b)

Figure 1. Molecular structures of 1,3,5,7-tetra-methyl-cyclo-tetrasiloxane (a) and bis (triethoxysilyl) ethane (BTESE) (b)

EXPERIMENTAL

Preparation of organic-inorganic hybrid silica polymer

Organic-inorganic hybrid silica polymer sol was prepared by hydrolysis and polymerization reactions of 1,3,5,7-tetramethylcyclotetrasiloxane (TMCTS) (Gelest) and/or bis (triethoxysilyl) ethane (BTESE) (Gelest).   The preparation procedures are as follows, for example: A specified amount of 1,3,5,7-tetramethylcyclotetrasiloxane (1.25g) was added to 2-propanol (50g) with a specified amount of distilled water (0.3 g).   After stirred for enough time at room temperature, the catalyst (NH$_4$OH) was added to the solution for hydrolysis and polymerization reactions.   Molecular weight of organic-inorganic hybrid polymer was measured by GPC method with Refractive Index Detector (RID) using GPC column (TSK gel α-2500, Tosoh) at 25°C.   Tetrahydrofuran (THF) was selected as solvent and flow rate was controlled at 0.8 cc·min^{-1} by HPLC pump (LC-10ADvp, Shimadzu).   The molecular weight of polymer was calibrated with polystyrene standards of molecular weights from 800-2,000,000 (Aldrich).   TMCTS gel powder was prepared by drying at 40-50°C and calcination at 300°C in air atmosphere for thermogravimetric analysis (TG).   The sample gel powder was heated up 1000°C under air flow rate of 30 cc·min^{-1}.

Fabrication of organic-inorganic hybrid silica membrane

Porous α-alumina tubes (porosity: 50%, average pore size: 1 μm, outside diameter: 10 mm, length: 100 mm) were used as the support for the silica-zirconia (Si/Zr=1/1) intermediate layer of pore size at around several nm. SiO_2-ZrO_2 (Si/Zr=1/1) intermediate layer was calcinated at 550°C for 15-30 min [18-22]. After formation of SiO_2-ZrO_2 intermediate layer, organic-inorganic hybrid silica polymer, which was diluted to about 0.5 wt%, was coated by contacting the module with a wet cloth (Asahikasei Co., Japan). Then, the membrane was calcinated at 300°C in air for 30 min. The coating and calcination procedures were repeated several times for formation of crack-free hybrid silica layer.

Single gas permeation measurement

Figure 2 shows schematic diagram of experimental apparatus for gas permeation measurement. Single gas of industrial grade (He, H_2, N_2 and SF_6) was fed on the membrane surface at atmospheric pressure, while the permeate side (downstream) was evacuated by a vacuum pump. The temperature of the permeation cell was kept at a given temperature between 100 and 300°C. The permeances were calculated from the observed pressure difference across the membrane and the permeation rate, which was obtained by a calibrated critical nozzle placed between the permeation cell and the vacuum pump. After measuring the gas permeances for a fresh membrane, the membrane was kept in humid atmosphere at 40°C and of 60% relative humidity for a specified period. Then the gas permeances were measured again in the same temperature range to evaluate the hydrothermal stability of hybrid silica membrane.

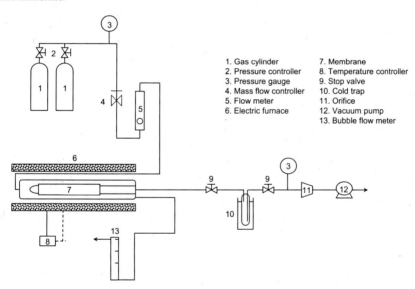

1. Gas cylinder	7. Membrane
2. Pressure controller	8. Temperature controller
3. Pressure gauge	9. Stop valve
4. Mass flow controller	10. Cold trap
5. Flow meter	11. Orifice
6. Electric furnace	12. Vacuum pump
	13. Bubble flow meter

Figure 2. Schematic diagram of experimental apparatus for gas permeation measurement

RESULTS AND DISCUSSION

Characterization of hybrid silica polymer derived by BTESE and TMCTS

Molecular weight of BTESE and TMCTS polymers were measured to know the reaction characteristics. Figure 3 shows molecular weight of BTESE and TMCTS polymer as a function of reaction time. BTESE polymer was prepared by hydrolysis and polymerization under acid condition (pH: 2-3), while TMCTS polymer under alkaline condition (pH: 8-9). It should be noted that reaction rate of TMCTS under acid condition was too low to control the molecular weight of TMCTS polymer. Molecular weight of BTESE polymer drastically increases with reaction time under hydrolysis and polymerization at 25°C and reaches to 10^6 g·mol^{-1} within 10 h. While molecular weight of TMCTS polymer slightly increases with reaction time at 25°C and approach to 10^3 g·mol^{-1}. With increasing the reaction temperature from 25 to 60°C, molecular weight starts to increase again and approach to 10^4 g·mol^{-1}. It is believed that the difference of these reaction behaviors depend on the number of hydrolyzed groups in the alkoxides (BTESE: 6, TMCTS: 4) and Si-(OEt) groups can be easily hydrolyzed compared to Si-H groups. In this work, molecular weight of BTESE polymer 10^5 g·mol^{-1} and that of TMCTS polymer 10^3 and 10^4 g·mol^{-1} was used as coating polymers for the preparation of hybrid silica membranes.

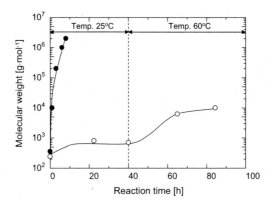

Figure 3. Molecular weight of BTESE (closed symbols) and TMCTS (open symbols) polymers as a function of reaction time (hydrolysis condition: Si/H$_2$O molar ratio: 3 for BTESE, 1 for TMCTS)

TG measurement was carried out to show the decomposition temperature of organic groups in TMCTS gel powder. Figure 4 shows TG curve of TMCTS gel powder in air atmosphere. There is no significant weight loss for TMCTS gel powder below 400°C. A weight loss of TMCTS gel powder gradually starts around at 400°C and approximately 30 % of weight loss was observed at 500°C. After that no weight loss was observed. A ideal weight loss can be calculated about 30 % if all CH$_3$ groups in TMCTS gel were decomposed, indicating the decomposition temperature of CH$_3$ groups are expected to 400-500°C. For BTESE derived hybrid silica polymer, Castricum et al. [15,16] reported that the decomposition of the (CH$_2$)$_2$ groups between 2 Si atoms started gradually above 315°C in air atmosphere. Organic-inorganic hybrid silica membranes, which were prepared in this work, are expected to contain organic groups in amorphous silica networks due to the fact that both hybrid silica membranes were calcined at 300°C in air atmosphere.

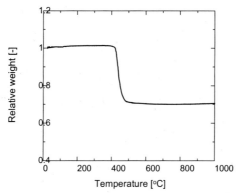

Figure 4. TG curve of TMCTS gel powder in air atmosphere (ramping rate: 5°C·min^{-1})

Pore size distribution of hybrid silica membranes derived by BTESE and TMCTS

The SEM image of the cross section of the hybrid silica membrane derived by BTESE is shown in Figure 5. As shown in this micrograph, crack-free continuous hybrid silica layer for selective hydrogen separation could be formed on silica-zirconia intermediate layer. The thickness of the active separation layer is clearly less than 0.5 μm.

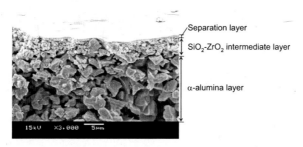

Figure 5. SEM image of hybrid silica membrane derived by BTESE

Figure 6 shows gas permeances for organic-inorganic hybrid silica membranes (BTESE, TMCTS) and silica membrane derived by TEOS [18] at 200°C as a function of kinetic diameter. Silica membrane derived by TEOS [18] shows relatively high hydrogen permeance (1 x 10^{-6} mol·m^{-2}·s^{-1}·Pa^{-1}) with selectivity of H$_2$ to N$_2$, CH$_4$ (H$_2$/N$_2$ permselectivity: 200, H$_2$/CH$_4$ permselectivity: ~1,000), indicating the average pore size of approximately 0.4 nm. On the other hand, hybrid silica membrane derived by BTESE shows one order high hydrogen permeance of 1 x 10^{-5} mol·m^{-2}·s^{-1}·Pa^{-1} with high selectivity of H$_2$ to SF$_6$ (H$_2$/SF$_6$ permselectivity: ~1,000) and low H$_2$ to N$_2$ permselectivity (~10). For TMCTS derived hybrid silica membrane, the permeance of H$_2$ is comparable with that for a silica membrane (TEOS). However, pore size distribution of hybrid silica membrane (TMCTS) is clearly different from that of a silica membrane (TEOS) and similar to that of a hybrid silica membrane (BTESE), i.e. high H_2/SF_6 permeance ratio (~1,000) with low H_2 to N_2 permselectivity (~10).

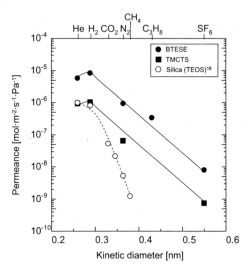

Figure 6. Gas permeances at 200°C for organic-inorganic hybrid silica membranes derived by BTESE and TMCTS and a silica membrane derived by TEOS [18] as a function of kinetic diameter

The difference of these pore size distributions can be attributed to the minimum unit of amorphous silica networks. Figure 7 shows schematic image of amorphous silica networks derived by TEOS (a), BTESE (b) and TMCTS (c). For amorphous silica networks derived by BTESE, Si-C-C-Si bonds can be the minimum for organic-inorganic hybrid silica networks because BTESE contains the organic groups between 2 silicon atoms, which can not be hydrolyzed in the process of hydrolysis and polymerization reaction. Compared to the silica networks using TEOS, which consists of one Si atom, the formation of a much looser structure is expected that can give high hydrogen permeability. For amorphous silica networks derived by TMCTS, 4-membered Si ring can be the minimum unit. In this case, the spaces formed by 4-membered Si ring can be the pores, through which hydrogen molecules can permeate, because space formed inside 4-membered Si ring is too small for hydrogen molecules to permeate [23]. The spaces formed by 4-membered Si ring, should be much larger than that by TEOS due to larger minimum unit of amorphous silica networks. N_2 adsorption measurement and Si-NMR analysis is a subject of future study to evaluate the spaces formed by 4-membered Si ring.

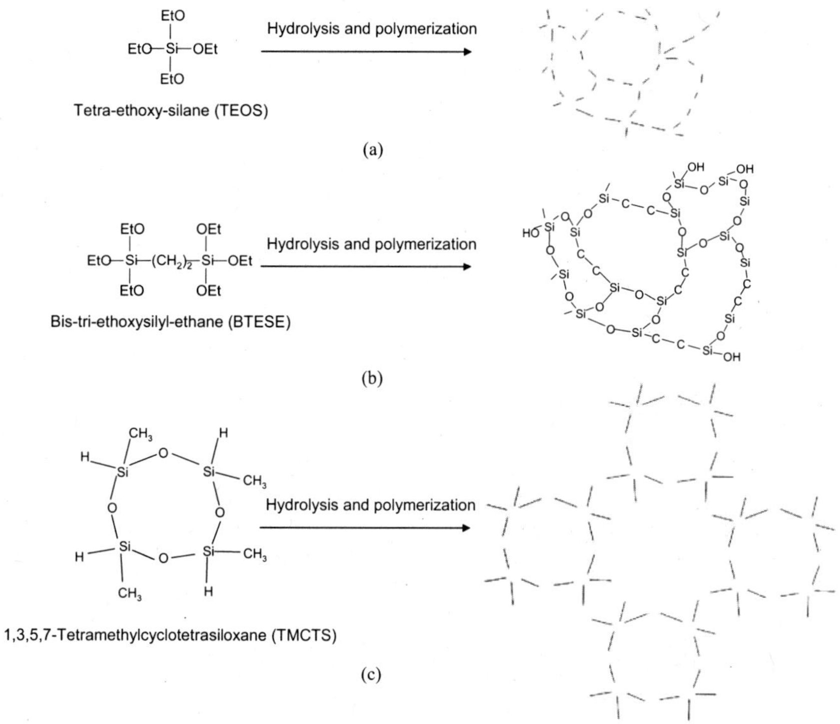

EtO
|
EtO—Si—OEt
|
EtO

Hydrolysis and polymerization ⟶

Tetra-ethoxy-silane (TEOS)

(a)

EtO   OEt
| |
EtO—Si—(CH₂)₂—Si—OEt
| |
EtO   OEt

Hydrolysis and polymerization ⟶

Bis-tri-ethoxysilyl-ethane (BTESE)

(b)

1,3,5,7-Tetramethylcyclotetrasiloxane (TMCTS)

Hydrolysis and polymerization ⟶

(c)

Figure 7. Schematic image of amorphous silica networks derived by TEOS (a), BTESE (b) and TMCTS (c)

There is a possibility that molecular weight of coating polymer can affect the gas permeability. In order to examine effect of molecular weight of coating polymer on gas permeation behavior, two types of hybrid silica membranes derived by TMCTS were prepared; type 1: coating with TMCTS polymer (M_w: 10,000) on SiO_2-ZrO_2 intermediate layer (pore size: ~1 nm) then TMCTS polymer (M_w: 1,000), type 2: coating with TMCTS polymer (M_w: 1,000) on SiO_2-ZrO_2 intermediate layer. It should be noted that total coating times of type 1 and type 2 membrane was 5 times. Pore size distribution of hybrid silica membranes (type 1, type 2) is shown in Figure 8. Every gas permeances for type 1 membrane increases about double compared to those for type 2 membrane and shows similar pore size distribution. When the polymer of small molecular weight was coated on SiO_2-ZrO_2 intermediate layer, penetration of coating polymer into SiO_2-ZrO_2 intermediate layer should be occurred, resulting in an increase of the resistance for gas permeation. For type 1 membrane, penetration of TMCTS polymer could be prevented by coating with polymer of large molecular weight on SiO_2-ZrO_2 intermediate layer before coating with polymer of small molecular weight.

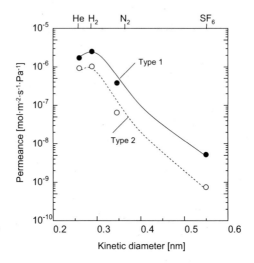

Figure 8. Gas permeances at 200°C for organic-inorganic hybrid silica membranes derived by TMCTS as a function of kinetic diameter (closed symbols on solid line: type 1 membrane, open symbols on broken line: type 2 membrane)

Hydrothermal stability of hybrid silica membrane derived by TMCTS

Hydrothermal stability of hybrid silica membrane was evaluated by measuring the gas permeation behavior before and after kept in humid atmosphere. Figure 9 (a) shows temperature dependency of gas permeances for a fresh silica membrane derived by TEOS in the temperature range 100-300°C. In the same figure are shown gas permeances measured after kept the membrane in humid atmosphere (40°C, 60%RH) for 30 days. The permeances of He, H_2 and N_2 slightly increase with increasing temperature, which is similar to activated permeation behavior. The permselectivity of H_2 to N_2 at 300°C is 150 and SF_6 permeance was less than detection limit ($<10^{-9}$ $mol \cdot m^{-2} \cdot s^{-1} \cdot Pa^{-1}$). After 30 days clear difference was observed for gas permeation behavior. The permeances of He and H_2 decrease about 60% and strongly depend on the temperature. When small gas molecules, especially at He and H_2, permeate through amorphous silica networks, the activation energy for gas permeation is needed to overcome the energy barrier imposed by the amorphous silica networks, i.e., larger activation energy for gas permeation indicates much smaller pore size of amorphous silica networks. The activation energy of He and H_2 permeation before and after kept in humid atmosphere was calculated by Arrhenius equation. Activation energy increases from 3.8 to 5.3 $kJ \cdot mol^{-1}$ for He-permeation and from 4.2 to 7.2 $kJ \cdot mol^{-1}$ for H_2-permeation, respectively. The permeance of N_2 also decreases about 1/5 of initial value and is almost close to detection limit. The change of gas permeation behavior for silica membrane (TEOS) after kept in humid atmosphere is due to the densification of amorphous silica networks. Water molecules can be adsorbed or condensed in the pores (silica networks) and formed the Si-OH groups during kept in humid atmosphere. By raising the temperature again, condensation of Si-OH groups can be occurred in the amorphous silica networks, resulting in much densified amorphous silica structure [3,9]. The densified amorphous structure can give much larger activation energy for permeation of He and H_2 molecules.

Figure 9 (b) shows temperature dependency of gas permeances for a fresh hybrid silica membrane derived by TMCTS before and after kept in humid atmosphere. The permeances of H_2 and He for hybrid silica membrane are independent of temperature, while those of N_2 and SF_6 slightly increase with decreasing temperature, which is similar to Knudsen type permeation. A hybrid silica membrane shows hydrogen permeance of 0.3×10^{-5} $mol \cdot m^{-2} \cdot s^{-1} \cdot Pa^{-1}$ with H_2 to SF_6 selectivity of 200 at $300°C$. After kept in humid atmosphere ($40°C$, $60\%RH$) for 30 days, the gas permeances (He, H_2, N_2, SF_6) show little change (increase of 5-10%) and no difference was observed for temperature dependency of H_2, He, N_2 and SF_6 permeances, suggesting the prevention of densification of amorphous silica networks. The improvement of hydrothermal stability for hybrid silica membrane can be explained by the presence of CH_3 groups in the silica networks, resulting in an increase in hydrophobicity, which leads to a decrease in the formation of Si-OH groups in a hydrothermal atmosphere [24].

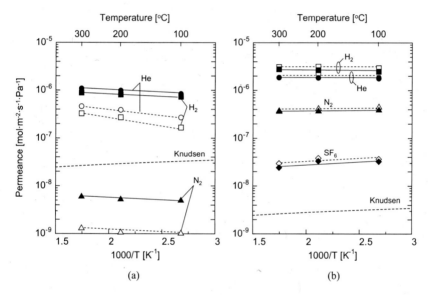

Figure 9. Temperature dependency of gas permeances for silica membrane derived by TEOS (a) and organic-inorganic hybrid silica membrane derived by TMCTS (b) (closed symbols on solid line: gas permeances for fresh membrane, open symbols on broken line: those for membrane after kept in humid atmosphere ($40°C$, $60\%RH$) for 30 days)

CONCLUSIONS

Sol-gel derived organic-inorganic hybrid silica membranes were prepared using 1,3,5,7-tetramethylcyclotetrasiloxane (TMCTS) and bis (triethoxysilyl) ethane (BTESE) for improvement of hydrogen permeability and hydrothermal stability of amorphous silica networks. Molecular weight of hybrid silica polymer was successfully controlled by reaction time and reaction temperature. The design of amorphous silica networks using an organic-inorganic hybrid structure for improvement of H_2 permeability could be successfully conducted using BTESE and TMCTS as a silica precursor. Hybrid silica membranes derived by BTESE and TMCTS showed high hydrogen permeance (0.3-1 x 10^{-5} mol·m^{-2}·s^{-1}·Pa^{-1}) with high selectivity of H_2 to SF_6 (H_2/SF_6 permselectivity: ~1,000) and low H_2 to N_2 permselectivity (~10). Penetration of coating polymer could be prevented by coating with polymer of large molecular weight on SiO_2-ZrO_2 intermediate layer before coating with polymer of small molecular weight, resulting in the improvement of hydrogen permeability for hybrid silica membrane. There was no significant gas permeance changes for TMCTS derived hybrid silica membrane before and after kept in humid atmosphere, indicating the possibility for hydrothermal stability of amorphous silica networks due to the presence of CH_3 groups in amorphous silica networks.

REFERENCES

[1]N.W. Ockwig, and T.M. Nenoff, Membranes for Hydrogen Separation, *Chem. Rev.*, **107**, 4078-110 (2007).

[2]T. Tsuru, Nano/Subnano-Tuning of Porous Ceramic Membranes for Molecular Separation, *J. Sol-Gel Sci. Technol.*, **46**, 349-61 (2008).

[3]G.R. Gavalas, C.E. Megris, and S.W. Nam, Deposition of H_2-Permselective SiO_2 Films, *Chem. Eng. Sci.*, **44**, 1829-35(1989).

[4]S. Yan, H. Maeda, K. Kusakabe, and S. Morooka, Hydrogen-Permselective SiO_2 Membrane Formed in Pores of Alumina Support Tube by Chemical Vapor Deposition with Tetraethylorthosilicate, *Ind. Eng. Chem. Res.*, **33**, 2096-101 (1994).

[5]S. Nakao, T. Suzuki, T. Sugawara, T. Tsuru, and S. Kimura, Preparation of Microporous Membranes by TEOS/O_3 CVD in the Opposing Reactants Geometry, *Micropor. Mesopor. Mater*, **37**, 145-52 (2000).

[6]D. Lee, and S.T. Oyama, Gas Permeation Characteristics of a Hydrogen Selective Supported Silica Membrane, *J. Memb. Sci.*, **210**, 291-306 (2002).

[7]R.J.R. Uhlhorn, M.H.B.J. Huisintveld, K. Keizer, and A.J. Burggraaf, High Permselectivities of Microporous Silica-Modified γ-Alumina Membranes, *J. Mater. Sci. Lett.*, **8**, 1135-38 (1989).

[8]S. Kitao, H. Kameda, and M. Asaeda, Gas Separation by Thin Porous Silica Membrane of Ultra Fine Pores at High Temperature, *Membrane(Maku)*, **15**, 222-27 (1990).

[9]M. Asaeda, and S. Yamasaki, Separation of Inorganic/Organic Gas Mixtures by Porous Silica Membranes, *Sep. Purif. Technol.*, **25**, 151-59 (2001).

[10]C.J. Brinker, R. Sehgal, S.L. Hietala, R.D. Deshpande, D.M. Smith, D. Loy, and C.S. Ashley, Sol-Gel Strategies for Controlled Porosity Inorganic Materials, *J. Membr. Sci.*, **94**, 85-102 (1994).

[11]N.K. Raman, and C.J. Brinker, Organic "Template" Approach to Molecular Sieving Silica Membranes, *J. Membr. Sci.*, **105**, 273-79 (1995).

[12]G. Cao, Y. Lu, L. Delattre, C.J. Brinker, and G.P. Lopez, Amorphous Silica Molecular Sieving Membranes by Sol-Gel Processing, *Adv. Mater.*, **8**, 588-91 (1996).

[13]K. Kusakabe, S. Sakamoto, T. Saie, and S. Morooka, Pore Structure of Silica Membranes Formed by a Sol-Gel Technique Using Tetraethoxysilane and Alkyltriethoxysilanes, *Sep. Purif. Technol.*, **16**, 139-46 (1999).

[14]G. Xomeritakis, S. Naik, C.M. Braunbarth, C.J. Cornelius, R. Pardey, and C.J. Brinker, Organic-Templated Silica Membranes I. Gas and Vapor Transport Properties, *J. Membr. Sci.*, **215**, 225-33 (2003).

[15]H.L. Castricum, A. Sah, R. Kreiter, D.H.A. Blank, J.F. Vente, and J.E. ten Elshof, Hybrid Ceramic Nanosieves: Stabilizing Nanopores with Organic Links, *Chem. Commun.*, 1103-05 (2008).

[16]H.L. Castricum,; A. Sah,; R. Kreiter,; D.H.A. Blank, J.F. Vente, and J.E. ten Elshof, Hydrothermally Stable Molecular Separation Membranes from Organically Linked Silica, *J. Mater. Chem.*, **18**, 2150-58 (2008).

[17]M. Kanezashi, K. Yada, T. Yoshioka, and T. Tsuru, Design of Silica Networks for Development of Highly Permeable Hydrogen Separation Membranes with Hydrothermal Stability, *J. Am. Chem. Soc.* in press.

[18]M. Asaeda, M. Kanezashi, T. Yoshioka, and T. Tsuru, Gas Permeation Characteristics and Stability of Composite Silica-Metal Oxide Membranes, *Mater. Res. Soc. Symp. Series*, **752**, 213-18 (2003).

[19]M. Kanezashi, T. Fujita, and M. Asaeda, Nickel-Doped Silica Membranes for Separation of Helium from Organic Gas Mixtures, *Sep. Sci. Technol.*, **40**, 225-38 (2005).

[20]M. Kanezashi, and M. Asaeda, Stability of H_2-Permselective Ni-Doped Silica Membranes in Steam at High Temperature, *J. Chem.Eng. Jpn*, **38**, 908-12 (2005).

[21]M. Kanezashi, and M. Asaeda, Hydrogen Permeation Characteristics and Stability of Ni-Doped Silica Membranes in Steam at High Temperature, *J.Memb. Sci.*, **271**, 86-93 (2006).

[22]R. Igi, T. Yoshioka, Y.H. Ikuhara, Y. Iwamoto, and T. Tsuru, Characterization of Co-Doped Silica for Improved Hydrothermal Stability and Application to Hydrogen Separation Membranes at High Temperatures, *J. Am. Ceram. Soc.*, **91**, 2975-81 (2008).

[23]P. Hacarlioglu, D. Lee, G.V. Gibbs, and S.T. Oyama, Activation Energy for Permeation of He and H_2 Through Silica Membranes: An Ab Initio Calculation Study, *J.Memb. Sci.*, **313**, 277-83 (2008).

[24]M.C. Duke, J.C.D. da Costa, D.D. Do, P.G. Gray, and G.Q. Lu, Hydrothermally Robust Molecular Sieving Silica for Wet Gas Separation, *Adv. Funct. Mater.*, **16** 1215-20 (2006).

COMPUTER SIMULATION OF HYDROGEN CAPACITY OF NANOPOROUS CARBON

V. Kartuzov, Y. Gogotsi[*], A. Kryklia
Institute for Problems in Materials Science,
Kyiv, 03680, Krzhizhanivsky 3, Ukraine,
* Drexel Nanotechnology Institute
Philadelphia, PA 19104, 3141 Chestnut St, USA

ABSTRACT

Key idea of work is computer imitation of variable geometrical structure of porous materials. Porous space is set as a 3D orthogonal net of similar channels in cluster of nanoporous carbon, such as carbide-derived carbon (CDC) or templated carbon. Cross-sections of channels are: circle, square, hexagon, and octagon. Models of original clusters are forming by molecular dynamics method. Model is realized in program system, which allows calculating: pore-size distribution, specific surface area, hybridization, porosity, radial distribution function, structure parameters and hydrogen capacity. System allows investigating temperature influence on parameters of nanoporous cluster. Simulated results are compared with experimental parameters and hydrogen capacity that have been measured for carbide derived carbons with a comparable pore size and pore volume.

INTRODUCTION

Nanoporous carbon is now more attracting researchers due to its wide spectrum of appliances. In work[1] original imitation of process of self-organization of nanoporous carbon during its forming by excluding of silicon atoms from structure of silicon carbide was performed. Technological process of forming of nanoporous carbon, chosen as basic for computer modeling described in work[2] (fig. 1).

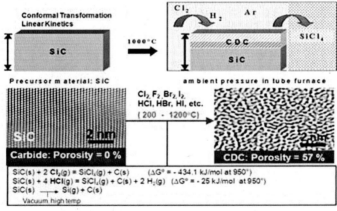

Figure 1. Technological process of forming of nanoporous carbon

Parameters of process: temperature: 400 – 1200 °C, normal pressure, linear dimensions of sample do not change.

Parameters of obtained sample: density 0,96 g/cm^3, size of pores from 0,5 to 1,5 nm, hydrogen capacity from 3 to 4 wt%, porosity 50 – 60% and 50 – 60 cm^3/g, narrow maximums of pore-size distributions, specific surface area ~ 2000 m^2/g (BET).

Nanoporous carbon is forming under chlorine flow and temperature 400 – 1200 °C and normal pressure above brick of solid SiC until full yield of $SiCl_4$ with excluding of chlorine residuals. Nanoporous carbon is forming on SiC substrate. Linear dimensions of sample do not change. For further research in work[3] computer imitation of porous structure as 3D orthogonal net of similar channels in cluster of nanoporous carbon was performed. This porous structure satisfies such experimental data: channel size corresponds to maximum of pore-size distribution; specific surface area corresponds to experimental one[4].

It was shown, that for such sizes of pores clusters of nanoporous carbon after molecular dynamics (MD) modeling with such pore geometry are stable. For decision of applied tasks, such as determination of maximum hydrogen capacity in sample and transport of molecules of given type in constructed structure, it was needed to continue research clusters with different pore sizes and forms. In this work objects of modeling are clusters with dimensions 6,54x4,36x4,01 nm (15x10x10 SiC unit cells) with number of carbon atoms 5400 and 6,54x6,54x5,67 nm (15x15x15 SiC unit cells) with number of carbon atoms 11700, pore diameters: 0,3, 0,6, 0,9 nm and distances between pore channels: 1,2, 1,5, 1,8 nm. Cross-sections of channels are: square, circle, hex- and octagon.

MODEL OF FORMING OF ORIGINAL SAMPLE

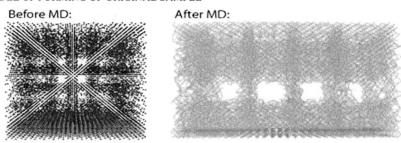

Figure 2. Different clusters before and after molecular dynamics simulation.

Structure of sample clusters of obtained nanoporous carbon was modeled by MD method, by XMD[5] program, with Tersoff potentials for carbon and silicon[6] (fig. 2). Original structure contains from SiC unit cells translated parallel to coordinate axes in resulted form of brick. Boundary conditions are set only to X and Y axes. Lower 10% atoms of sample are fixed during MD as substrate; atoms of silicon were not removed from one. All other silicon atoms were removed. In not-substrate space of sample 3D net of orthogonal pore channels was built. No atoms are located within pore channels. Carbon atoms form channels are uniformly relocated in other sample space. As result cluster of 12000 SiC atoms became cluster with 5400 C and 27000 SiC → 11700 C. Channel diameter, cross-sections, and distances between channels vary for every sample (fig. 3).

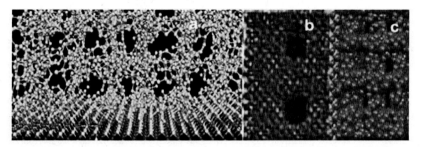

Figure 3. Model cluster with pores. Cross section: a–circle, b–octagon, c–rectangle.

PARAMETERS OF MODELING
- Pore size, specific surface area (SSA) calculated by Monte-Carlo method – 30 000 sampling points, van-der-vaals radius of carbon – 0,17 nm
- As there are no boundary conditions by Z axis, during MD simulation some atoms might get out of cluster's Z-border, their relative quantity – $\leq 0,87\%$. That's why we consider cluster as stable
- For each cluster 30 000 steps of MD simulation were taken (fig. 4) with purpose to minimize total system energy, for MD steps after 30 000: $\frac{\Delta E}{E} \leq 0,02\%$.

- After MD simulation substrate was cut and results are presented for pure nanoporous carbon

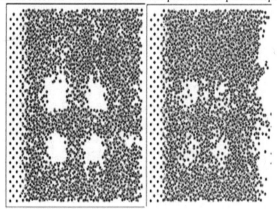

Figure 4. First and 30 000 steps of MD modeling.

RESULTS AND DISCUSSION

Data for different cross-sections very less than 2%, so results are presented for circle cross-section. It may be said that dependence for temperature of 300 K is given by construction, since at this temperature original structure of sample practically was not changed during MD modeling. (6, 12) means cluster with channel's diameter 0,6 nm and distances between pore channels 0,12 nm.

Data for cluster of 5400 carbon atoms (fig. 5 – 10).

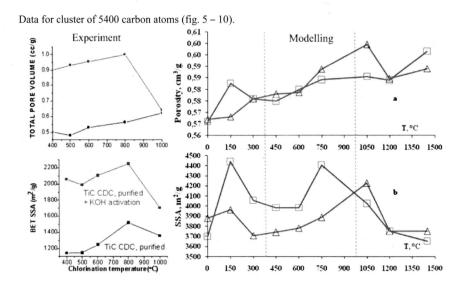

Figure 5. Dependence of porosity (a) and specific surface area (b) on temperature. Square marker – (6, 18), triangle – (6, 12).

On left side of (fig. 5) are shown dependences from work[8] for nanoporous carbon obtained by same technology from TiC. Dependences type in TiC and SiC cases are similar. Specific surface area (SSA) shown on ordinate axis is measured by BET method[9].

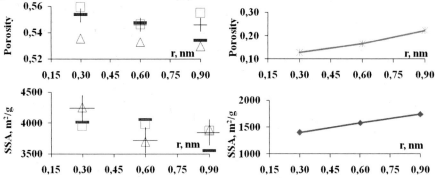

Figure 6. Dependences of porosity and SSA on pore size for distances between channels 1,8 nm. Triangle marker – 300, linear – 600, square – 900, plus – 1200 K. On the right – graphs of calculated data for channels.

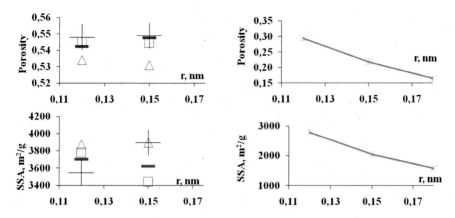

Figure 7. Dependence of porosity and SSA on distance between channels on pore size 0,6 nm. Triangle marker – 300, linear – 600, square – 900, plus – 1200 K. On the right – graphs of calculated data for channels.

Figure 8. Quota of neighbor's number in 0,16 nm sphere.

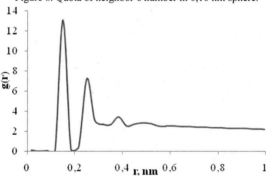

Figure 9. Radial distribution function g(r) for (6, 12).

From graph of g(r) it can be said that cluster structure is amorphous. For determine exact bond length, g(r) was built at segment 0 – 0,3 nm (fig. 10).

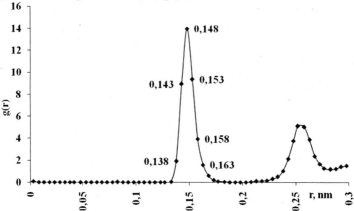

Figure 10. Radial distribution function g(r) at segment 0–0,3 nm

Table I. Carbon-carbon bond lengths.

Link type	sp^3–sp^3	sp^3–sp^2	sp^3–sp	sp^2–sp	sp^2–sp^2	sp–sp
Length, nm	0,154	0,150	0,146	0,143	0,135	0,120

It is arguable from comparing table I data and g(r) graph (fig. 10), that in sample there are all types of links, except sp-sp link, and also that numbers of 0,154 nm (diamond) and 0,143 nm (graphite) links are roughly equal. Also on sample there are links longer than 0,154 nm. What does it mean is not clear, as links of such lengths are registered only in organic compounds[10, 11]. With aim for results compare fig. 11 is given. First part (fig. 11a), experimental, describes graphitic amorphous carbon (a-C)[12]. Second (fig. 11b) describes model amorphous carbon[13].

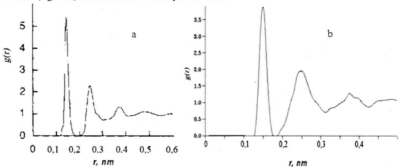

Figure 11. Functions g(r) for graphitic amorphous carbon (experimental) (a) and for model amorphous carbon (b).

Data for cluster of 11700 carbon atoms (fig. 12 – 15)

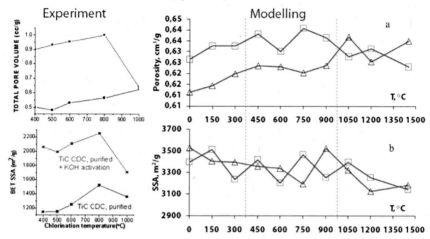

Figure 12. Dependence of porosity (a) and specific surface area (b) on temperature. Square marker – (6, 18), triangle – (6, 12).

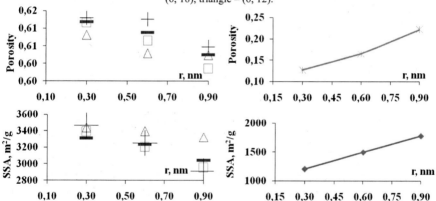

Figure 13. Dependences of porosity and SSA on pore size for distances between channels 1,8 nm. Triangle marker – 300, linear – 600, square – 900, plus – 1200 K. On the right – graphs of calculated data for channels.

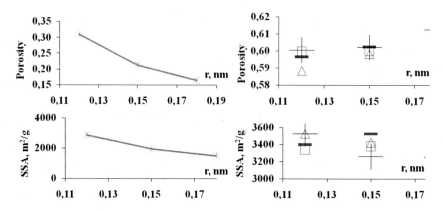

Figure 14. Dependence of porosity and SSA on distance between channels on pore size 0,6 nm. Triangle marker – 300, linear – 600, square – 900, plus – 1200 K. On the left – graphs of calculated data for channels.

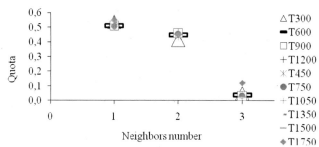

Figure 15. Quota of neighbor's number in 0,16 nm sphere.

Radial distribution function for cluster with 11700 carbon atoms is exactly equal to one for cluster with 5400 carbon atoms.

Table II. Statistics for clusters.

	Experiment	Cluster of 5400 carbon atoms	Cluster of 11700 Carbon atoms
Density, g/cm^3	0,96	0,94	1,04
Average by clusters minimal distances, nm for 600 and 900 K for 300 and 1200 K		0,1492 and 0,1491 0,1500 and 0,1503	practically identical

Fractality data (fig. 16)

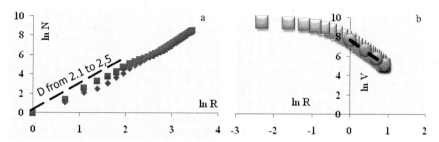

Figure 16. Dependence of number of atoms (N) on measure radius (R), blue marker – cluster with 5400, red – with 11700 carbon atoms (a); dependence of volume (V) on sample sphere radius (R), for cluster with 5400 carbon atoms (b).

Hydrogen capacity (fig. 17, table III)

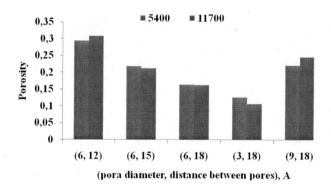

Figure 17. Dependence of porosity on parameters of geometry of porous space.

Table III. Hydrogen capacity per finding method for (6, 12)

Number of carbon atoms	Measure	Hydrogen capacity finding method		
		Volume: Monte-Carlo / theoretical for channel space	Sampling balls radius 0,4 A	Sampling balls aside pore walls
5400	%	54 / 29	2,3	2,3
	wt%	>100	15,6	15,6
11700	%	60 / 31	1,3	1,3
	wt%	>100	8,5	8,2

CONCLUSIONS

Chosen model rather adequate reflects experimental results on values of porosity, density and SSA. Dependences of porosity and SSA on temperature are similar by nature with experimental.

For 600 and 900 K average minimal distances – 0,1492 and 0,1491 nm respectively – are characteristic for nanoporous carbon, since these temperatures are conform to experimental temperatures of formation. For 300 and 1200 K values of average minimal distances are greater and closer one to another – 0,1500 and 0,1503 nm respectively. Since at 300 K equilibrium state after MD simulation practically identical to original state – temperature is low for structure alteration. And temperature of 1200 K is rather high for increasing average distances during MD modeling. This fact might influence on character of SSA dependences on pore size: pairs for 600 and 900 K and for 300 and 1200K are of similar type, this question demand on further research.

Constructed clusters of nanoporous carbon with given pore geometry have radial distribution functions identical to experimental amorphous carbon and modeled one given in literature.

Fractal structure is not found. Hydrogen capacity of cluster is determined.

REFERENCES

1. V. Kartuzov, O. Kryklia, Program System for Modeling of Process of Obtaining Nanoporous Carbon // *Math. model. and comput. exper. in materials science*, **8**, 69—77 (2006).
2. G. Yushin, Y. Gogotsi, A. Nikitin, Carbide derived carbon // *Nano-materials Handbook*, 237-280 (2006).
3. V. Kartuzov, O. Kryklia, Computer Simulation of Pore Structure of Nanoporous Carbon // *Math. model. and comput. exper. in materials science*, **9**, 69—77 (2007).
4. R. Dash, J. Chimola, G. Yushin et. al., Titanium Carbide-Derived Nanoporous Carbon for Energy-Related Applications // *Carbon*, **44**, 2489—2497 (2006).
5. http://xmd.sf.net
6. J. Tersoff, Modeling Solid-state Chemistry: Interatomic Potentials for Multicomponent Systems // *Phys. Rev. B.*, **39 (8)**, 5566-5568 (1989).
7. N. Metropolis, S. Ulam, The Monte Carlo Method // *J. of the Americ. Statis. Association.*, **44**, 335 (1949).
8. J. Fischer, Y. Gogotsi, T. Yildirim, Carbide-Derived Carbons with Tunable Porosity Optimized for Hydrogen Storage // *DOE Hydrogen program*, (2008).
9. S. Brunauer, P. Emmett, E. Teller, Adsorption of Gases in Multimolecular Layers // *J. Amer. Chem. Soc.*, **60**, 309 (1938).
10. Naphthocyclobutenes and Benzodicyclobutadienes: Synthesis in the Solid State and Anomalies in the Bond Lengths // *Fumio Toda European J. of Organic Chemistry*, **8**, 1377-1386 (2000).
11. http://en.wikipedia.org/wiki/Bond_length#cite_note-1
12. C. Wang, K. Ho, C. Chan, Tight-binding molecular-dynamics study of amorphous carbon // *Phys. Rev. Lett.*, **70**, 611 (1993).
13. C. Stewart, http://cmt.dur.ac.uk/sjc/thesis/thesis/node75.html — 1996

NANOSTRUCTURED ALUMINA COATINGS FORMED BY A DISSOLUTION/PRECIPITATION PROCESS USING AlN POWDER HYDROLYSIS

Andraz Kocjan*[1], Kristoffer Krnel[1], Peter Jevnikar[2], Tomaz Kosmac[1]

[1]Engineering Ceramics Department, Jozef Stefan Institute, Ljubljana, Slovenia
[2]Faculty of Medicine, University of Ljubljana, Ljubljana, Slovenia

ABSTRACT

The hydrolysis of aluminum nitride (AlN) powder was exploited in the preparation of nanostructured coatings on a sintered yttria-stabilized tetragonal zirconia (Y-TZP) ceramic substrate, with the aim to enhance the adhesion of dental cements to the ceramic core structure. The nanocrystalline coating formed using this method consists of γ-AlOOH (boehmite) in the form of 6-nm-thick and 240-nm-long interconnected lamellas. During a subsequent heat treatment in the temperature range from 600 °C to 1200 °C this coating was transformed to various transient aluminas without any noticeable change in the morphology, but its bonding to the substrate was significantly improved. By varying the precipitating conditions, such as the temperature and the time, it was also possible to synthesize coatings differing in composition and morphology, resulting in nano- or micron-sized particles. This nanosized alumina coating has the potential to improve the adhesion of composite dental cements to the sintered Y-TZP ceramic.

INTRODUCTION

The reaction of AlN powder with water has been known for a long time. In the presence of water, AlN will decompose, forming aluminum hydroxide and ammonia. The hydrolysis of AlN powder was mainly studied in order to prevent the reaction, since AlN ceramics are used as a substrate material for power circuits and as a packaging material for integrated circuits.[1] Only a few authors have investigated the mechanisms of aluminum hydroxides formation during the reaction of AlN powder with water.[2-5] Bowen et al.[2] investigated the hydrolysis at room temperature and proposed the following reaction scheme:

$$AlN + 2H_2O \rightarrow AlOOH_{amorph} + NH_3 \tag{1}$$
$$NH_3 + H_2O \rightarrow NH_4^+OH^- \tag{2}$$
$$AlOOH_{amorph} + H_2O \rightarrow Al(OH)_3 \tag{3}$$

The AlN powder first reacts with water to form amorphous aluminum oxyhydroxide (pseudoboehmite phase, γ-AlOOH), which later re-crystallizes as aluminum 3-hydroxide (bayerite, α-Al(OH)$_3$) by the dissolution/precipitation process, according to reaction (3).[6,7] The kinetics of AlN hydrolysis was described using an unreacted-core model, proposed by Levenspiel,[8] and the chemical reaction at the product-layer/unreacted-core interface was proposed to be the rate-controlling step during the initial stage of the reaction.[2,5]

At elevated temperatures, the starting temperature and especially the ageing time in the mother liquor, i.e., the time of the hydrolysis, strongly influence the reaction products and their morphology.[3,4,5,9]

Svedberg et al.[3] studied the corrosion of AlN in aqueous solutions at various constant pH values (5, 8, 11, 14) heated to 85 °C for 1 h. In all pH regimes the pseudoboehmite and bayerite/gibbsite phases, in various proportions, were detected with XRD.

Fukumoto et al.[4] investigated the hydrolysis behavior of AlN powder in water at room temperature and elevated temperatures up to 100 °C. According to these authors the hydrolysis behavior changes at 78 °C: below this temperature, crystalline bayerite is the predominant phase, while above 78 °C, crystalline boehmite is formed.

In our recent work we also found that, at temperatures higher than room temperature, the first crystalline product is boehmite, and with a prolonged ageing time the bayerite conversion takes place by the dissolution of pseudoboehmite and the recrystallization of bayerite.[9]

More recently, the AlN powder hydrolysis at elevated temperatures has been exploited in the Hydrolysis Assisted Solidification (HAS) forming process, in which the hydrolysis reactions provoke the solidification of the aqueous ceramic suspension in an impermeable mould. In this process a few percent of AlN powder is added to the aqueous slurry of the host ceramic powder. After homogenization, the suspension is cast into a closed, nonporous mould, where the hydrolysis is thermally activated. Due to the water consumption, the formation of aluminum hydroxides on the surface of the parent ceramic particles and the change in the ionic strength of the suspension during the hydrolysis of AlN powder, the viscosity of the host slurry is increased to such an extent that a saturated solid body is formed.[10-12] The precipitation of the aluminum hydroxides on the surface of the host ceramic particles in the ceramic suspension during the solidification in the HAS process also indicated that the formation of nanostructured aluminum hydroxide coatings on hydrophilic surfaces with the use of AlN powder hydrolysis is a feasible process.

In this work the preparation of nanostructured alumina coatings by exploiting the AlN powder hydrolysis is presented. The influence of precipitation conditions, such as the time and temperature, on the morphology of the precipitated boehmite coatings was studied and the crystal structure of the coatings before and after the heat treatment was analyzed. In addition, the potential of this coating to improve the adhesion of dental cements to the sintered Y-TZP ceramic, depending on the surface treatment, was investigated. This is because Y-TZP is widely used as a core material in restorative dentistry, due to its high strength, biocompatibility and favorable optical properties. However, the adhesion of the dental cement to the Y-TZP surface is poor.[13] For this reason sandblasting is commercially used to increase the surface roughness of the ceramics and thereafter improve the adhesion of the dental cement.[14,15] As an alternative, chemical modification of the zirconia surface, i.e., silanization and others, is being investigated to improve the binding between the dental cement and the zirconia surface.[16,17]

EXPERIMENTAL

The AlN powder used in the experimental work was AlN Grade C powder (H.C. Starck, Berlin, Germany) with a nominal particle size of 1.2 μm, an oxygen content of 2.2 wt.%, and a specific surface area of 3.2 m^2/g. The hydrolysis tests were carried out in dilute suspensions containing 3 wt% of AlN in deionized water. In these tests the water was preheated with an electric heater under constant stirring to the desired temperature (40 °C, 50 °C, 60 °C, 70 °C, 80 °C and 90 °C) and then the AlN powder was added to the water. The pH and the temperature were measured versus time using a combined glass-electrode/Pt 1000 pH meter (Metrohm 827) equipped with an accurate thermoelement.

The crystalline phases of the hydrolysis products were characterized using X-ray diffraction (XRD), scanning electron microscopy (SEM) and transmission electron microscopy (TEM).

The zirconia substrates were produced from commercially available ready-to-press TZ-3YSB-E (Tosoh Corp, Tokyo, Japan) powder containing 3 mol.% yttria and 0.25 wt.% alumina in the solid solution to hinder the ageing process. The discs were fabricated by uni-axial dry pressing, and sintering at 1500 °C for 2 hours in air. One side of each disc was ground and polished with 6-μm and 3-μm diamond pastes using a standard metallographic procedure.

In the preparation of aluminate coatings, the zirconia disc was inserted into preheated deionized water and the AlN powder was added under constant stirring and heating. The immersion time ranged from a few minutes to several hours. Afterwards, the substrate was removed from the slurry, dried in air at 100 °C for 30 minutes and subjected to a heat treatment in dry air for 1 hour at 900 °C, 1200 °C and 1300 °C. The heating rate was 10 °C/min.

For the characterization of the coated Y-TZP substrates SEM and TEM analyses were employed.

The adhesion of the coating to the Y-TZP substrate before and after the heat treatment was evaluated using a simple "Scotch tape" test. In addition, the shear bond strength of a commonly used composite dental cement to uncoated and coated Y-TZP substrates, differing in the surface pre-treatment, was determined according to the ISO 6782 (2008) standard.

RESULTS and DISCUSSION

The hydrolysis of AlN powder

The variation of pH versus time of the AlN-powder water suspension for various starting temperatures is presented in Figure 1. The starting pH of the deionized water ranged from 4 to 7, because of the adsorption of CO_2 from the air. When the AlN powder was dispersed in water preheated to 40 °C, 50 °C and 70 °C, the starting pH value initially decreased even more (revealed by the inset graph in Figure 1.), presumably due to the uptake of additional CO_2 during the vigorous stirring needed to homogenize the suspension. The higher starting pH values for the higher starting temperatures correspond to the solubility of the CO_2 in water, i.e., the solubility of the CO_2 in the water is decreasing with the increasing water temperature. After the addition of the AlN powder to the preheated water the pH started to increase and reached an equilibrium pH value within a few minutes to several hours, depending on the starting temperature. A higher starting temperature evidently speeds up the reaction of the AlN powder with water, also lowering the equilibrium pH value, presumably due to the decreasing solubility of ammonia with the increasing temperature.[9]

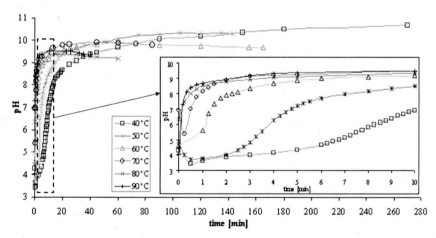

Figure 1: pH versus time for the hydrolysis of a 3 wt.% AlN suspension in deionized water at various starting temperatures.

Since the reaction of AlN powder with water is an exothermic process,[18] we also monitored the temperature increase versus time for the AlN powder suspension for various starting temperatures. The results are presented in Figure 2. For the starting hydrolysis temperatures of 50 °C, 60 °C and 70 °C, the temperature increase (ΔT), i.e., the difference between the highest measured temperature and the starting temperature, is about 18 °C. The ΔT value for the AlN powder hydrolyzed at 80 °C is 15 °C, and at 90 °C it is only 8 °C, because the temperature reached the boiling point of the suspension. The time needed to reach the maximum ΔT value (ΔT_{MAX}; the black lines perpendicular to the ΔT curve) for the starting temperatures of 50 °C, 60 °C and 70 °C was 49 min, 27 min and 13 min, respectively. For the 80 °C and 90 °C curves, the ΔT_{MAX} (the dashed black lines perpendicular to the ΔT curve) values were estimated to be 9 min and 7 min, respectively. After reaching ΔT_{MAX}, the

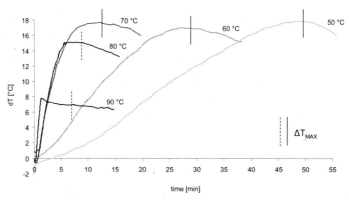

Figure 2: dT versus time for the hydrolysis of a 3 wt% AlN powder suspension in deionized water at 50 °C, 60 °C, 70 °C, 80 °C and 90 °C.

slurries were filtered, washed with isopropanol and dried for subsequent XRD analyses. In all cases only boehmite and unreacted AlN were indentified from the corresponding XRD pattern. With even longer ageing times, peaks of bayerite were also detected in the XRD pattern, in agreement with previously reported studies.[3,4,5,9] A typical SEM micrograph of AlN powder hydrolyzed at 50 °C for 2 hours is presented in Figure 3, showing small, agglomerated, nanostructured boehmite (AlOOH) lamellas and large, elongated bayerite (Al(OH)$_3$) crystals. Bayerite particles usually occur as somatoids, which are defined as bodies of uniform shape that are not enclosed by crystal faces. These shapes resemble hour glasses, cones, or spindles.[6] On the other hand, the crystallites of lamellar boehmite have at least two growth dimensions, ranging from 5 nm to 50 nm. The thickness of the crystallites constitutes the third dimension, generally ranging from about 2 nm to 10 nm.[19]

From the above results it can be assumed that at ΔT_{MAX} the extensive boehmite formation is terminated, and with prolonged hydrolysis times bayerite will be formed at a slower reaction rate. The vigorous reaction during the initial stages of hydrolysis, i.e., prior to the ΔT_{MAX} being reached, should be accompanied by a rapid increase in the concentration of dissolved aluminum (poly)cations in the suspension, which favor the nucleation of boehmite.[20,21] With an appropriate amount of AlN powder in the suspension it is likely that the degree of supersaturation will be sufficiently high to provoke a heterogeneous nucleation on the surface of an immersed hydrophilic substrate. This assumption was experimentally confirmed by monitoring the boehmite formation on a polished Y-TZP substrate immersed in the 3 wt.% AlN powder suspension at 70 °C for various periods of time.

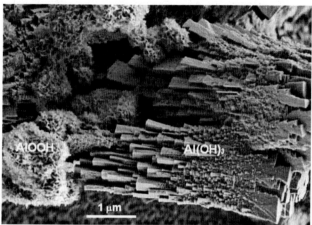

Figure 3: SEM micrograph of AlN powder after hydrolysis for 2 hours at 50 °C, showing large, elongated bayerite (Al(OH)$_3$) crystals and small, agglomerated, nanostructured boehmite (AlOOH) lamellas.

Coating morphologies

The SEM micrographs in Figure 4 reveal that the nucleation of the boehmite lamellas after soaking the substrate for 1 minute (Figure 4a) had already begun. After 3 minutes, the lamellas became interlocked (Figure 4b), and eventually, after 15 min of immersion, the

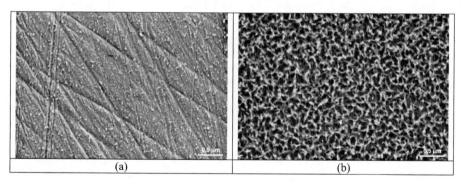

| (a) | (b) |

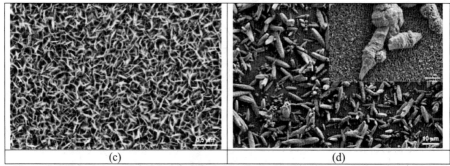

(c)	(d)

Figure 4: SEM micrographs of the precipitated boehmite and bayerite coatings on the polished Y-TZP surface using AlN powder hydrolysis at 70 °C for (a) 1 min, (b) 3 min, (c) 15 min and (d) 4 h.

perpendicular growth of the nanostructured boehmite lamellas was complete (Figure 4c), as predicted from the ΔT versus time diagram (Figure 3). With longer reaction times, large, elongated, micron-sized bayerite particles started to grow on the coated Y-TZP substrate, whereas the boehmite coating remained unchanged (Figure 4d).

The boehmite coatings prepared at other starting temperatures and the corresponding ΔT_{MAX} times, as estimated from Figure 3, are shown in Figure 5. The micrographs indicate that the thickness of the lamellas, which was relatively insensitive to the starting temperature, was around 6 nm, which was later confirmed by the TEM analysis. The thickness of the boehmite coatings, as estimated from the SEM micrographs of tilted cross-sections, was about 240 nm, and was also not influenced by the starting temperature. In contrast, the surface density of the lamellas, i.e., the number of lamellas per unit area, was found to be temperature dependent and decreased with higher hydrolysis temperatures. At the same time, their width and distinctness were increased with higher temperatures.

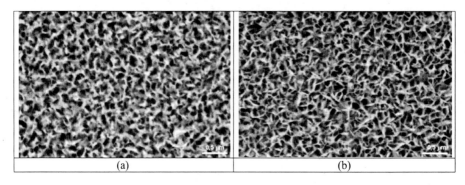

(a)	(b)

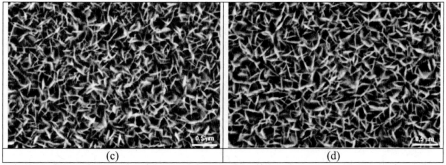

| (c) | (d) |

Figure 5: SEM micrographs of the precipitated boehmite coatings on the Y-TZP surface using AlN powder hydrolysis at (a) 50 °C for 50 min, (b) 60 °C for 30 min, (c) 80 °C for 10 min and (d) 90 °C for 7 min.

The effects of heat treatment

    After drying, the adhesion of the boehmite coating to the Y-TZP surface was poor and it could be easily peeled off during the "Scotch tape" test. In order to improve the adhesion, a heat treatment of the coated substrates was performed, during which the topotactic transformation of the boehmite to various transient aluminas occurred. SEM micrographs and the corresponding results of the TEM analyses of the boehmite precipitates followed by a heat treatment are shown in Figure 6. The morphology of the coating remained basically unchanged up to 900 °C (Figure 6b), whereas at higher temperatures the lamellas started to coarsen and sinter (Figure 6c), until at 1300 °C (Figure 6d) they practically disappeared, forming a thin discontinuous film. The structure of the as-precipitated polycrystalline lamellas corresponds to boehmite (orthorhombic, γ-AlOOH, Figure 6a). The corresponding TEM analysis of the boehmite precipitates after heat treatment revealed that at 900 °C delta alumina (tetragonal, δ-Al$_2$O$_3$) is formed, at 1000 °C to 1200 °C it transforms to theta alumina (monoclinic, θ-Al$_2$O$_3$), whereas at 1300 °C alpha alumina thin film (rhomboedric, α-Al$_2$O$_3$) is formed.

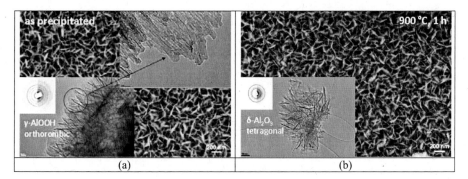

| (a) | (b) |

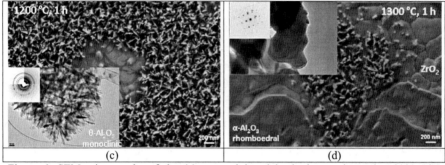

Figure 6: SEM micrographs of the (a) as-precipitated boehmite coating using 3 % AlN powder hydrolysis at 90 °C for 7 min, (b) coating after heat treatment at 900 °C for 1 h, (c) coating after heat treatment at 1200 °C for 1 h and (d) coating after heat treatment at 1300 °C for 1 h on the Y-TZP surface. The inset pictures represent the corresponding TEM and SAED analysis of the precipitates.

Bond strengths with dental cement

In order to explore the potential of this coating for establishing a stronger bond between the ceramic core material and the luting agent in restorative dentistry, the adhesion of dental cements to the Y-TZP ceramics was studied using the shear-bond test, following the ISO 6872 (2008) standard. The adhesion of a modern composite cement with an active phosphate monomer (MDP) on the differently prepared surfaces, i.e., as-sintered, polished and sandblasted, of Y-TZP ceramic was investigated and the results obtained with commercial MDP cement (Rely X Unicem, 3M ESPE, USA) are presented. As shown in Fig. 7, this coating indeed has the potential to substantially enhance the adhesive bond by a factor of 2–4, depending on the mechanical pre-treatment of the substrate.

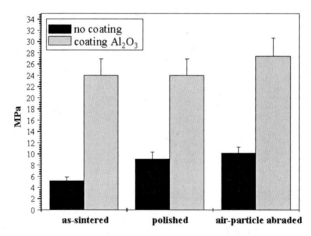

Figure 7: Shear bond strength of the composite dental cement Rely X Unicem to the surface of the Y-TZP.

CONCLUSIONS

The hydrolysis of AlN powder dispersed in deionized water at elevated temperatures is accompanied by a rapid increase in pH and temperature. A vigorous reaction during the initial stages of hydrolysis, i.e., prior the ΔT_{MAX} value, provokes the heterogeneous nucleation of boehmite onto the surface of the immersed Y-TZP substrate. This can be exploited for the formation of nanostructured boehmite coatings. Irrespective of the starting temperature, the coating consisted of 6-nm-thick and 240-nm-long interconnected polycrystalline γ-AlOOH lamellas. The surface density and the width of these lamellas depend on the initial temperature of the aqueous AlN slurry. During a subsequent heat treatment up to 900 °C, these coatings are transformed to a transient alumina without any noticeable change in the morphology, but their bonding to the substrate after the heat treatment is improved. This nanosized alumina coating has the potential to significantly improve the adhesion of composite dental cements to the sintered Y-TZP ceramic.

REFERENCES

[1]L. M. Sheppard, Aluminum nitride: a versatile but challenging material. *Am. Ceram. Soc. Bull.*, **69**, 1801–1812 (1990).
[2]P. Bowen, J.G. Highfield, A. Mocellin, and T.A. Ring, Degradation of Aluminum Nitride Powder in an Aqueous Environment, *J. Am. Ceram. Soc.*, **73** [3], 724-728 (1990).
[3]L. M. Svedberg, K. C. Arndt, and M. J. Cima, Corrosion of aluminum nitride (AlN) in aqueous cleaning solutions, *J. Am. Ceram. Soc.*, **83**, 41–46 (2000).
[4]S. Fukumoto, T. Hookabe, and H. Tsubakino, Hydrolysis behavior of aluminum nitride in various solutions, *J. Mater. Sci.*, **35**, 2743–2748 (2000).
[5]K. Krnel, G. Drazic, and T. Kosmac, Degradation of AlN powder in aqueous environments, *J. Mater. Res.*, **19**, 1157–1163 (2004).
[6]K. Wefers, and C. Misra, *Oxides and Hydroxides of Aluminum*, Technical Paper No. 19 (revised 1987) available from Alcoa, Pittsburg, PA.
[7]T. Graziani, and A. Belosi, Degradation of Dense AlN Materials in Aqueous Environments, *Mater. Chem. Phys.,* **35**, 43-48 (1993).
[8]O. Levenspiel, *Chemical Reaction*, Engineering (2nd ed.). John Wiley & Sons, New York, pp. 357–408 (1972).
[9]A. Kocjan, K. Krnel, and T. Kosmac, The influence of temperature and time on the AlN powder hydrolysis reaction products, *J. Eur. Ceram. Soc.,* **28**, 1003-1008 (2008).
[10]T. Kosmac, S. Novak, and M. Sajko, Hydrolysis-assisted solidification (HAS): a new setting concept for ceramic net-shaping, *J. Eur. Ceram. Soc.*, **17**, 427–432 (1997).
[11]T. Kosmac, S. Novak, and K. Krnel, Hydrolysis assisted solidification process and its use in ceramic wet forming, *Z. Metallkd.*, **92**, 150–157 (2001).
[12]T. Kosmac, The densification and microstructure of Y-TZP ceramics formed using the hydrolysis-assisted solidification process, *J. Am. Ceram. Soc.*, **88**, 1444–1447 (2005).
[13]C. Piconi, and G. Maccauro, Zirconia as a ceramic biomaterial, *Biomaterials*, **20**, 1-25 (1999).
[14]M. Wolfart, F. Lehman, S. Wolfart, and M. Kern Durability of the resin bond strength to zirconia ceramic after using different surface conditioning methods. *Dent Mater* **23**, 45-50 (2007).
[15]M. Uo, G. Sjogren, A. Sundh, M. Goto, ynd F. Watari, Effect of surface condition of dental zirconia ceramic Denzir on bonding. *Dent Mater J*, **25**, 1-6 (2006);
[16]M. Blatz, G. Chiche, S. Holst, and A. Sadan, Influence of surface treatment and simulated aging on bond strengths of luting agents to zirconia. *Quint Int*, **38**, 745-53 (2007).
[17]J. P. Matinlinna, T. Heikkinen, M. Ozcan, L. V. J. Lassila, and P. K. Vallittu, Evaluation of resin adhesion to zirconia ceramic using some organosilanes, *Dental Mat*, **22**, 824-831 (2006)

[18]W.M. Mobley, *Colloidal Properties, Processing and Characterization of Aluminum Nitride Suspensions*, Ph.D. Thesis, Alfred University, Alfred, New York (1996), p. 110.
[19]P. Euzen, P. Raybaud, X. Krokidis, H. Toulhoat, J. L. Le Loarer, J. P. Jolivet, Froidefond, "Alumina"; pp. 1591–677 in Hand Book of Porous Solids, Edited by F. Schuth, K. S. W. Sing, and J. Weitkamp. Wiley, Chichester, 2002.
[20]J. P. Jolivet, *Metal Oxide Chemistry and Synthesis—From Solution to Solid State*, Wiley, Chichester, (2000).
[21]M. Henry, J. P. Jolivet, and J. Livage, *Aqueous Chemistry of Metal Cations: Hydrolysis, Condensation, and Complexation*, Struct. Bonding, **77**, 153–206 (1992).

POROUS FeCr-ZrO$_2$(7Y$_2$O$_3$) CERMETS PRODUCED BY EBPVD

B.A. Movchan, F.D. Lemkey and L.M. Nerodenko
International Center for Electron Beam Technology
E.O. Paton Electric Welding Institute, Kiev, 03150, Ukraine

ABSTRACT

Porous cermets were produced by simultaneous independent evaporation of Fe,20 wt.% Cr and partially stabilized ZrO$_2$ + 7Y$_2$O$_3$ ingots in vacuum and subsequent vapor phase condensation at rates of 18-20 µm/min on steel with a substrate temperature interval of 350-1050°C. During deposition the surface of the condensate was bombarded by argon ions. The thickness of the condensate (sheet) after separation from the substrate was 0.3-0.5 mm. Fe,20 wt.% Cr condensates containing 5-38 wt.% ZrO2 (7Y$_2$O$_3$) were examined. It was established that argon ion treatment at substrate deposition temperature intervals of 780-1050°C produced a porous microstructure. Maximum values of total porosity in condensates containing 15-25 wt.% ZrO2 (7Y$_2$O$_3$) were found to equal 20-28%. The average pore size was 0.5-0.8 µm and annealing at 1000°C for 2 hours increased both the average size and total porosity. The estimated values of mechanical properties are presented and possible applications are considered.

The possibility of producing condensates with a controllable porosity by evaporation and subsequent condensation of the vapor phase has been shown for many materials and described in reviews, for instance [1, 2]. Results of these investigations reveal that one of the variants of porous structure formation is addition of the second phase of refractory compounds into an inorganic matrix.

Physical essence of the mechanism of pore formation at deposition of two-phase condensates consists is the so-called "shadowing" effect. Faster growing particles of the "second phase" form topological relief on the condensation surface, the individual sections of which cut off the incoming rectilinear vapor flow from the evaporator. In this way, shadow areas are created, which are transformed into pores.

The purpose of this investigation was to study the features of formation of the porous structure of two-phase condensates with a metal matrix of Fe-20%Cr(wt%) and refractory second phase of zirconium dioxide,ZrO2, partially stabilized by yttrium oxide (7wt.%Y2O3). Selection of material for investigations was determined by the possibility of using ferrite Fe,Cr alloys with a porous structure for various applications including solid oxide fuel cell electrical interconnect support structures [3].

The initial material for producing the condensates were Fe,Cr alloy ingots of 50 mm diameter and compacted sintered cylinders of yitria and zirconium dioxide powders. Electron beam evaporation of both materials was conducted in vacuum of 10^{-2} Pa from two separate electron beam sources with subsequent condensation of the mixed vapor flow on a low-carbon steel substrate with temperature gradient T$_s$ in the range of 350 -1050°C. Two variants of condensate structure were studied: without treatment and with treatment (bombardment) by Ar ions [4]. A magnetron type ion source located in the vacuum chamber was used at a voltage of 2.2 kV, current of 220 mA, and density of ion current on the substrate of ~ 1 mA/cm ^{2}.

For easy separation of the condensate from the substrate, a separation layer of

CaF2 of ~ 5 μm thickness was first deposited on the substrate surface. The porous condensate deposition rate was equal to ~ 18 – 20 μm/min. Thickness of the separated condensates was 0.3 – 0.5 mm.

Metallographic examination of the structure and analysis of the condensate composition were performed using scanning electron microscopy. The level of total porosity was determined by a numerical method of analysis of the condensate cross-sectional microstructure using a computer program Image Pro Plus. Calculations were performed based on identification of the studied phase boundary contrast. Sample's structures were studied after condensation and after annealing in vacuum at 1000°C for two hours.

Microhardness was determined in Polyvar-Met instrument at the load of 20 – 50 gm. Susceptibility to brittle fracture of the condensates was evaluated at testing by three point bending with a loading rate of 0.15 mm/min.

Fig. 1, a and b show structures of condensates deposited at temperature Ts = 530°C, after deposition and after annealing at 1000°C. As observed from the presented structures, porosity after deposition has a clear orientation with predominant localizing of pores along the boundaries of columnar crystallites. During annealing at 1000°C a certain spheroidization of pores takes place with increase of their dimensions and decrease of columnar orientation. Percentage of total porosity increases at the expense of coalescence and growth of fine (< 0.1-0.2 μm) pores: approximately 5-6% before annealing and 10-12% after annealing. These structures are characteristic for condensates produced within a temperature interval Ts = 500 – 700°C.

At increased Ts>700°C the pore size is observed to grow as a result of greater oxide particle size and a more uniform pore distribution in the crystallite volume. In this temperature range 1000° C annealing of condensates with 27 – 38 wt.% ZrO2 is accompanied by a practically two times increase of total porosity: from 11-15% to 25% as shown in Fig. 2.

A dashed line illustrated in Fig. 3 depicts the dependencies of total porosity of Fe, Cr condensates containing 10 and 27wt.%ZrO2 (7Y2O3) after deposition in the temperature range of 530-1050°C. A smooth increase of porosity is observed with temperature rise. As was shown above, annealing of such condensates at 1000°C for two hours increases the total porosity of the condensates by approximately two times.

Experiments were also conducted to study the influence on porosity level from argon ion bombardment during deposition in the range of Ts = 530 – 1050°C. Ion sputtering, heating and coarsening of the condensation surface are capable of enhancing the "shadowing effect". The result are reflected in the solid curve in Fig. 3.

Dependence of porosity of the condensates subjected to ion treatment during deposition, on Ts is of a pronounced non-monotonic nature. At deposition temperatures below 700°C, the condensates produced without application of ion treatment, have higher values of porosity compared to those treated by Ar ions. In the range of Ts – 700 – 800°C an abrupt increase of porosity in condensates subjected to ion treatment is observed. Maximum levels of porosity (~28%) were obtained at Ts = 800°C.

It should be noted, in particular, that the above porosity level of 25%, as was earlier discussed, can be obtained by annealing samples at 1000°C for two hours, i.e. Ar ion treatment during condensation is approximately equivalent to high-temperature annealing.

Results of determination of microhardness, Hμ, given in Fig.4, reflect the combined contribution of the porous matrix and hard ceramic phase of ZrO2. At condensation temperature of 530 – 775°C the porosity of the condensates with 5-10%ZrO2 is minimum (~5%). In this temperature range ZrO2 particles have a predominant effect on Hμ level by the mechanism of dispersion strengthening of the matrix. Increase of deposition temperature is accompanied by weakening of the strengthening action of ZrO2 particles due to increase of particle size and simultaneous increase of porosity because of intensification of the shadowing effect.

Mechanical bend testing at 20°C revealed a brittle fracture susceptibility of the condensates. Bend angle to fracture below 20° was demonstrated by condensates with 5–10% ZrO2. The other condensates failed in the brittle mode.

CONCLUSIONS

1. Porosity of Fe,Cr-ZrO2 (7Y2O3) condensates of 0.3-0.5 mm thickness is controllable in the range of 5-28% using the following parameters:
- temperature of substrate surface on which deposition is performed,
- quantity and dimensions of the oxide phase,
- bombardment by Ar ions during deposition,
- high-temperature annealing at 1000°C after deposition
-

2. Condensates in the form of a sheet separated from the substrate or coatings on substrates can be accepted as structural elements for solid oxide fuel cells (SOFC),filters, catalyst carriers, etc.

REFERENCES

1. L.S. Palatnik, P.G. Cheremskij, M.Ya. Fux, Pores in films. Energoizdat, Moscow, 1982 (in Russian)
2. B.A. Movchan, F.D. Lemkey. Some approaches to producing microporous materials and coatings by EB PVD. Surface and Coatings Technology. 165 (2003). P.90-100.
3. I. Villaral, C. Jacobson, A. Leming, Y. Matus, S. Visco, L. De Jonghe. Metal-Supported Solid Oxide Fuel Cells /Electrical and Solid-State Letters. – 2003. - 6 (9). – P.178 -179.
4. Sputtering by Particle Bombardment III. Edited by Behrisch and K. Wittmaak. Springer-Verlag, Berlin, Heidelberg,1991, 550p.

a) as deposited

b) as annealed at 1000°C

Fig 1 Structure of Fe,Cr-ZrO$_2$ condensate with 15 wt% ZrO$_2$ at T$_s$=530°C

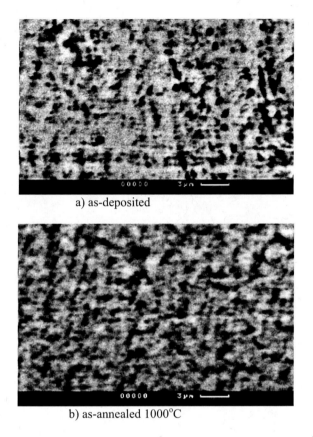

a) as-deposited

b) as-annealed 1000°C

Fig. 2 Structure of Fe,Cr, - ZrO$_2$ condensates with 27 wt% ZrO$_2$ at T$_S$ = 975°C

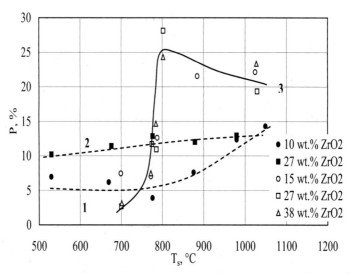

Fig. 3 Porosity of Fe,Cr-ZrO₂ condensates depending on substrate deposition temperature, ZrO₂ concentration and Ar ion bombardment. 1,2 – without Ar ion bombardment, 3 – with Ar ion bombardment.

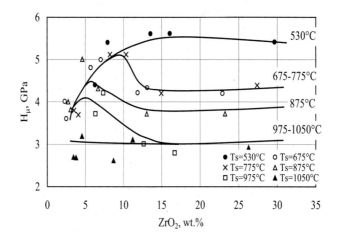

Fig. 4 Microstructure of Fe,Cr-ZrO₂ condensates depending on ZrO₂ content and deposition temperature, T_s

USE OF CERAMIC MICROFIBERS TO GENERATE A HIGH POROSITY CROSS-LINKED
MICROSTRUCTURE IN EXTRUDED HONEYCOMBS

James J. Liu, Rachel A. Dahl, Tim Gordon, Bilal Zuberi
GEO2 Technologies
12-R Cabot Road
Woburn, MA 01801.

ABSTRACT

Honeycomb ceramic bodies with a novel cross-linked microstructure were generated by the extrusion of ceramic fiber-based pastes. The high fiber content in the paste leads to differences in rheology as compared to traditional powder-based pastes. The cross-linked micro-structure forms in-situ during sintering and is influenced by the fiber-diameter, fiber length and the organic constituents. The sinterability is highly dependent on the chemistry, sintering conditions and chemical reactions during sintering. In addition to standard characterization techniques, 3-D X-ray nano-Computed Tomography was utilized to analyze the three-dimensional porous microstructure. The honeycombs exhibit a uniform pore-structure, and textured wall surface with a preferred alignment of the fibers along the principle extrusion axis. The microstructure leads to high strength at high porosity, improved permeability, and high nano-particle filtration efficiency for emission control applications. The microstructure has been produced for mullite, silicon carbide, cordierite and alumina compositions.

INTRODUCTION

Advanced porous ceramic materials are commonly utilized in auto emission control systems as substrates for catalytic converters and for diesel engines, as particulate/NOx filters.[1-5] Extruded porous honeycomb ceramic bodies are of particular use in diesel particulate filter (DPF) applications due to their ability to withstand high temperatures and for their resistance to chemical corrosion. Porous honeycomb ceramic DPF substrates provide high specific surface area for filtration and support for catalytic reactions and, at the same time, are structurally stable at the high operating temperatures associated with an automotive engine environment.

As the emission control regulations have tightened, there is a need for improved DPF materials in order to achieve high particulate matter (PM) filtering efficiency and meet diesel emission regulations as well as to realize low back pressure and reduce power loss. A high strength and high porosity (>60%) and large pore size is desirable for low back pressure which has not been possible to achieve with powder-based ceramic materials.[6,7] Advantages of these high porosity filters include better emission control performance, cost reduction due to a reduction in the amount of precious metal catalyst required, and integration of multiple catalyst functionalities onto a single substrate. The tradeoff for improved performance is a decrease in the filtering efficiency and mechanical robustness of the filter.

A successful material for DPF applications must have low backpressure, high filtration efficiency (especially for nano-particles), thermo-mechanical durability, and the ability to be manufactured at scale at low cost. Several processing improvements and new materials have been developed to satisfy the required characteristics. For example, both cordierite and silicon carbide DPFs have been produced with optimized pore size distributions as well as various honeycomb cell geometries, but none of these advanced materials have met all of the criteria listed above.[8-10] Although several new filter materials have been introduced such as mullite[9] and aluminum titanate[10] they have

267

not achieved high porosity or have been found to be compromised. Despite recent progress in developing higher porosity filter materials, substantial improvement is necessary in order to provide viable substrates for the complex control systems necessary to meet the current stringent emissions requirements.[11] In addition to new filter materials, optimizing the microstructure of existing materials is expected to result in substrates that satisfy all the previously-mentioned requirements. However, it is difficult to accomplish a microstructure as designed because of its complexities.

A vast family of random microstructures is possible, ranging from dispersions with varying degrees of clustering to complex interpenetrating connected multiphase media, including porous media. Disordered materials exhibit a broad spectrum of rich and complex microstructures. In this work we demonstrate a family of novel, microstructurally-ordered materials – the fibrous cross-linked microstructure (CLM) honeycomb ceramic materials. There are significant differences between fibrous CLM materials and fibrous materials, such as those with pleated structures, documented in previous literature[12].

Honeycomb filters produced with the fibrous CLM ceramic materials possess the advantages of a high porosity honeycomb macro-structure and the high performance mechanical properties of the fibrous microstructure.

The aim of this study is to characterize the fibrous CLM ceramic honeycomb materials and their emissions control performance advantages.

EXPERIMENTAL PROCEDURES

Honeycomb Fibrous CLM ceramic honeycombs were extruded from ceramic paste which was prepared by mixing raw powders with >50 wt% of fiber and the addition of binders, pore formers and deionized water. For CLM-mullite and CLM-cordierite, polycrystalline Al_2O_3-SiO_2 fiber was used, CLM SiC was formed through reaction of Si and graphite fiber. Starting fibers were chopped fiber with diameter greater than 3 micron and length range from 0.25 inch to 2 inch. These constituents were mixed in a high-shear sigma-blade mixer for ~60 minutes until a ceramic paste was formed during mixing fibers were chopped to more uniform length. The pastes were then de-aired, formed into a billet and extruded at about 10 m/min into a 36 mm square bar using a honeycomb die in a hydraulic ram extruder. Extruder parameters were adjusted to accommodate the varying rheological parameters of each set of pastes. The extruded parts were dried with a Radio-Frequency (RF) oven and sintered. After sintering, the approximate final cell density is 300 cpsi (cells per square inch) and 12 mil (0.012 inch) wall thickness.

Sample preparation

The extruded ceramic filters were cored into 21.6x 25.4 mm samples for porosity and crush strength measurements. Bars were also cut out from the sintered parts to measure the Young's modulus and modulus of rupture. The bars were 3 cells by 6 cells or roughly 5 mm by 10 mm and 90 mm in length.

Porosity and pore Size

Porosity of the substrate was measured using both standard water absorption and mercury intrusion porosimetry techniques (Micromeritics, AutoPore IV 9500 V1.07). The porosity and pore size distribution are plotted from the mercury porosimetry results.

Compressive strength

The axial compressive strength was measured from the cored samples on an Admet universal testing machine with a crosshead speed of 22.2N /second. Compressive strength was measured in the directions parallel, perpendicular and diagonal to the axis of extrusion.

Young's Modulus and Modulus of Rupture

The Young's modulus of the filter materials was measured using sonic resonance on a Dynamic Elastic Properties Analyzer. Four point bending was used to measure the modulus of rupture of the materials using a crosshead speed of 0.5 mm/min in accordance with ASTM standard ASTM C 1674 – 08 [13]. The MOR data are reported as the average of at least 20 tested samples.

Coefficient of thermal expansion

The coefficient of thermal expansion was measured at Orton Ceramics on an Orton dilatometer, model 1000D. The standard heating rate of 3 degrees Celsius per minute was used. The reported values are for the temperature range of 200 to 800 degrees Celsius.

Chemical stability

Sections of ceramic filters were exposed to chemical/ash components at various temperatures and times to determine the chemical stability of fibrous CLM ceramics. Portions of filter samples were immersed in concentrated solutions of Zinc Nitrate, Cerium Nitrate, Sodium Hydroxide, Sulfuric Acid, engine oil, Sodium Chloride, and Potassium Nitrate. The filter samples were then heated to 1000°C, 1100°C, 1300°C and 1400°C for 1 hour and 5 hours. SEM images of the chemical and ash exposed filter samples were also examined.

Pressure drop and filtration efficiency

Filter pressure drop performance was conducted in a lab test device at ambient temperature with air and soot. Filters were 50x153 mm. Soot was loaded into samples via a diesel engine. After a specific and controlled amount of soot was loaded inside the filter, the filter was removed, weighed, installed on a cold flow pressure drop rig and tested for pressure drop as a function of flow rate. The filter was then incrementally loaded with greater amounts of soot and the process was repeated until attainting the desired soot loading level. Pressure drop was typically plotted as a function of soot loading for the highest measured flow rate.

Filtration efficiency was also measured on the same lab testing device. Both weight and particle count efficiency were measured at the same time. The filter was weighed before and after the test to determine the filtration efficiency at different times. The filtration efficiency of the entire system was also determined by comparing the upstream and downstream numbers of artificial soot particles while the filter was in a quasi stationary test operation. A micro-diluter was coupled with a condensation particle counter. These instruments were used in conjunction to dilute the aerosol flow upstream and measure the particles present upstream and downstream of a particulate filter.

Scanning electron microscope

The microstructure was analyzed using scanning electron microscopy performed at Alfred University. The operating voltage used was 20 kV.

X-ray computed tomography

The final substrate was scanned for internal defects and flaws using X-ray computed tomography (CT) with a BIR ACTISTM 800/450/225 system, Lincolnshire IL. The X-ray source was a FeinFocus 225kV, and a Perkin Elmer XRD 160 Si detector was used. Typically 1000 slices were viewed with a slice thickness of 0.2 mm. The resulting pixel resolution was 0.2 mm.

RESULTS AND DISCUSSION

Microstructure and 3-D microstructure

Several specific oxide and non-oxide chemistries (including mullite, silicon carbide, cordierite, and alumina) have been extruded into honeycombs with the novel CLM, as shown in Figure 1. This microstructure produces a combination of high strength at high porosity and high material permeability for a given pore size distribution. The unique material properties of CLM substrates are due to the cross-links between fibers and the crystal and grain structure induced through chemistry and sintering. Fibers are arranged in a 3-D interconnected matrix during mixing and extrusion. Fibers preferentially align along the extrusion direction on a macro scale. Fibers are cross-linked through sintering and in situ chemical reactions. Scanning electron micrographs (SEM) of sintered composite honeycombs are shown in Figure 2. These images indicate that the cross-linked microstructure ranges from fine crystalline mullite fiber to large grain size fibrous silicon carbide. Grain growth is inhibited for the mullite CLM and growth is guided along a certain direction for CLM silicon carbide.

Figure 1. Extruded CLM Ceramics.

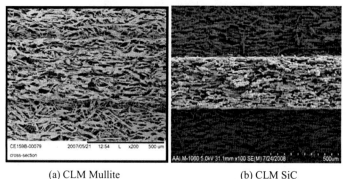

(a) CLM Mullite   (b) CLM SiC

Figure 2. SEM of CLM in (a) mullite and (b) SiC filters, showing fiber alignment in the extrusion direction.

This novel fibrous CLM was also examined with three-dimensional visualization techniques. Several techniques are available for investigating 3-D structure over a range of size scales. In this study, industrial X-ray computed tomography provided image data which was reconstructed to produce 3-D views of the honeycomb channels and to detect macroscopic extrusion and sintering flaws, such as internal voids and cracks. In addition, X-ray nano-CT was used to investigate the CLM and demonstrate its unique characteristics in 3-D at high resolution.

An example 2-D image from an industrial CT scan is shown in Figure 3, where it can be seen that the honeycomb channel walls inside the filter are defect-free. Typically >85% yield rate is achieved for most CLM ceramics. Most of the defects are due to extrusion problems, but some occur during drying and sintering. 3-D X-ray nano-CT images of both a CLM fibrous material and a particle-based material are shown in Figure 4. The characteristic high porosity of CLM ceramics (as compared to particle-based ceramic materials), coupled with the increased tortuousity and interconnectivity of pore spaces are evident in Figure 4. High tortuosity increases filtration efficiency and high connectivity is responsible for substantial improvement in mechanical properties of DPFs constructed from CLM materials.

The CLM pore structure is different from traditional particle-based honeycomb ceramics. The anisotropy of CLM materials is a function of the unique microstructure, and this microstructure also results in orientation-enhanced filtration efficiencies and increased fracture toughness. Detailed quantitative results from the X-ray nano CT analysis of several CLM honeycomb ceramics will be published separately. Another publication will detail the fracture toughness enhancement effect of CLM materials.

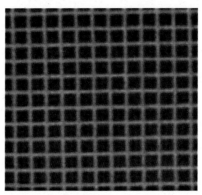

Figure 3. 2-D industrial CT scan of mullite CLM (300 cpsi /12 mil)

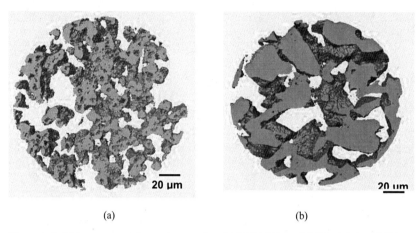

(a)   (b)

Figure 4. 3-D X-ray computed nanotomography of (a) CLM SiC and (b) Particle based SiC.

Anisotropic properties and orientation effects

The anisotropy result should not be surprising given the inherent structure of fibers itself, the unusual textured structure and the complex chemistry of extruded fibrous CLM ceramics. All extruded honeycomb ceramics possess anisotropic mechanical properties for its 2-D honeycomb channel. As documented in the literature, [13, 14, 15] the mechanical properties of honeycomb ceramics vary significantly in different directions (a, parallel to extrusion direction and b, perpendicular to extrusion direction and c, 45 degree of extrusion direction). Similarly, CLM honeycomb ceramics have similar anisotropic mechanical properties as powder-based honeycomb ceramics, due to the honeycomb channel structure layout. In a micro-scale, the fibrous structure introduces a preferred-orientation microstructure, i.e. fibers are preferably aligned along the extrusion direction, due to the severe plastic characteristics of the paste. In 3-D X-ray computed nano tomography, test results also indicates a higher anisotropic of CLM ceramics compared to particle-based honeycomb ceramics. This preferred-orientation has significant an impact not just on compressive strength but also other properties, such as fracture toughness and thermal expansion.

The preferred-orientation behavior is controlled by several factors, including fiber content, size, distribution, and morphology. The extrusion process results in the fibers being preferentially oriented along the extrusion axis.

Porosity and pore size distribution

Porosity and pore size distribution are important parameters for the catalyst coating and mechanical properties, while cell structure, i.e. wall thickness and cell density are important for back pressure and heat mass. It is more difficult to achieve a high porosity with high strength than variation of cell density structure. If fibrous structure is used, it is possible to extend the porosity range with various designed cell structure at the same time, while maintaining adequate mechanical strength. High porosity is one of the key characteristics of fibrous structured materials. The porosity of a fibrous structure is created by open space between fibers. A highly porous fibrous structure is utilized in many applications, such as fiber insulation and fiber filtration. The extruded fibrous CLM ceramics are also inconsistent with other high porosity material as shown in Table I and the previous SEM and nano CT scan figures. The results indicate that extruded fibrous CLM ceramics have porosity greater than 60% and can reach up to 75% and maintain equivalent strength. The porosity of extruded fibrous CLM ceramics depends on several factors, such as fiber diameter, pore-former, chemical reactions and sintering. Figure 5 shows the pore size distributions of CLM materials and particle-based filter materials. In Figure 6, various porosity and pore size distributions of mullite CLM ceramics are produced. It is indicated that the porosity and pore structure can be controlled by chemistry and sintering of CLM ceramics. This is particularly important for applications that require high washcoat and catalyst loadings.

Table I. Summary of physical properties of CLM and commercially available ceramic materials.

	Approximate Cell Geometry (cpsi/t)	Porosity (%)	Pore Diameter (μm) *	CTE (parallel) ($\times 10^{-6}$/C)	CTE (perpendicular) ($\times 10^{-6}$/C)	Axial Compression (MPa)	MOR (MPa)	E-MOD (GPa)	Surface Area** (m^2/g)
Fibrous CLM Mullite	300/12	67%	16.9	4.6	4.3	4.2	5.5	4.9	0.187
Fibrous CLM SiC	300/12	66%	17.1	4.1	4.3	5.7	6.3	12.2	N/A
Fibrous CLM Cordierite	200/20	59%	15.8	0.5	2.1	7.0	2.1	4.3	N/A
Commercial Particulate SiC	300/12	49%	20.9	4.6	5.6	9.8	9.4	13.3	0.0894
Commercial Particulate Cordierite	200/12	48%	13.2	0.5	0.9	8.1	2.2	4.8	0.1722

* Pore diameter is median based on volume.
**: Specific Surface Area of the filters (SSA, in m^2/g), measured by BET

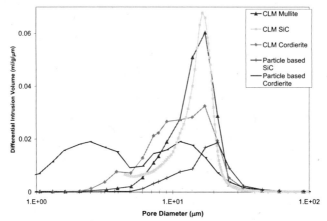

Figure 5. Pore size distribution of CLM and particle-based ceramics.

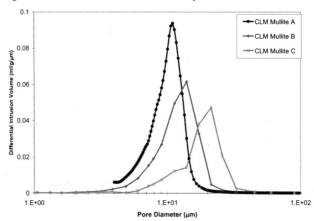

Figure 6. Pore size distribution of various CLM mullite ceramics

Thermal expansion coefficient

The thermal expansion coefficient parallel and perpendicular to the extrusion direction of various fibrous extruded honeycomb ceramics is measured and compared with commercially available DPF products. Experimental results of the linear dimensional change between 200 and 800 degrees Celsius of extruded fibrous honeycomb ceramics are listed in the previous Table 1.

It is well known that there is often a large difference in the relative thermal expansions along different crystallographic directions in anisotropic material. The thermal-expansion anisotropy of crystals is important, in part, because of the effect of this anisotropy on local stresses, porosity, and strength of polycrystalline bodies made from these crystals. It is also confirmed that fiber materials also have anisotropic CTE, and it is dependent on crystalline size. Among the previous studies on the thermal expansion behavior of composites, very little attention has been paid to the extrusion-induced anisotropy caused by preferred orientation of fibers.

Results of this study suggest that the anisotropic nature of fibrous CLM also exists with regard to thermal expansion. Anisotropic thermal expansion of fibrous honeycomb ceramics is caused by extrusion induced preferred orientation alignment of fiber. A profound influence of orientation effects on extruded fibrous CLM cordierite is noted for the cordierite material as shown in Table 1. This effect is similar to preferred oriented platelet structure[14,15,16] as discussed in the literature of extruded particle-based cordierite honeycomb body. However, the anisotropic preferred orientation effect is much closer to the theoretical calculation as discussed by Bruno et al.[14]. In this study, the anisotropic CTE is less noted for isotropic beta silicon carbide due to its cubic structure and large crystalline size.

Mechanical Properties

The fibrous CLM has much better mechanical properties compared to particle-based microstructures, because of the inherent characteristic of fiber. Table I lists the mechanical properties of fibrous CLM honeycomb ceramics. They indicate that the fibrous CLM materials have adequate mechanical strength at high porosity. The porosity effects on mechanical properties are also shown in Figures 7, 8 and 9. The relationship between mechanical properties and porosity can be analyzed with various analytical models [3,17,18]. In preliminary results, there is a correlation between fracture strength and porosity for CLM mullite, and this correlation may be different from porous particle-based honeycomb ceramics. Accurate prediction of the fracture strength of porous CLM ceramics requires more adequate information for comparison. It is necessary to point out that the differential results of theoretical predication models may arise from irregular shapes, varied size and random distribution of the pores and others.

For similar porosity and cell density, CLM ceramics have higher mechanical properties compare to particle-based ceramics. The improved mechanical performance of CLM ceramics may be due to the unique microstructure and chemistry. The microstructural evolution of CLM ceramic materials with increasing porosity is related to 3-D connectivity. Connectivity represents the contact area between fibers or fibrous grains. The increasing connectivity results in increased mechanical strength as discussed in previous section, however, the bonding strength between the fibers is important for increased load bearing. The minimum contact solid area (load-bearing area) which is the actual sintered or the bond area is significantly different between particle-based and fiber-based materials.

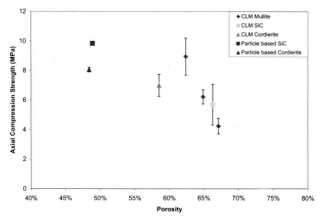

Figure 7. Comparison of compressive strength vs porosity of CLM ceramic and particle-based ceramics.

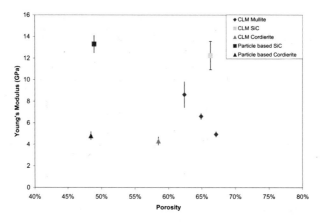

Figure 8. Comparison of Young's Modulus vs porosity of CLM ceramic and particle-based ceramics.

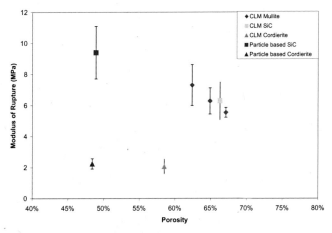

Figure 9. Comparison of Modulus of Rupture vs porosity of CLM ceramic and particle-based ceramics.

Back pressure and filtration efficiency

In DPF applications, the most important performance attributes are pressure drop, filtration efficiency, and filter durability. It is particularly true for applications requiring high catalyst loadings.

Fibrous CLM honeycomb ceramics have significantly better soot loaded back pressure while maintaining high filtration efficiency and filter durability. Figures 10 and 11 show a significantly lower back pressure and the high filtration efficiency of fibrous CLM honeycomb ceramics compared to traditional powder-based honeycomb ceramic. This excellent pressure drop performance and filtration efficiency can be ascribed to the improvement of total porosity, pore characteristics and microstructures as discussed above. Detailed engine testing results along with catalytic conversion performance are published in previous publication[19].

As demonstrated here, new high porosity CLM filters have substantially improved soot loaded pressure drop performance over particle-based materials. Fibrous CLM pressure drop performance is not accompanied with poor filtration efficiency and low filter durability as is typically seen with most of high porosity particle-base ceramic materials. As shown in Table 1 and Figure 11, with similar cell density, high porosity (66%) CLM SiC has better soot filtration efficiency compared to particle-based SiC (49%). The improved filtration efficiency may be attributed to the smaller pore size of CLM SiC, but the majority of benefit is attributed to the cross-linked microstructure as discussed above.

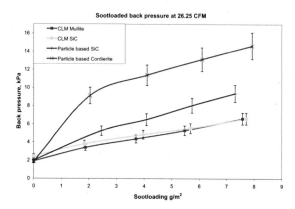

Figure 10. Soot loaded backpressure of CLM ceramic and particle-based ceramics,

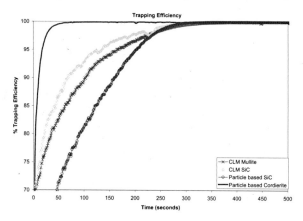

Figure 11. Filtration efficiency of CLM ceramic and particle-based ceramics.

Chemical Stability/Ash Resistance

It is necessary to examine the chemical stability of CLM ceramics for its increased connectivity and complex chemistry. No significant difference in chemical resistance is noted between fibrous CLM and particle-based ceramics. A DPF is exposed to a variety of chemical species[20] during the catalyst coating process or in operation that could interact unfavorably with the filter material. These chemicals could originate from the diesel fuel itself, such as the sulfur content in the fuel, from the ash deposited from burned lubricant oil, or from wear and tear of the engine components itself.

No visual or mechanical degradation, adhesion, pitting or melting of the fibrous CLM ceramics (mullite and silicon carbide material) was observed below 1000°C. SEM images of the reference and ash exposed mullite filter samples are shown in Figure 12. Additionally, no degradation of strength (measured using MOR) was observed for this material. Similar measurements were also conducted by exposing the filter to a mixture of commonly known ash constituents (e.g. CaO, P_2O_5, ZnO, MgO, Fe_2O_3, Na_2O) at temperatures between 1000°C to 1400°C for 1 hour. Around 1100°C ash adhesion and initial ash sintering was observed on the filter wall surface, without degradation of the filter structure. Pitting of the ceramic occurred at temperatures exceeding 1300°C.

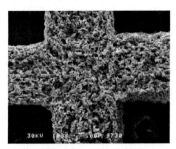

(a). Control CLM Mullite   (b). Ash exposed CLM Mullite

Figure 12. SEM of chemical exposure study showing no degradation.

ACKNOWLEDGMENTS

We wish to thank contributing team members at GEO2 Technologies, including Adam Wallen, Art O'Dea, Kyle Smith, Leonard Newton, Noah Loren, and Rob Lachenauer.

REFERENCES

1. T. V. Johnson, "Diesel Emission Control in Review", SAE Technical Paper No 2006-01-0233 (2007).

2. A.G. Konstandopoulos, and J. H. Johnson, "Wall- Flow Diesel Particulate Filters-Their Pressure Drop and Collection Efficiency", SAE Tech. Paper No. 890405 (1989)

3. A. Schaefer-Sindlingera, I. Lappas, C.D. Vogt, T. Ito, H. Kurachi, M. Makino, and A. Takahashi., "Efficient Material Design for Diesel Particulate Filters", Topics in Catalysis Vols. 42–43, May 2007.

4. J. Adler, "Ceramic Diesel Particulate Filters", International Journal of Applied Ceramic Technology 2 (6), 429–439 (2005).

5. "Retrofitting Emission Controls On Diesel-Powered Vehicles" Manufacturers of Emission Controls Association, April, (2006).

6. N. Miyakawa, H. Maeno, H. Takahasi, "Characteristics and Evaluation of Porous silicon Nitride DPF," SAE Technical Paper No. 2003-01-0386, (2003).

7. A. J. Pyzik and G. G. Li, "New Design of a Ceramic Filter for Diesel Emission Control Applications", Int. J. Appl. Ceram. Technol., 2 [6] 440–451 (2005)

8. R. K. Miller, W. C. Haberkamp, K. M. Badeau and Z. G. Liu, R. C. Shirk and T. Wood., "Design, Development and Performance of a Composite Diesel Particulate Filter", SAE Paper 2002-01-0323 (2002)

9. H. S. Hashimoto, Y. Miyairi, T. Hamanaka, R. Matsubara, T. Harada and S. Miwa, "SiC and Cordierite Diesel Particulate Filters Designed for Low Pressure Drop and Catalyzed, Uncatalyzed Systems", SAE Paper 2002-01-0322 (2002).

10. S. Ogunwumi and P. Tepesch, "Aluminum-Titanate Based Materials for Diesel Particulate Filters," 28th International Cocoa Beach Conference on Advanced Ceramics & Composites (2004)

11. B. Zuberi, "Multi-functional substantially fibrous mullite filtration substrates and devices", US 2006/0120937A1 (2006).

12. T. Sakaguchi, A. Ohgushi, S. Suzuki, H. Kita, K. Ohsumi, T. Suzuki and H. Kawamura "Development of High Durability Diesel Particulate Filter by using SiC Fiber" SAE Paper 1999-01-0436 (1999).

13. ASTM C 1674 – 08 Standard Test Method for Flexural Strength of Advanced Ceramics with Engineered Porosity (Honeycomb Cellular Channels) at Ambient Temperatures.

14. G. Bruno, S. Vogel, "Calculation of the Average Coefficient of Thermal Expansion in Oriented Cordierite Polycrystals," Journal of the American Ceramic Society, 91, 2646-2652 [8] (2008).

15. M. Lachman, R. D. Bagley, and R. M. Lewis, ''Thermal Expansion of Extruded Cordierite Ceramics,'' Ceram. Bull., 60 [2] 202–5 (1981).

16. D. L. Evans, G. R. Fischer, J. E. Geiger, and F. W. Martin, ''Thermal Expansions and Chemical Modifications of Cordierite,'' J. Am. Ceram. Soc., 63 [11–12] 629–34 (1980).

17. F. W. Nyongesa, and B.O. Aduda, "Fracture Strength of Porous Ceramics: Stress Concentration Vs Minimum Solid Area Models" Science and Engineering Series Vol. 5, No. 2, pp. 19-27 (2004).

18. I. Mikioh, H.Yuichi, O. Mitsunori,, "The Latest Technology of , Controlling Micro-Pore in Cordierite Diesel Particulate Filter for DPNR System", SAE Paper 2004-01-2028 (2004).

19. B. Zuberi, J. J. Liu, S. C. Pillai, J. G. Weinstein, A. G. Konstandopoulos, S. Lorentzou, and C. Pagoura, "Advanced High Porosity Ceramic Honeycomb Wall Flow Filters" SAE Paper 2008-01-0623 (2008).

20. A. Sappok and V. Wong, V.., "Detailed Chemical and Physical Characterization of Ash Species in Diesel Exhaust Entering Aftertreatment Systems", SAE Paper 2007- 01-0318 (2007).

POROUS β-Si₃N₄ CERAMICS PREPARED WITH FUGITIVE GRAPHITE FILLER

POROUS β-Si$_3$N$_4$ CERAMICS PREPARED WITH FUGITIVE GRAPHITE FILLER

Probal Chanda and Kevin P. Plucknett
Materials Engineering Program, Department of Process Engineering and Applied Science,
Dalhousie University, 1360 Barrington Street, Halifax, Nova Scotia, B3J 1Z1, CANADA

Liliana B. Garrido
CONICET, Centro de Technologia de Recursos Minerales y Cerámica (CETMIC, CIC-
CONICET-UNLP), Cam. Centenario y 506, C.C.49 (B 1897 ZCA) M.B. Gonnet. Pcia. De
Buenos Aires, ARGENTINA

Luis A. Genova
IPEN Instituto de Pesquisas Energéticas e Nucleares, CCTM Centro de Ciência e Technologia de
Materiais, Cidade Universitária, Travessa R 400, 05508-900 São Paulo, BRAZIL

ABSTRACT
Porous silicon nitride (Si$_3$N$_4$) ceramics have been prepared by pressureless-sintering of α-Si$_3$N$_4$ pellets with single Y$_2$O$_3$ additions (5 wt. %). In order to vary the porosity, fugitive graphite particle filler has been used, with the nominal graphite content varied up to 10 weight percent (on top of the baseline composition). After the preparation of isostatically pressed pellets, removal of the graphite filler is achieved using a low temperature oxidation treatment. Optimization of the oxidation treatment was performed using thermogravimetric analysis. After oxidation, samples were pressureless-sintered in a nitrogen atmosphere at temperatures between 1500 and 1750°C, for two hours. Post sinter evaluation of the materials was then conducted using scanning electron microscopy and x-ray diffraction, while pore size was determined for selected samples using mercury intrusion porosimetry. Strengths of up to 178 MPa were determined using ball-on-ring biaxial testing, which were subsequently related to the maximum estimated flaw size in the sintered material.

INTRODUCTION
There is growing interest in the development of high performance porous ceramics for a variety of applications, including use as hot gas and molten metal filters, bio-prosthetic implants, bioreactor supports, and potentially lightweight structural materials. Among the various materials that have been assessed, silicon nitride (Si$_3$N$_4$) based ceramics possess a desirable combination of mechanical and thermal properties for many of these applications, due to the formation of a highly anisotropic and interlocking grain structure.[1,2] In order to prepare materials with controlled nano-/micro-porosity, it is possible to select sintering aids that promote the α- to β-Si$_3$N$_4$ transformation, but have only a limited effect on densification.[3] Suitable sintering aids include oxides of the lanthanide and Group III elements.[4] Alternatively, it is also possible to use a low volume fraction of multiple sintering aids, where the additive selection can be chosen to promote the growth of high aspect ratio β-Si$_3$N$_4$.[5,6]

Combined compositional and microstructural design can also be undertaken, using β-Si$_3$N$_4$ seeding and/or bimodal α-Si$_3$N$_4$ powders to prolong the α- to β-Si$_3$N$_4$ transformation period. This latter approach has led to the development of materials with mean β-Si$_3$N$_4$ aspect ratios exceeding 12:1, and biaxial flexure strengths up to ~300 MPa, while still retaining 30 vol. % porosity (D. Gould and K. P. Plucknett, manuscript in preparation). It can be seen that following these methods, the sintered materials can be expected to retain good mechanical properties, provided that an interlocking β-Si₃N₄ grain structure is formed, while still retaining a moderately high volume fraction of interconnected porosity.

In the present work, single yttrium oxide (Y_2O_3) sintering additions have been used to produce porous β-Si₃N₄ ceramics, where the final porosity content has been varied through the addition of fugitive graphite filler particles. The graphite particles can be simply removed from the processed preforms, prior to sintering, through the use of a low temperature oxidation treatment. In the present case, optimization of the oxidation treatment has been achieved through the use of thermogravimetric analysis (TGA). The effects of filler content on the microstructural development and mechanical behavior have been assessed for nominal initial graphite filler contents up to 10 wt. %.

EXPERIMENTAL PROCEDURES

All samples in the current work were prepared using Ube SN E-10 α-Si₃N₄ powder (Ube Industries, New York, NY), together with additions of 5 wt. % Y_2O_3 (99.99 %; Metall Rare Earth, China). In order to vary the porosity content, graphite powder (-300 mesh; Alfa Aesar, Ward Hill, MA) was added as a fugitive filler, with the amount varied from 0 to 10 wt. % *on top of* the baseline Si₃N₄/Y_2O_3 composition (i.e. up to 110 wt. % total). 50 g baseline batches (plus graphite content) of the appropriate composition were prepared by ball-milling for 24 h in isopropyl alcohol using TZP media, followed by drying and sieving through a 75 μm mesh stainless steel sieve. The powders were then uniaxially pressed (at ~31 MPa) into 31 mm diameter x 4 mm thick discs, followed by vacuum bagging and cold-isostatic pressing at a pressure of ~170 MPa.

A low temperature oxidation pre-treatment was then utilized to remove the graphite filler, with appropriate oxidation conditions determined through the use of TGA (TA Instruments Q600 SDT, New Castle, DE). This approach involved the use of both constant heating rate tests (RT to 1000°C at 10°C/min) and also isothermal experiments at temperatures between 575 and 650°C. Sintering was subsequently conducted with the oxidized samples sited within a graphite crucible, in a powder bed comprised of 50 wt. % BN and 50 wt. % Si₃N₄. A static nitrogen atmosphere (99.998 purity; Air Liquide Canada, Inc., Dartmouth, Canada) was used to prevent Si₃N₄ decomposition during sintering. The sintering furnace (Materials Research Furnaces, Inc., NH) was first evacuated, then back-filled with nitrogen, evacuated a second time and then heated under vacuum to 750°C at 10°C/min, at which point the furnace was again back-filled with nitrogen and then heated to the final sintering temperature at 10°C/min. Sintering was performed at temperatures between 1500 and 1750°C, with a hold time at temperature of 2 hours.

After sintering, the densities were measured by immersion in mercury. Scanning electron microscopy (SEM; S-4700, Hitachi Industries, Tokyo, Japan) was performed on carbon coated fracture surfaces, while phase identification was conducted using X-ray diffraction (XRD; D-8 Advance, Bruker AXS, Inc., Madison, WI), with the α- to β-Si₃N₄ ratio calculated using the method of Yeheskel and Gefen.[7] Pore size distribution was determined using mercury porosimetry (Poremaster PM33-7, Quantachrome Instruments, FL). The strength of as-sintered samples (i.e. with no further surface preparation) was determined by biaxial flexure testing, using a ball-on-ring geometry following ASTM C1499. For strength determination, the materials Poisson ratio was determined using the approach of Arnold et al,[8] with a value of 0.23 assumed for dense Si₃N₄.

RESULTS AND DISCUSSION

Preform oxidation characterization

In order to determine suitable conditions for oxidative removal of the graphite filler content, prior to sintering, TGA was used to analyze weight loss for both constant heating rate (i.e. RT to 1000°C) and isothermal conditions. Figure 1 demonstrates the response of samples, with varying graphite contents, heated to 1000°C at a rate of 10°C/min in following dry air. It is clear that for samples containing graphite additions, weight loss occurs above approximately 600°C at this heating rate, and is' essentially complete above ~850°C. The onset of weight loss is confirmed to occur above 550°C when observing the derivative curve (Figure 2). The total weight loss in each case corresponds closely with the theoretical graphite content in the sample. There is minimal observed weight gain for the 0 wt. % graphite sample, even up to 1000°C, as might be expected.

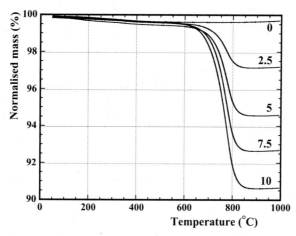

Figure 1. TGA observation of mass change as a function of temperature for samples prepared with varying graphite filler contents. All samples heated at 10°C/min in flowing dry air (100 mL/min).

As a consequence of these observations, isothermal experiments were subsequently performed on the 5 wt. % graphite samples at temperatures between 575 and 650°C for a period of 24 hours (Figure 3). In this instance the atmosphere was switched from nitrogen, during heating to temperature, to dry air at temperature to avoid oxidation prior to the isothermal hold period. It is apparent that the weight loss is essentially complete after 24 hours at 600°C, but is still incomplete after the same duration at 575°C. As a consequence, subsequent static oxidation treatments for graphite removal from all samples were performed at 600°C for a period of 48 hours. In each case, the final weight loss corresponded closely with the initial graphite content of the preform. It is notable that an experimental artifact is apparent near the start of the isothermal hold, arising from a buoyancy effect when changing the sample gas from nitrogen to air. Based on these studies, an isothermal heat treatment at 600°C for 48 hours was subsequently chosen for graphite removal from all samples.

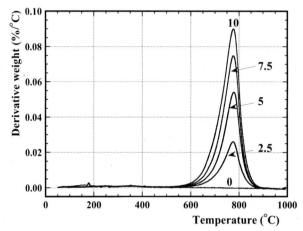

Figure 2. The derivative curves relating to data presented in Figure 1, demonstrating the consistent onset of weight loss, through graphite oxidation, at ~550°C for all compositions (heated at 10°C/min in flowing air).

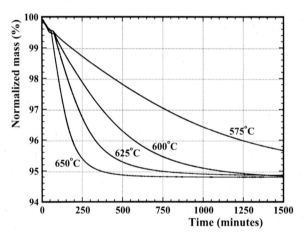

Figure 3. Mass change of 5 wt. % graphite filler samples as a function of time, for isothermal holds at temperatures between 575 and 650°C.

Microstructural development

The effects of graphite content upon green density after oxidation at 600°C for 48 hours are shown in Figure 4. It is apparent that the retained porosity can be readily varied between 50 and 60 vol. % using this approach. Oxidation marked a visual macroscopic change in appearance from grey to white in all the preforms, which was confirmed to be consistent through the thickness as well. Figure 4 also demonstrates the effect of sintering temperature (in this instance sintering at 1750°C for 2 hours) on densification behavior. It is clear that all samples show evidence of densification, regardless of initial graphite content. However, it is also clear that the samples with the highest original graphite filler content have the lowest sintered densities, confirming the trend observed for the green bodies.

The progression of the α- to β-Si₃N₄ phase transformation was followed using XRD. Samples sintered at 1700 and 1750°C showed complete transformation from α- to β-Si₃N₄, while samples sintered at lower temperatures showed residual α-Si₃N₄, with the amount decreasing with increasing sintering temperature. It was also notable in the intermediate temperature range (i.e. 1550-1650°C) that samples with higher graphite content showed a decreased extent of transformation at a constant sintering temperature. This general observation has also been made for similar materials processed in air (K.P. Plucknett, manuscript in preparation). In both cases, this effect can be attributed to increased diffusion distances, due to the lower initial green density of the high graphite containing preforms. This is particularly noticeable for the present compositions due to the high liquid viscosity that can be expected with the Y-Si-O-N oxynitride glass phase present at the sintering temperature (i.e. above ~1550°C).[3]

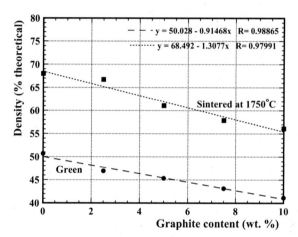

Figure 4. The effects of initial graphite filler content on both green density (after oxidation at 600°C for 48 hours) and sintered density (sintered at 1750°C for 2 hours in nitrogen). Typical standard deviation errors were less than 0.5 % of the measured density.

The microstructure of the sintered materials showed a typical progression from equiaxial at lower temperatures, to a highly anisotropic structure at the highest temperatures, indicative of β-Si₃N₄ formation (as confirmed by the XRD observations). Qualitatively it was observed that the β-Si₃N₄ aspect ratio increased with increasing initial graphite content, potentially due to a reduction of growth hindrance in the more open structure. However, this was complicated by the retention of large pores in the highest graphite containing samples, and requires further investigation; it was apparent that very high aspect ratio grains could be formed in such pores (up to ~12:1). The retained pore size distribution for selected samples is shown in Figure 5, and highlights a generally narrow pore size distribution that broadens with increasing initial graphite filler content. In each case a primarily monomodal pore size distribution is apparent. The exception to this is the highest graphite content sample, where a low volume fraction of pores up to 100 μm were apparent, which arose from agglomeration of graphite particles during drying of the powder (visible as dark spots in the pressed powder preforms prior to oxidation treatment).

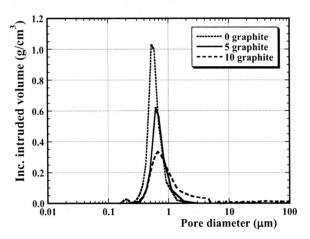

Figure 5. Intruded mercury pore size distribution for samples sintered at 1750°C for 2 hours with 0, 5 and 10 wt. % initial graphite contents.

Mechanical behavior

The effect of initial graphite content on the biaxial flexure strength of samples sintered at 1750°C for 2 hours is shown in Figure 6. Strengths of up to ~178 MPa are achieved for the samples prepared without graphite filler, which decreases slightly to ~168 MPa for samples prepared with up to 5 wt. % filler. However, a significant decrease in strength was observed for higher filler content samples. This was attributed to the formation of graphite aggregates during drying of the powder mixtures, prior to sintering, that were not subsequently eliminated. This aggregate formation and retention was also apparent from mercury porosimetry evaluation, as noted in the previous section. In order to assess the approximate flaw size in each case, the Griffith equation was employed,[9] with a toughness value of ~0.78 MPa·m$^{1/2}$ used, based on a previous study of porous β-Si₃N₄ ceramics, with a similar baseline composition (5 wt. % Y₂O₃) and microstructure.[10] While the porosity content is varied in the present study, there is limited data available on the toughness of isotropic, porous β-Si₃N₄. Ohji has demonstrated a slightly higher toughness of ~ 2 MPa·m$^{1/2}$, for materials with 5 wt. % Yb₂O₃ sintered at 1750°C,[11] which would increase the apparent flaw size in the present work. Estimated critical flaw sizes for each sample are shown in Table 1. It is clear that the estimated flaw size is relatively small for materials prepared with up to 5 vol. % graphite content, and likely relates to the interaction of sub-micron pores. However, for higher graphite content samples the critical flaw size was dramatically increased, which corresponds to the presence of sizeable graphite aggregates as described in the previous section.

It is clear that in most cases the estimated flaw size exceeds the apparent maximum pore size observed. However, it must be remembered that the nature of these materials is that they contain continuous porosity, and therefore it is possible that there will be significant interaction between neighboring porous regions. In addition, as noted earlier, for the case of the 10 wt. % graphite sample, larger pores exceeding 100 μm were measured during porosimetry. A major issue when processing these materials with initially coarse graphite particles is the potential retention of larger particles after milling and, perhaps more importantly, the tendency of the graphite to aggregate during drying. While not assessed in the present study, this problem has been subsequently alleviated through the use of finer graphite particles (7-10 μm) in more recent experimental work (K.P. Plucknett, manuscript in preparation).

Graphite addition (wt. %)	0	2.5	5	7.5	10
Porosity (vol. %)	32	34	39	43	44
Critical flaw size (μm)	24	25	27	198	361

Table 1. The effect of graphite filler content on the nominal critical flaw size determined for porous β-Si₃N₄ ceramics.

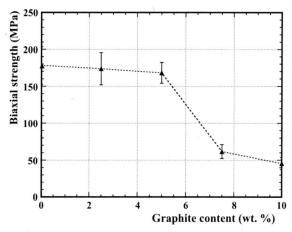

Figure 6. The effect of initial graphite filler content on the biaxial flexure strength of porous β-Si₃N₄ samples sintered at 1750°C for 2 hours.

SUMMARY

The present work has focused on the use of graphite fugitive fillers to prepare porous β-Si₃N₄ ceramics with varying amounts of porosity. The use of TGA has shown that the optimal temperature range for graphite removal of the green bodies is 600-650°C, such that Si₃N₄ oxidation is minimized. As anticipated, sintering between 1500 and 1750C results in increasing densification and α- to β-Si₃N₄ transformation, although a lower extent of transformation was observed at intermediate temperatures in samples with higher initial graphite content (and hence greater diffusion distances). The final porosity was effectively varied from ~32 to 44 vol. %, with increasing initial graphite content, in materials sintered at 1750°C. It was apparent that increasing the extent of porosity in the green body, and hence the final sintered material, resulted in an increase in the observed β-Si₃N₄ aspect ratio. This can be attributed to a reduction in the growth hindrance in the more open structure of the high graphite content samples. Mercury porosimetry confirmed a gradual change from a narrow pore size distribution in zero graphite content samples (mean ~600 nm), to a much broader one in 10 wt. % graphite samples. Graphite aggregation during processing even resulted in some measured pores up to ~100 μm in size. This increased pore size, and the potential presence of interacting pore clusters led to low strengths (< 70 MPa) in the highest graphite containing samples, while those with graphite contents up to 5 vol. % exhibited relatively high strengths (168-178 MPa) in comparison.

ACKNOWLEDGEMENTS
The authors would like to acknowledge NSERC (Canada), CONICET (Argentina) and CNPq (Brazil) for provision of research funding through the Inter-American Research in Materials (CIAM) program. We also acknowledge the support of the Canada Foundation for Innovation, the Atlantic Innovation Fund, and other partners who helped fund the Facilities for Materials Characterisation, managed by the Dalhousie University Institute for Materials Research, who provided access to the FE-SEM. Mr. Brian Liekens (Dept. of Civil and Resource Engineering, Dalhousie University) is also thanked for performing the mercury porosimetry measurements. Mr. Mervin Quinlan is gratefully acknowledged for assistance with the sintering work conducted during this study.

REFERENCES
1. J.-F. Yang, Z.-Y. Deng and T. Ohji, 'Fabrication and Characterisation of Porous Silicon Nitride Ceramics Using Yb₂O₃ as Sintering Additive,' *J. Eur. Ceram. Soc.*, **23** [2] 371-78 (2003).
2. Y. Inagaki, Y. Shigegaki, M. Ando and T. Ohji, 'Synthesis and Evaluation of Anisotropic Porous Silicon Nitride,' *J. Eur. Ceram. Soc.*, **24** [2] 197-200 (2004).
3. S. Hampshire and K.H. Jack, pp. 225-30 in *Progress in Nitrogen Ceramics*, Ed. F.L. Riley, Martinus Nijhoff Publishers, The Hague, The Netherlands, 1983.
4. M. Quinlan, D. Heard, L. Garrido, L. Genova and K.P. Plucknett, 'Sintering Behaviour and Microstructure Development of Porous Silicon Nitride Ceramics in the Si-RE-O-N Quaternary Systems (RE = La, Nd, Sm, Y, Yb),' *Ceram. Eng. Sci. Proc.*, **28** [9] 41-48 (2007).
5. M. Quinlan, L. Garrido, L. Genova and K.P. Plucknett, 'Compositional Design of Porous β-Si₃N₄ Prepared by Pressureless-Sintering Compositions in the Si-Y-Mg-(Ca)-O-N System,' *Ceram. Eng. Sci. Proc.*, **28** [9] 49-56 (2007).
6. K.P. Plucknett, M. Quinlan, L. Garrido and L. Genova, 'Microstructural Development in Porous β-Si₃N₄ Ceramics Prepared with Low Volume RE₂O₃-MgO-(CaO) Additions (RE = La, Nd, Y, Yb),' *Mater. Sci. Eng. A*, **489** [1-2] 337-50 (2008).
7. O. Yeheskel and Y. Gefen, 'The Effect of the α-Phase on the Elastic Properties of Si₃N₄,' *Mater. Sci. Eng.*, **71** [1-2] 95-99 (1985).

8. M. Arnold, A.R. Boccaccini and G. Ondracek, 'Prediction of the Poisson's Ratio of Porous Materials, J. Mater. Sci., 31 [6] 1643-46 (1996).
9. B. Lawn, *Fracture of Brittle Solids-Second Edition*, Cambridge University Press, Cambridge, UK, 1993.
10. K.P. Plucknett and M.H. Lewis, 'Microstructure and Fracture Toughness of Si₃N₄ Based Ceramics Subjected to Pre-Sinter Heat Treatments,' *J. Mater. Sci. Lett.*, **17** [23] 1987-90 (1998).
11. T. Ohji, 'Microstructural Design and Mechanical Properties of Porous Silicon Nitride Ceramics,' *Mater. Sci. Eng. A*, **498** [1-2] 5-11 (2008).

DATA RELIABILITY FOR HONEYCOMB POROUS MATERIAL FLEXURAL TESTING

Randall J. Stafford
Cummins Inc.
Columbus, IN, USA

Stephen T. Gonczy
Gateway Materials Technology
Mount Prospect, IL, USA

ABSTRACT
A new standard test method to evaluate the flexural strength of honeycomb porous ceramic materials has been published by ASTM International as ASTM C1674[1]. An interlaboratory study was completed where the test method was exercised to develop an understanding of the data variation possible from the method. Cordierite in two different honeycomb architectures and two different specimen sizes was evaluated using over 400 flexure test specimens. Broad, world wide participation in the interlaboratory study (9 laboratories) made this a robust data set for analysis.

The key conclusions of the study were 1. humidity control in the test environment is critical to producing comparable data; 2. the equations for calculation of strength are accurate and produce reasonable values which are comparable between laboratories; and 3. the effect of lineal cell count for nominal beam strength values reported previously[2] is possible but not statistically confirmed in this limited data set.

INTRODUCTION
Honeycomb substrate materials have been used in aftertreatment devices since 1975 when they were introduced for automotive catalyst supports[3]. The use of honeycomb substrates expanded to diesel applications[4] with the continuing tightening of worldwide emission regulations for nitrous oxides and particulate matter. As the applications of the material have expanded, so has the expectation for the performance and durability. The strength has been a key factor in providing critical information on durability and has been measured by manufacturers[5, 6, 7] and used in development of empirical models to predict survival.[8, 9] More demanding applications and the advancement of technology have resulted in finite element models which require high quality data.

The original flexural strength testing on honeycomb ceramics used the available test methods for structural ceramics MIL-STD 1942A and then ASTM C1161[10]. The specimen geometry and test parameters were modified by each laboratory to meet their needs in conducting flexural strength tests by these methods. The influences of test technique, specimen alignment, fixture articulation and specimen preparation were difficult to compare when tests were conducted using different materials, cell geometries and specimen sizes. This was the impetus for creating and publishing a standard flexure test method on honeycomb ceramics. Following publication, a measure of the data precision of the method was needed. This was planned as an interlaboratory study. The work reported below uses the new standard test method, ASTM C1674, to evaluate the data reliability and robustness of the method in the interlaboratory study.

EXPERIMENTAL

Materials
Ten (10) commercial designation C558* cordierite particulate filter substrates and twenty (20) commercial designation Celcor** cordierite catalyst substrates were provided for specimen preparation. The C558 filters were 228 mm (9 inch) diameter by 178 mm (7 inch) long with a 46.5

cell/cm^2 (300 cell/in^2) cell density and a wall thickness of 0.305 mm (0.012 inch). The Celcor catalysts were 144 mm (5.66 inch) diameter by 152 mm (6 inch) long with a 93 cell/cm^2 (600 cell/in^2) cell density and a wall thickness of 0.102 mm (0.004 inch).

Specimen Preparation

Three groups of flexure test specimens were prepared for testing and the materials, specimen cross sections and test methods are shown in Table I. Each of the specimens was cut to size using a fine toothed (20 tpi) hacksaw blade. The faces of the specimens were sanded to remove most of the residual walls using 220 grit paper on a laboratory benchtop. All specimens were prepared at Cummins Inc. to remove the specimen preparation as a variable in the testing program.

Table I. Specimen Description

Material Designation	C1674, Test Method	Specimen Cross Section	Linear Cell Count	Manufacturer
Cordierite 1A	Method B	12 mm x 25 mm	9 cells x 17 cells	NGK, C558
Cordierite 1B	Method A	22 mm x 25 mm	15 cells x 17 cells	NGK, C558
Cordierite 2A	Method B	12 mm x 25 mm	12 cells x 24 cells	Corning, Celcor

Test Procedure

The flexure test specimens from each set were randomly divided into groups of twenty for distribution to the nine participating laboratories.

The basic test parameters used for all specimens at each laboratory tested in accordance with ASTM C1674, Test Method A and Test Method B were:

- Crosshead rate of 0.01 mm/second
- Load and support bearing diameters between 10 mm and 25 mm
- Fully articulating flexural fixtures

Cordierite 1A and Cordierite 2A specimens (12 x 25 x 100mm) were tested at each laboratory in accordance with ASTM C1674 Method B under the additional condition:

- Four point ¼ point flexure geometry with a 90 mm outer span and a 45 mm inner span. (This gives a 7:1 span/depth ratio to minimize shearing)

Cordierite 1B specimens (22 x 25 x 155 mm) were tested at each laboratory in accordance with ASTM C1674 Method A under the additional condition:

- Four point ¼ point flexure geometry with a 130 mm outer span and a 65 mm inner span. (This gives a 6:1 span/depth ratio to minimize shearing)

The specimen mass, dimensions, and breaking force were measured and recorded and the nominal beam strength and wall fracture strength were calculated for each specimen in accordance with the equations in ASTM C1674. The wall thickness and cell pitch (cell opening + one wall) were measured optically at Cummins for use in the results calculations for wall fracture strength.

Round Robin Organization

The round robin test program was conducted with nine laboratories representing material manufacturers, intermediate suppliers, product end users, academic institutions and national laboratories.*** The testing was conducted at sites in Europe, Japan and USA to provide a broad spectrum of test input for validation of the test method. The tests were conducted between October and December 2008.

RESULTS

Cordierite 1A was distributed to 9 laboratories and 158 specimens were tested with 155 specimen breaks inside the inner load span and 3 specimen breaks outside the load span. Breaks outside the load span were censored and not used in the data analysis. The mean nominal beam strength and mean wall fracture strength were calculated from the measurements of the specimens and are shown in Figure 1 for each of the laboratories.

Cordierite 1B was distributed to 8 laboratories and 136 specimens were tested with 127 specimen breaks inside the inner load span, 7 specimen breaks outside the load span and 2 specimens with incorrect cell counts. Specimens with breaks outside the load span and incorrect cell counts were censored and not used in the data analysis. The mean nominal beam strength and mean wall fracture strength were calculated from the measurements of the specimens and are shown in Figure 2 for each of the laboratories. Specimens censored for incorrect cell count had either one more or one less cell in the thickness or width than the rest of the specimens tested. While this will not affect the wall fracture strength, there is a difference in the nominal beam strength, therefore to limit the number of variables under consideration, these data points were removed.

Cordierite 2A was distributed to 9 laboratories and 161 specimens were tested with 147 specimen breaks inside the load span, 12 specimen breaks outside the load span and 2 specimens with incorrect cell counts. Specimens with breaks outside the load span and incorrect cell counts were censored and not used in the data analysis. The mean nominal beam strength and mean wall fracture strength were calculated from the measurements of the specimens and are shown in Figure 3, showing

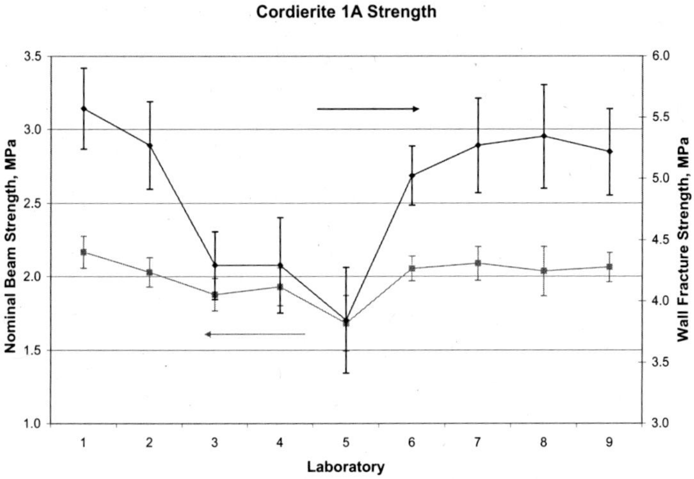

Cordierite 1A Strength

mean values of each data set with ± 1 standard deviation error bars.
Figure 1. Cordierite 1A - Nominal Beam Strength and Wall Fracture Strength by Laboratory

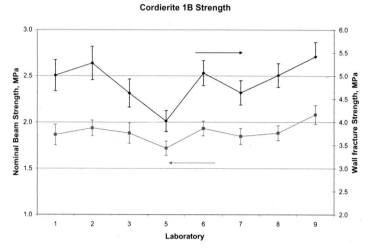

Figure 2. Cordierite 1B - Nominal Beam Strength and Wall Fracture Strength by Laboratory

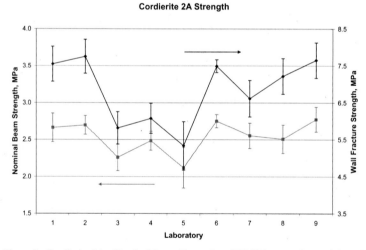

Figure 3. Cordierite 2A - Nominal Beam Strength and Wall Fracture Strength by Laboratory

DISCUSSION

The results for all three material sample sets show significant differences in strength results between some of the laboratories. Since the temperature, humidity and bearing diameters were varied between the laboratories, these were possible factors of the strength differences. In addition to the test parameter variations, analysis of the dimensional measurements (width and thickness at the fracture point) showed major variation between the laboratories as shown for width measurements in Figure 4A. Further contact with the laboratories revealed that some measurements included the height of the residual wall stubs. These specimens were re-measured for width and thickness (remeasured widths shown in Figure 4B) and the strengths recalculated as shown in Figures 5, 6 and 7 (the mean value of each data set is plotted with ± 1 standard deviation error bars).

Specimen dimensional measurement accuracy is critical to the stress value. Previous work[11] has found that specimen measurements including the thickness of the wall stubs (on the order of 0.15 mm) can reduce the calculated stress value by 8%. If the specimen measurements at some laboratories included the wall stub and other laboratories did not, then there is inherently more scatter in the calculated nominal beam and wall fracture strengths. The strength reduction variation found between the calculation using the initial dimension measurements and the remeasured dimensions was between 2.0% and 8.3% for the different material sample sets.

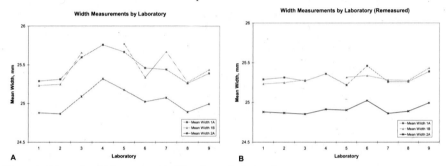

Figure 4. Dimensional Measurements of Specimens by Laboratory

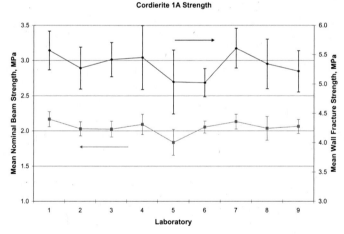

Figure 5. Cordierite 1A – Recalculated Nominal Beam Strength and Wall Fracture Strength by Laboratory

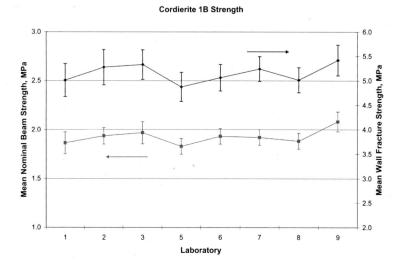

Figure 6.   Cordierite 1B – Recalculated Nominal Beam Strength and Wall Fracture Strength by Laboratory

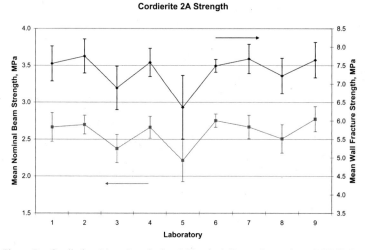

Figure 7.   Cordierite 2A – Recalculated Nominal Beam Strength and Wall Fracture Strength by Laboratory

The nominal beam and wall fracture strength values from the remeasured specimens were used for analysis of the interaction effects of the temperature, bearing diameter, and humidity. Statistical analysis, by ANOVA, of each material sample set showed that bearing diameter and temperature were not significant factors in the strength differences between laboratories. From the ANOVA analysis, humidity was projected to be a significant interaction factor for Cordierite 1A (weak interaction) and Cordierite 2A (strong interaction).

The humidity effect on the nominal beam strength for Cordierites 1A and 2A are shown in Figures 8 and 9, where the mean value of each data set is plotted with ± 1 standard deviation error bars. The mean values and coefficient of variation for each material sample set are shown in Table II. The coefficient of variation value for Cordierite 2A is affected by the lower strength at high humidity test conditions.

Table II. Ambient Nominal Beam Strength and Wall Fracture Strength for Cordierite 1A, 1B, 2A

	Nominal Beam Strength		Wall Fracture Strength	
Cordierite	Mean, MPa	Coeff. Variation, %	Mean, MPa	Coeff. Variation, %
1A	2.05	4.5	5.33	4.0
1B	1.93	4.0	5.15	3.7
2A	2.59	7.3	7.35	6.2

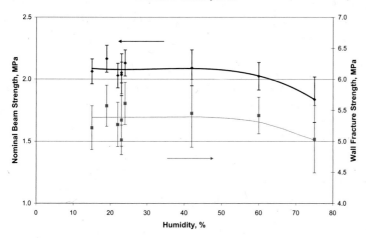

Figure 8. Strength and Humidity Dependence for Cordierite 1A

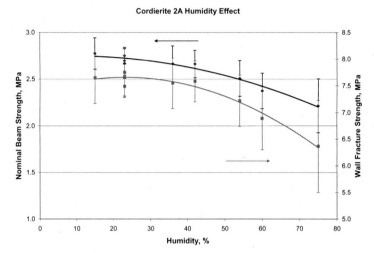

Figure 9. Strength and Humidity Dependence for Cordierite 2A

These figures indicate a possible humidity effect that deserves further investigation and shows that humidity (in storage or during testing) may be a critical variable for cordierite mechanical properties.

Comparison of Cordierite 1A to Cordierite 1B test results is possible as they are the same basic structure (300 cpsi, 0.012 inch wall thickness) and composition with the only difference in lineal cell count in the thickness (A =9 cells, B = 15 cells). Previous work by Webb[2] and Gulati[12] has shown there is an overprediction of the Nominal Beam Strength for small lineal cell counts. In this study the mean nominal beam strength for Cordierite 1A with the smaller thickness is 6% higher strength than the mean nominal beam strength for Cordierite 1B, which agrees with the previous work. When the nominal beam strength for Cordierites 1A and 1B is converted to wall fracture strength, the values for Cordierite 1A and Cordierite 1B are different by only 3%. (The rigorous statistical validity in this data comparison is affected by the between-laboratory variability introduced by the humidity differences and the apparent humidity effect only on the smaller Cordierite 1A and 2A specimens.) These results indicate the validity of the wall fracture strength equation to utilize moment of inertia calculations to convert structure strength into a material strength for better comparison and use in finite element models.

Cordierite 2A can not be directly compared to Cordierites 1A and 1B as it has a different macro and microscopic structure. The calculation results for Cordierite 2A do show a similar relationship between nominal beam strength and wall fracture strength, as observed for Cordierites 1A and 1B.

The larger specimen geometry (Cordierite 1B) has a lower coefficient of variation (4.0%), compared to the small specimen geometry [Cordierites 1A (CoV = 4.5%) and 2A (CoV = 7.3%)]. This is most likely based on the higher precision possible in measuring dimensions and aligning fixtures for the larger specimen geometry.

CONCLUSIONS

Overall, ASTM C1674 can be used for determination of nominal beam strength and wall fracture strength in honeycomb ceramics to provide comparable data between sites.

Use of the standard ASTM C1674 test method produced a large data set for each material which showed that the method was robust. The worldwide participation was essential in generating interest and developing confidence in use of the method.

The nominal beam strength calculation is straight forward and easily completed. The determination of the wall fracture strength is also a simple calculation, when the cell pitch and wall thickness are characterized. These two strengths showed a general relationship factor where the wall fracture strength $\cong 2.5$ x nominal beam strength, for specimens with this cell architecture.

Care should be taken when measuring the specimens to ensure that residual wall stubs on the specimen surface are not included in the dimensional measurements.

The comparison of lineal cell count showed that there is a difference in the nominal beam strength due to differences in cell count, with lower cell counts having higher strength. However, the effects of humidity on the data between the laboratories for a single material was greater than the shift between the two specimen geometries so this difference cannot be statistically confirmed.

In this study the presence of water in the atmosphere appears to have an effect on the strength of the cordierite materials, with strength apparently decreasing as humidity levels increase. The test method should be reviewed for inclusion of a warning on the effect of humidity for cordierite and any other ceramics with known environmental sensitivity, especially if the data is to be used for comparison between laboratories and different materials.

FOOTNOTES

* NGK Automotive Ceramics USA Inc, Novi, MI
** Corning Inc, Corning, NY
*** Caterpillar Inc, Mossville, IL; Corning Inc, Corning, NY; Cummins Inc., Columbus, IN; Dow Chemical Co., Midland, MI; Johnson Matthey, Wayne, PA; NGK Automotive Ceramics USA Inc, Novi, MI; Oak Ridge National Laboratory, Oak Ridge, TN; Robert Bosch Gmbh, Stuttgart, Germany; Università di Padova, Padova, Italy

ACKNOWLEDGEMENTS

We appreciate the support of NGK Ceramics and Corning Inc and providing the test material that was instrumental in conducting this round robin program. Also, the diligence and persistence of Leigh Rogoski and Ken Rogoski at Cummins Inc. in preparing, packaging and shipping the specimens to the different laboratories for testing.

REFERENCES

[1] ASTM C1674, Standard Test Method for Flexural Strength of Advanced Ceramics with Engineered Porosity (Honeycomb Cellular Channels) at Ambient Temperature, Annual Book of ASTM Standards, Vol 15.01, ASTM International, West Conshohocken, PA, 2008
[2] Webb, J. E., Widjaja, S., Helfinstine, J. D., Strength Size Effects in Cellular Ceramic Structures, *Ceramic Engineering and Science Proceeding – Mechanical Properties and Performance of Engineering Ceramics II*, Vol. 27, Issue 2, 521-531 (2006).
[3] Then, P. M., Day, P,, The Catalytic Converter Ceramic Substrate – An Astonishing and Enduring Invention, *Interceram (Germany)*, Vol. 49, No. 1, 20-23 (2000)
[4] Adler, J., Ceramic Diesel Particulate Filters, *International Journal of Applied Ceramic Technology*, Vol. 2, Issue 6, 429-439 (2005)

[5]Bagley, R. D., Doman, R. C., Duke, D. A., McNally, R. W., Multicellular ceramics as catalyst supports for controlling automotive emissions, SAE Paper 730274 (1973).
[6]Helfinstine, J. D., Gulati, S., T., High Temperature Fatigue in Ceramic Honeycomb Catalyst Supports, *SAE Paper 852100* (1985).
[7]Gulati, S. T., Sweet, R. D., Strength and Deformation Behavior of Cordierite Substrates from 70° to 2550°F, *SAE Paper 900268* (1990).
[8]Howitt, J. S., Thin Wall Ceramics as Monolithic Catalyst Supports, *SAE Paper 80082* (1980).
[9]Gulati, S. T., Long-Term Durability of Ceramic Honeycombs for Automotive Emissions Control, *SAE Paper 850130* (1985).
[10]ASTM C1161, Standard Test Method for Flexural Strength of Advanced Ceramics at Ambient Temperature, Annual Book of ASTM Standards, Vol 15.01, ASTM International, West Conshohocken, PA, 2008
[11]Stafford, R. J., Wang, R., Effect of Test Span on Flexural Strength Testing of Cordierite Honeycomb Ceramic, Proceedings of the 31st International Conference on Advanced Ceramics and Composites, Jonathon Salem and Dongming Zhu, ed., The American Ceramic Society, Westerville, OH, 2007
[12]Gulati, S. T., Reddy, K. P., Size Effects on the Strength of Ceramic Catalyst Supports, *SAE Paper 922333* (1992).

ALUMINUM SILICATE AEROGELS WITH HIGH TEMPERATURE STABILITY

Roxana Trifu, Wendell Rhine, Irene Melnikova, Shannon White
Aspen Aerogels, Inc.
Northborough, MA, USA

Frances Hurwitz
NASA Glenn Research Center
Cleveland, OH, USA

ABSTRACT
Aluminum silicate aerogels were prepared from tetraethylorthosilicate and two types of alumina precursors: $AlCl_3 \cdot 6H_2O$ and boehmite. Silica and alumina aerogels have been prepared as well and compared with the aluminum silicate samples. The monolithic and fiber reinforced aerogels were dried with supercritical CO_2. The samples were characterized by TGA/DSC, BET nitrogen adsorption and thermal conductivity analyses. Some samples were also heated in a furnace at various temperatures between 300 and 1300 °C. These heat treated samples were analyzed by XRD and FESEM and the density, shrinkage and thermal conductivities of the samples were determined. Both aluminum chloride and boehmite aerogels were diphasic and formed well crystallized mullite after heating at 1300 °C and 1400 °C in air according to XRD analysis. Aerogels prepared from boehmite exhibited superior thermal stability compared to their aluminum chloride counterparts as determined by thermal conductivity, TGA, shrinkage and weight loss data recorded after heat treatments up to 1100 °C.

INTRODUCTION
NASA's push to hypersonic flight regimes requires novel materials that are lightweight as well as thermally and structurally efficient for airframes and thermal protection systems. The materials required must increase safety and decrease system weight while maintaining their performance throughout the lifetime of the system. A critical component of the system is the thermal protection system, required to maintain internal temperatures compatible with the airframe. Currently available thermal protection system (TPS) designs and materials are not capable of providing the level of protection required by NASA without a significant increase in TPS weight and volume. Therefore, NASA needs new TPS concepts for hypersonic vehicles that will provide the highest level of thermal performance and can also be structurally integrated with the airframe rather than just add parasitic weight. The desired insulation materials must be oxidatively and thermally stable and not degrade over the lifetime of the vehicle.

The development of porous insulation materials that are resistant to sintering at high temperature offers many technical challenges. The sol-gel method is an excellent method for preparing insulative aerogels and also provides an approach for tailoring the compositions of aerogel materials on a molecular scale. Silica aerogels have limited thermal stability because they begin sintering at temperatures above 600 °C. One of the most widely studied sinter resistant ceramic material is mullite ($3Al_2O_3 \cdot 2SiO_2$) which actually is a solid solution that ranges to include $2Al_2O_3 \cdot SiO_2$ and is the only crystalline phase in the binary alumina – silica system.[1,2,3,4] Previous studies have proven that doping silica aerogels with alumina is a successful method for improving the sinter resistance of silica aerogels.[5,6,7]

The sol-gel method has shown promise for preparing porous materials at relatively low temperatures. The synthesis of mullite by the sol-gel method has been extensively studied, and it is well known that its formation depends on the starting materials and the preparation methods. Mullite precursors can be characterized as single phase or diphasic which is determined by the scale of silica-alumina mixing. The diphasic gels are a colloidal mixture of crystalline alumina phases and amorphous silica which transform to mullite upon heating above 1200 °C.[8,9,10] On the other hand, single phase gels are homogeneous on a molecular scale and transform to mullite at temperatures below 1000 °C.[8] Al-Jasha et al.[11] investigated preparation methods using tetraethylorthosilicate (TEOS) and aluminum chloride (AlCl₃) and determined the effect of organic and inorganic additives on mullite formation. Hoffman et al.[8] described several methods for making single phase gels from TEOS and Al(NO₃)₃ and diphasic gels from TEOS and boehmite, and showed that changes in the Al:Si and ethanol:precursor ratios can change DTA behavior. Okada and Otsuka[12] classified single-phase gels prepared from Al(NO₃)₃ and TEOS according to the rate of hydrolysis and demonstrated that reaction paths can be quite different depending on the hydrolysis conditions.

For this effort, we prepared both monolithic and fiber reinforced aluminum silicate aerogels by two routes. The first route was based on an approach developed by Clapsaddle et al.[13] and Baumann et al.[14] and used aluminum chloride as the alumina source, and the second route used boehmite as the alumina source and TEOS as the silica precursor.[7,8] The gels prepared were extracted with supercritical CO_2 to prepare nanoporous coherent aerogel structures.

EXPERIMENTAL METHODS
Preparation of Aluminum Silicate Aerogels from Aluminum Chloride and TEOS
Alumina silica aerogels with various alumina to silica ratios were prepared using $AlCl_3 \cdot 6H_2O$ and TEOS (tetraethylorthosilicate) precursors. Table 1 lists the compositions prepared by this approach. The aluminum chloride was dissolved in water and the solution added to TEOS dissolved in ethanol. The pH of the sol at this point was 1.5-2. Solutions were mixed to hydrolyze the TEOS, cooled to 5 °C in an ice bath and then cold (refrigerated) propylene oxide was added. The gelation time was monitored immediately after addition of propylene oxide (PO) to the mixture. Each sol was cast as 2" diameter monoliths and on 4" x 4" Quartzel™ fiber batting to make fiber reinforced coupons. Once gelled, the batting reinforced gels were aged in ethanol for two days at room temperature during which time the solvent was exchanged three times. Then the samples were aged in ethanol at 55 °C for 65 hours prior to extraction of solvent with supercritical CO_2. The aluminum silicate sols with Al:Si molar ratio of 3:1 were used to infiltrate both Nextel™ 440 fabric and Nextel 720 chopped ceramic fiber reinforcements.

Table 1. Alumina silicate sols prepared with variable alumina content.

Formulation	Al1Si0	Al8Si1	Al4Si1	Al3Si1	Al1Si1
Al:Si (mole:mole)	1:0	8:1	4:1	3:1	1:1
PO/Al molar ratio	6.4	7.4	8.2	8.8	13.1
Hydrolysis Time (h)	1	1	1	1	1
Gelation Time (min)	36	26	23	23	23

Preparation of Aluminum Silicate Aerogels from Boehmite and TEOS
Dispersible boehmite offers an inexpensive, safe source of alumina, which eliminates the use of toxic propylene oxide. Dispal 23N4-80 with 90nm dispersed particle size and Disperal P2 with 20nm particle size obtained from Sasol were dispersed in water or water/ethanol mixtures. The boehmite dispersions were mixed with prehydrolyzed TEOS sols at various Al:Si ratios. After stirring the mixed sol for 1 hour, base or acid catalysts were added to the sol to obtain aluminum silicate gels (Table 2 and Table 3). The boehmite doped sols were cast as both 2" diameter monoliths and 4" x 4" fiber reinforced coupons using Quartzel fiber battings, Nextel 440 fabrics and Nextel 720 chopped fibers. The gels were aged in ammonia solutions at 55 °C for 1 day, followed by two ethanol exchanges. The gels were dried by extraction with supercritical CO_2.

Table 2. Formulations of alumina silica aerogels based on Disperal P2.

Formulation	Al3Si1A	Al4.5Si1B	Al1Si0A
Target density, g/cm³	0.082	0.073	0.067
Al:Si, mole:mole	3:1	4.5:1	1:0
Diperal P2, g	10	10	10
H₂O, mL	100	90	100
Catalyst	Nitric acid	Ammonia	Nitric acid

Table 3. Formulations of aluminum silicate aerogels based on Dispal 23N4.

Sample No.	Alumina content (%)	Al:Si mole:mole	Target density (g/cc)	Catalyst
Al0Si1	0	0:1	0.06	Ammonia
Al0.02Si1	2	0.02:1	0.06	Ammonia
Al0.10Si1	10	0.10:1	0.06	Ammonia
Al0.23Si1	20	0.23:1	0.06	Ammonia
Al3.6Si1	76.4	3.6:1	0.06	Ammonia

CHARACTERIZATION METHODS

The as prepared samples were characterized by various techniques. Thermal analyses were performed at Aspen at 20 °C/min heating rate in air using a Netzsch STA 449C Jupiter TGA/DSC (thermogravimetric analysis/differential scanning calorimetry) and at NASA Glenn Research Center (GRC) using a Netzsch Model STA 409C at 5 °C/min in helium. Helium was used to assist in separating thermal decomposition from oxidation events. Surface areas and pore size distributions were determined by nitrogen adsorption analysis using a Micromeritics ASAP 2010 at Aspen and Micromeritics Instruments Model ASAP2020 at NASA GRC. Some samples were also heated in a furnace for 1h at various temperatures between 300 °C and 1300 °C. These heat treated samples were analyzed at NASA by XRD (x-ray diffraction) using a Philips APD (Automated Powder Diffractometer) for Cu K_α or a Bruker D8 Discover with GADDS (General Area Diffraction Detection System) for Cr K_α and by SEM using a Hitachi S4700 FESEM (Field Emission Scanning Electron Microscopy). The density, shrinkage and thermal conductivities of the samples were determined. Elemental analysis of the aerogels was determined by ICP (inductively coupled plasma) analysis using a Varian Vista Pro in the axial configuration. All of the samples prepared from $AlCl_3$ and boehmite had close to the expected composition except for the $Al_{3.6}Si$ formulation which gave $Al_{3.07}Si$.

Thermal Conductivity

Calibrated hot plate (CHP) devices fabricated by Aspen Aerogels were used to measure the room temperature thermal conductivity of small, flat samples at 37.8 °C (100 °F). The CHP method is based on the principle underlying ASTM E 1225, *Standard Test Method for Thermal Conductivity of Solids by Means of the Guarded-Comparative-Longitudinal Heat Flow Technique*. The method is based on a comparison of temperature drops across the test specimen and a reference sample, which are placed in contact with each other and sandwiched between hot and cold plates to create a heat flux through the stack. Typical lateral dimensions of plates and samples are 125mm (5 inches) square, with thicknesses ranging from 3 to 15mm. Due to the large aspect ratio, the precise lateral dimensions of the plates and samples are not critical, and since calculations are based on temperature measurements made at the center of the specimen/reference stack, results are largely insensitive to edge effects.

RESULTS

TGA and DTA of aluminum silicate aerogels are shown in Figure 1 and Figure 2. The TGA results indicated that aerogels prepared from boehmite lose less weight than those prepared from AlCl$_3$. The aerogels prepared from AlCl$_3$ lose about 35% of their weight when heated to 1100 °C, while those prepared from Dispal 23N4 boehmite precursor lose only 18% with heating to 1400 °C. In both cases, the weight loss due to volatilization of water and decomposition of organics is essentially completed by 500 °C. The higher weight loss recorded for the AlCl$_3$ aerogels is due to loss of residual organic groups formed during the reaction between AlCl$_3$ and propylene oxide.

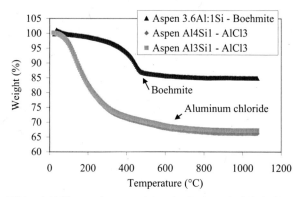

Figure 1. TGAs of Al-Si aerogels prepared from boehmite and AlCl$_3$ in flowing helium

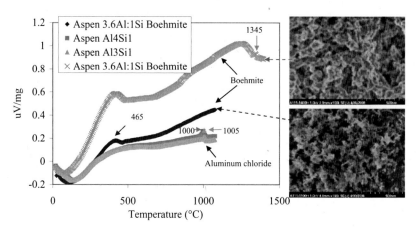

Figure 2. Differential thermal analysis and FESEM images of boehmite and aluminum chloride derived aluminum silicate aerogels.

DTA analysis at NASA GRC showed exothermic peaks which occurred at 1000-1005 °C for Al4Si1 and Al3Si1 aerogels prepared from aluminum chloride (Figure 2). When the Al3Si1 aerogel was heated in air at 20 °C/min, the exothermic peak occurred at 1025 °C. The DTA for the boehmite derived aluminum silicate aerogels showed an exothermic event at 1345 °C when heated at 5 °C/min in helium and at 1379 °C when heated at 20 °C/min in air. Any residual carbon formed during heating in helium is expected to hinder mullite formation;[15] therefore, the observed differences in temperature are attributed to the different heating rates rather than the atmosphere. Based on previously published papers, the exotherms could be due to crystallization of mullite, or the spinel or transitional alumina phases.[8,16] Comparison of FESEM images obtained after heating the boehmite derived aerogel to 1100 °C and 1400 °C clearly show a coarser microstructure at the higher temperature, indicating that the exotherms recorded by thermal analyses above 1300 °C are due to crystallization of the aerogels prepared form boehmite. The phases observed after heating the aerogels were determined by XRD and are discussed in the next section.

XRD Analysis of Aerogels

Mullite crystallizes in an orthorhombic space group (Pbam) and the size of the unit cell and lattice parameters varies with the processing temperature.[17] The a axis for mullite formed at lower temperatures (~1000 °C) is larger than for mullite formed at higher temperatures. Since the a axis is almost the same as the b axis when there is poor splitting of the 120 and 210 reflections, mullite formed at low temperatures is referred to as pseudotetragonal mullite. Mullite formed at high temperatures shows clear splitting of the 120 and 210 reflections and is orthorhombic mullite.[18]

The aerogel samples prepared at Aspen Aerogels were analyzed by XRD. Pure α-alumina was observed for an alumina aerogel after heating at 1300 °C, as shown in Table 4. Although we expected the aluminum silicate samples prepared from AlCl$_3$ to be single phase gels, none of them crystallized to mullite after heating to 1100 °C. In fact, no mullite formed in sample ASP 103 after heating to 1100 °C, although sample ASP 104 contained peaks due to weakly crystalline mullite. The two stage formation of mullite suggests that the "single phase gels" are not truly single phase but are diphasic and are a mixture of polymeric and colloidal gels which transform first to an alumina transitional phase and then to mullite at temperatures above 1200 °C.[16]

Table 4. Summary of XRD results for aerogel prepared from AlCl$_3$.

GRC Sample ID	Precursor Formulation	Temp (°C)	Atm	Exposure	Phases
ASP101	AlCl$_3$ Al1Si0	1300	Air	Furnace 1h	Corundum, trace of α- cristobalite
ASP103	AlCl$_3$ Al4Si1	1100	He	DTA	Weakly crystalline transition alumina (γ-Al$_2$O$_3$)
ASP104	AlCl$_3$ Al3Si1	1100	He	DTA	Weakly crystalline mullite and transition alumina
ASP104	AlCl$_3$ Al3Si1	1300	Air	Furnace 1h	Well crystallized mullite

The XRD pattern for sample ASP 104 that was heated to 1300 °C for 2 hours in air is shown in Figure 3. The XRD results indicate that well crystallized mullite was formed. There was good agreement between the observed d-spacings and those reported for PDF 15-776. The splitting between the 120 and 220 reflections is poor and the two peaks are not resolved indicating that the mullite formed form TEOS and AlCl$_3$ could be referred to as pseudotetragonal mullite.

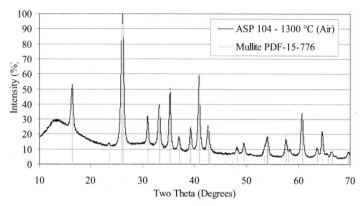

Figure 3. XRD of ASP 104 after heating to 1300 °C in air for 2 hours (Cu Kα).

Samples of diphasic aerogels prepared from boehmite were also analyzed by XRD, and the results are summarized in Table 5. The sample heated to 1100 °C was weakly crystalline and some peaks due to mullite were present. Diffraction peaks for a transitional alumina phase also were present which could be gamma (γ), delta (δ), or theta (θ) alumina. None of the transition alumina phases have diffraction peaks that exactly match the observed diffraction pattern.[19] Most likely the phases present are not pure alumina and contain both alumina and silica which alters the d-spacings.[20]

Table 5. Summary of XRD results for aerogel derived from boehmite.

Sample ID	Precursor	Temp	Atm	Exposure	Phases
ASP 111	Boehmite Al3.6Si1	1100	He	DTA	Weakly crystalline mullite; and gamma alumina
ASP 111	Boehmite Al3.6Si1	1400	He	DTA	Mullite
ASP 111	Boehmite Al3.6Si1	1300	Air	Furnace 1h	Mullite and δ, θ alumina

After heating in air at 1300 °C, the mullite diffraction peaks did not increase much in intensity, and the sample still contained phases other than mullite. The best fit seems to indicate that the transitional phase is a mixture of δ and θ alumina as shown in Figure 4. After heating to 1400 °C in helium, the only phase present is mullite, as indicated by the XRD pattern shown in Figure 5. The aerogel prepared from boehmite and silica required heating at 1400 °C to convert the aerogel to mullite. The splitting between the 120 and 220 reflections is poor and the two peaks are not resolved indicating that the mullite formed using boehmite as the alumina source could also be referred to as pseudotetragonal mullite. Although the mullitization temperature is high, these results are expected for diphasic aerogels. The aerogels prepared from aluminum chloride form mullite at lower temperatures than those prepared from boehmite, indicating that the gels prepared from aluminum chloride are more homogeneous than those prepared from boehmite. The shift to higher crystallization temperatures in the boehmite derived Al-Si aerogels actually is helpful in delaying shrinkage due to crystallization and improving their thermal stability.

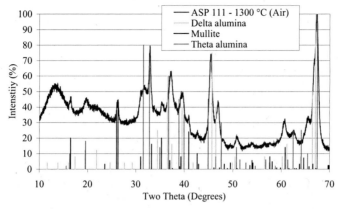

Figure 4. XRD of ASP-111 Al3.6Si1 heated to 1300 °C in air for 2h (Cu Kα).

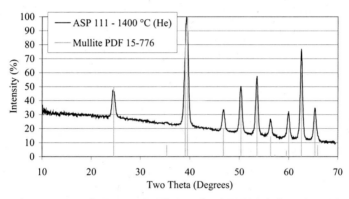

Figure 5. XRD of ASP 111 Al3.6Si1 heated to 1400 °C in helium (Cr Kα).

Pore Size Distributions and Surface Areas of the Aluminum Silicate Aerogels

Surface areas and pore sizes differ substantially in aluminum chloride and boehmite derived Al-Si aerogels as a result of the different precursors and other preparation variables, such as TEOS prehydrolysis, different gelation catalysts and pH.[21] The N_2 adsorption pore size distributions, pore volumes and surface areas were determined before and after heating the aluminum silicate aerogels. According to nitrogen adsorption porosimetry analysis, an aerogel monolith of Al3Si1 composition prepared from aluminum chloride precursor showed a narrow pore size distribution with the majority of pores between 5 and 25 nm. After heating this aluminum silicate aerogel to 1000 °C, the surface area decreased from 822m^2/g to 425m^2/g and the average pore size and pore volume decreased (Figure 6). Pores larger than 100nm may be present after heating but are not detected by N_2 adsorption methods. The observed decrease in surface area, pore size, and pore volume are consistent with densification of the aerogel by dehydroxylation of the surface and sintering.

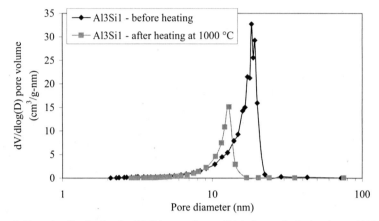

Figure 6. Pore size distribution for Al3Si1 aerogel monolith before and after heating at 1000 °C.

BET analysis after step heating in air of an aluminum silicate aerogel of mullite composition reinforced with Quartzel fibers is shown in Table 6. The data listed in Table 6 indicates that there was a continuous decrease of surface area and pore size with increasing heat treatment temperature, suggesting pore collapse and sintering. The decrease in pore volume and corresponding increase in density explains the increase of the sample's thermal conductivity from 20.2 to 40.5 mW/m-K upon heating to 1100 °C.

Table 6. BET analysis of Al3Si1 aerogel derived from AlCl₃ reinforced with Quartzel.*

Temperature Heat treatment	Initial	500 °C	700 °C	1000 °C	1100 °C
Surface Area (m^2/g)	913.5	460.7	405.0	171.0	161.3
Pore volume (cm^3/g)	5.9	2.9	2.4	0.8	0.7
Ave. Pore diameter (nm)	25.8	25.2	23.5	17.8	18.3

* Powder was extracted from the Quartzel fiber reinforced coupon

An aluminum silicate aerogel of Al3.6Si1 composition prepared from the boehmite precursor Dispal 23N4 was analyzed for surface area and porosimetry before and after heating in air (Table 7). BET analysis of the same sample determined by NASA showed that the surface area of the aerogel was 210 m^2/g and had a pore volume of only 0.376 cm^3/g. The pore volume of this sample was expected to be about 10 cm^3/g based on a bulk density of 0.096 g/cm^3 and a skeletal density of 2.8 g/cm^3. While the initial surface area was low, it decreased only slightly upon heating up to 1100 °C. After heating for 1h at 1300 °C in air, there was a significant decrease in surface area. This decrease in surface area was accompanied by an increase in the pore size and a sharp decrease in the pore volume, suggesting pore collapse caused by the beginning of mullite formation and sintering.

Table 7. BET Analysis of Al3.6Si1 boehmite aerogel.*

Heat Treatment Temperature (°C)	Initial		After 1100 °C	After 1300 °C
	Aspen Aerogels	NASA GRC		
Surface Area (m²/g)	218.2	210	194.2	32.3
Pore volume (cm³/g)	0.517	0.376	0.657	0.242
Ave. Pore diameter (nm)	9.5	8.48	13.5	29.9

*Powder was extracted from the Quartzel fiber reinforced coupon

Properties of Aerogel Composites Before and after Heat Treatment

The fiber reinforced aerogels were prepared using Dispal 23N4 as the alumina source with the formulations listed in Table 3. Thermal conductivities (TC) and densities for these aerogels are listed in Table 8, and the samples prepared are illustrated in Figure 7. With increasing boehmite doping, the aerogels became less translucent and turned opaque. The monolithic aerogel with highest alumina content was cracked (Figure 7), but the monolithic aerogels for the other compositions were crack free. The lowest thermal conductivity was obtained for the 100% silica aerogel, as expected. XY shrinkage of aerogel monoliths during processing was reduced in half by doping the silica sol with only 2% alumina.

Table 8. Alumina silica aerogels prepared from boehmite (Dispal 23N4).

Sample	Alumina content (%)	Al:Si mole:mole	TC coupon (mW/m-K)	Density coupon (g/cm³)	X-Y shrinkage Monolith (%)
Al0Si1	0	0:1	11.9	0.119	7.3
Al0.2Si1	2	0.02:1	13.7	0.099	3.5
Al0.10Si1	10	0.10:1	14.0	0.107	3.1
Al0.23Si1	20	0.23:1	13.7	0.110	3.5
Al3.6Si1	76.4	3.6:1	17.5	0.096	--

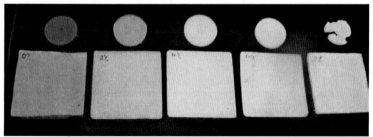

Figure 7. Alumina silica aerogels prepared from boehmite Dispal 23N4 (increasing alumina content from left to right).

After heating these aerogels at 1000 °C and 1100 °C the alumina rich samples with a mullite composition showed lower shrinkage and lower thermal conductivities than the silica rich aerogels (Figure 8). The Al3.6Si1 aerogel containing 76.4% alumina prepared from boehmite shrank less than the silica aerogel by 60%, and its thermal conductivity was 40% lower compared to the TC for the silica aerogel after the heat treatment as shown in Figure 8. By doping the silica matrix with alumina, the sintering temperature is increased by increasing the viscosity of the aluminum silicate, and in addition, the aluminum sites adjacent to silicon atoms hinder condensation of silanol groups which also reduces shrinkage and sintering.[6]

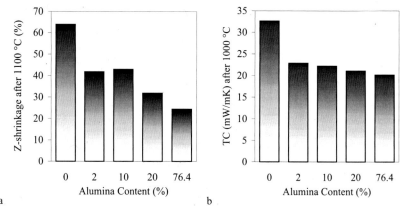

a   b

Figure 8. Plots of a) Z-shrinkage and b) TCs after heat treatment at 1000 °C versus the alumina content in aluminum silicate aerogels prepared form Dispal 23N4.

The thermal performance and stability of heat treated silica, aluminum silicate and alumina aerogels was investigated by comparing their thermal conductivities (Figure 9). Thermal conductivities of the aluminum silicate aerogels of mullite composition (Al3Si1 and Al3.6Si1) derived from Dispal 23N4 remained essentially unchanged after heat treatment up to 1100 °C, in contrast with those of silica and alumina aerogels which increased after heating at temperatures above 700 °C.

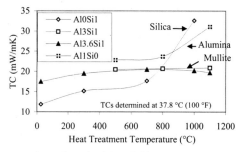

Figure 9. Room temperature thermal conductivies of silica, aluminum silicate and alumina aerogels after heat treatment in air.

Before heating, the thermal conductivities of the Quartzel fiber reinforced aerogels prepared from Disperal P2 showed little dependence on the alumina content or other preparation variables. For example, the thermal conductivity of samples prepared as described in Table 2 was 19 – 21 mW/m-K, at densities varying from 0.08 to 0.11 g/cc (Table 9). Usually, the surface area of aluminum silicate aerogels decreases and TC increases with increasing alumina content, but this was not the case for boehmite derived aluminum silicate aerogels. It is likely that the surface area and pore volume of aerogels prepared from boehmite show less variation with the preparation variables than those prepared from $AlCl_3$; however, more BET and mercury porosimetry analyses are required to elucidate the pore size distribution of these aerogels.

Table 9. TCs and densities of Quartzel fiber reinforced boehmite derived aluminum silicate aerogels.

Disperal P2 (Particle size: 20 nm)			
Formulation	Al3Si1A	Al4.5Si1B	Al1Si0A
TC, (mW/m-K)	18.7	18.9	20.6
Density, (g/cc)	0.115	0.092	0.081

The Quartzel fiber reinforced coupons listed in Table 9 were evaluated after heating at various temperatures. While X-Y shrinkage was negligible, shrinkage through the thickness (Z-shrinkage) increased most after heating at temperatures over 800 °C, reaching 20% for the aluminum silicate sample and 30% for the aluminum aerogel after the 1100 °C heat treatment. The aluminum silicate samples showed slightly better performance than the 100% alumina aerogel, and had lower shrinkage and thermal conductivities over the entire range of heat treatments (Figure 10). Regardless of the composition, the fiber reinforced boehmite derived aluminum silicate aerogels showed good stability with temperatures up to 1100 °C; the thermal conductivity values of these samples were below 25mW/m-K after the heat treatment.

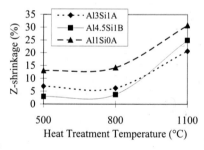
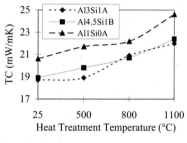

a   b

Figure 10. a) Z-Shrinkage and b) room temperature TCs of aluminum Quartzel fiber reinforced silicate aerogels prepared from boehmite Disperal P2 after heating in air.

For the aluminum silicate aerogels derived from aluminum chloride the weight loss generally increased with increasing temperature and with silica content. The fiber reinforcement inhibited linear shrinkage on the X-Y axis which was on the order of a few percent. Shrinkage on Z-axis was much larger and it generally reached over 25% after heating at 1100 °C (Figure 11). Silica rich aerogels shrank more than alumina rich aerogels. For example, increasing the alumina content in aerogels from 33% to 80% resulted in reducing the Z-shrinkage in half, from 50% to 25%.

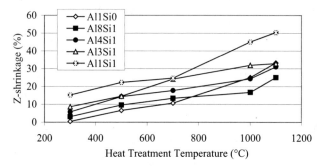

Figure 11. Shrinkage of Quartzel fiber reinforced aluminum silicate aerogels derived from aluminum chloride versus heat treatment temperature.

The thermal conductivity (TC) of aerogels prepared from the aluminum chloride precursor increased with increasing the heat treatment temperature, especially after heating at 700 °C (Figure 12). Although before heat treatment the thermal conductivity of aerogels showed a clear dependence on the alumina content, with higher values for alumina-rich samples, after heating at 1100 °C the thermal performance of the aerogels prepared from the aluminum chloride precursor was similar (TC of about 40 mW/m-K to 43 mW/m-K).

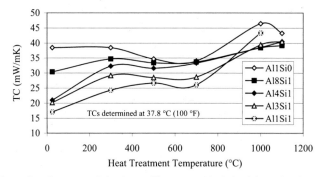

Figure 12. Thermal performance of aluminum silicate aerogels derived from aluminum chloride as a function of the heat treatment temperature.

The performance and stability after heating of Quartzel reinforced aerogels of mullite composition prepared from aluminum chloride and boehmite were compared. Aerogels with Al:Si molar ratio of 3:1 prepared from the boehmite precursor showed better performance and stability after heat treatments than their aluminum chloride counterparts. Although the Al3Si1 aerogel prepared from AlCl$_3$ shrank significantly above 500 °C, the boehmite derived aerogel remained stable up to 800 °C (Figure 13, a). As discussed above, shrinkage of Al-Si aerogels was correlated to viscous sintering and to the efficiency with which alumina increases the viscosity and prevents densification and pore collapse. It is likely that boehmite particles are more effective in increasing the viscosity and hindering densification of Al-Si aerogels than AlCl$_3$ which forms more homogeneous gels.

The boehmite derived aerogel showed almost no change in thermal conductivity after heating up to 1100 °C, while a much higher thermal conductivity increase was observed for the aerogel prepared from $AlCl_3$ (Figure 13, b). As the data listed in Table 6 indicates, the surface area and pore volume of Al-Si aerogels prepared from $AlCl_3$ decrease as a function of the heat treatment temperature, and as a result the thermal conduction increases in these heat treated aerogels. In contrast, the surface area and pore volume of the boehmite derived aerogels remain unchanged with heating up to 1100 °C as indicated by the data listed in Table 7. As a result their thermal conductivities do not increase and remain low (Table 7).

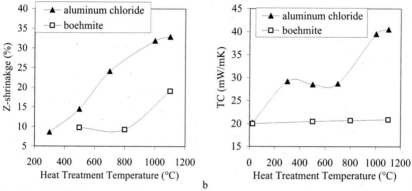

a   b

Figure 13. a) Z-Shrinkage and b) Room temperature TCs for aerogels with Al:Si molar ratio of 3:1 prepared from boehmite and aluminum chloride.

Characterization of Aerogels Reinforced with High Temperature Nextel Fibers

Quartzel batting used in the reinforcement of aerogels has a maximum service temperature in air of 1050 °C. Other reinforcements with higher temperature resistance, such as Nextel 440 and 720 ceramic fibers have also been investigated (
Figure **14**).

Figure 14. Photos of aluminum silicate aerogels reinforced with a) Nextel 440 fabric and b) Nextel 720 chopped fiber.

The thermal performance of aluminum silicate aerogels with Al:Si molar ratio of 3:1 reinforced with Nextel was investigated after heating them in air at several temperatures (Table 10). Regardless of the preparation method, an outstanding insulating effect was achieved by infiltration of Nextel fiber with aerogel. The thermal conductivity of the Nextel fabric before infiltration was 107 mW/m-K, at a density of 0.95 g/cc. A thin layer of aerogel was able to reduce the thermal conductivity of the Nextel reinforcement by 75-90% and the density by more than half.

The boehmite derived aerogel reinforced with Nextel 440 fabric exhibited a very low thermal conductivity of 12mW/m-K after heat treatment in air at 500 °C and weight loss of only 5% after heating at 1100 °C for one hour. The thin flexible aerogels reinforced with Nextel fabric outperformed those reinforced with chopped Nextel short fibers, however they had higher densities.

Table 10. Densities, TC and weight loss of Nextel reinforced boehmite aerogels heated to various temperatures.

Precursor	Nextel fiber type	Density g/cm³	TC, mW/m-K			Weight loss, %		
			500°C	800°C	1100°C	500°C	800°C	1100°C
AlCl₃	720, ¼" chopped	0.250	35.9	40.4	40.3	6.90	9.49	9.14
	440, 1 layer	0.435	24.4	31.2	29.9	5.69	7.28	7.68
Boehmite Disperal P2	720, ¼" chopped	0.256	21.6	21.5	22.0	5.80	6.87	7.84
	440, 1 layer	0.458	12.3	12.2	17.7	2.99	3.50	5.39

Consistent with previous findings, boehmite aerogels reinforced with Nextel fibers showed better thermal performance than the composites prepared using aluminum chloride as the alumina source. As shown in Figure 15, the aerogels prepared from boehmite showed excellent thermal stability with heating up to 1100 °C; their thermal conductivities changed very little with the heat treatment. However, the data listed in Table 11 indicates that the composites suffer significant changes after heating at 1300 °C which were reflected by an abrupt increase in their thermal conductivities and densities suggesting that structural changes occurred during heating. Aluminum chloride aerogels cracked extensively after heating at 1300 °C and were not tested.

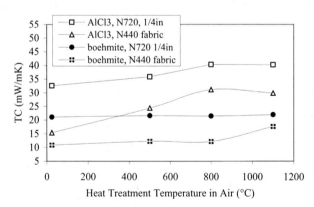

Figure 15. Room temperature thermal conductivity of Nextel reinforced aerogels after heat treatments in air at various temperatures.

Table 11. Effect of heat treatment on the thermal conductivity and density of boehmite derived aluminum silicate aerogels reinforced with Nextel fibers.

Temperature (°C)	Nextel 720, ¼ in chopped		Nextel 440 fabric	
	TC (mW/m-K)	Density (g/cc)	TC (mW/m-K)	Density (g/cc)
25	21.1	0.253	10.9	0.462
1100	22.0	0.268	17.7	0.490
1300	41.3	0.386	31.2	0.592

CONCLUSIONS

Aluminum silicate aerogels have been prepared from TEOS and two different types of alumina precursors: $AlCl_3 \cdot 6H_2O$ and boehmite. The preparation route, as well as the fiber reinforcements in the form of Quartzel, Nextel fabric and chopped fibers played an important role in determining the properties of aerogels such as thermal conductivity, density, weight loss and shrinkage. Alumina content had a strong effect on the physical properties of aluminum silicate aerogels and their thermal stability. In Quartzel reinforced aerogels derived from $AlCl_3$ shrinkage through the Z-axis was reduced in half by increasing the alumina concentration from 33% to 80%.

Thermal conductivity of aluminum silicate aerogels of mullite composition prepared from the boehmite precursor remained essentially unchanged after heat treatment to 1100 °C, in contrast with that of silica and alumina aerogels which suffered thermal performance loss, especially after heat treatment at 700 °C in air. Diphasic aluminum silicate aerogels of Al3:Si1 composition derived from boehmite precursors outperformed their counterparts prepared from aluminum chloride showing lower shrinkage, weight loss and better thermal stability after heat treatment in air up to 1100 °C. Thermal conductivity of Nextel reinforced aluminum silicate aerogels prepared from the boehmite precursor was half that of the aluminum chloride derived aerogels after heat treatments. This result correlates well with the decrease of the surface area of mullite aerogel derived from aluminum chloride from 913 m^2/g to 161 m^2/g with step heating in air up to 1100 °C, while the surface area of boehmite aerogel remained essentially unchanged after heat treatments.

After heating at 1300 °C the thermal conductivity of the boehmite aerogel increased abruptly suggesting that significant structural changes start to take place. Thermal analyses conducted at NASA GRC (Aspen) demonstrate that crystallization events start to occur at about 1000 °C (1025 °C) for aluminum silicate aerogels prepared from $AlCl_3$ and at 1345 °C (1379 °C) for those prepared from the boehmite precursor. Although weakly crystalline mullite was detected after heat treatment at 1100 °C in case of the aluminum chloride derived aerogel, well crystallized mullite was formed only after heat treatment at 1300 °C according to XRD analysis. The data suggests that the aluminum chloride aerogels do not form as a perfectly homogeneous single phase material but rather as diphasic gels. Although mullitization temperatures for boehmite derived Al-Si aerogels were high according to XRD analysis, 1400 °C in air, they are in agreement with formation of diphasic aluminum silicate aerogels as was expected from using colloidal boehmite as the alumina precursor.

REFERENCES

1   I.A. Aksay and J.A. Pask, Stable and Metastable Equilibria in System Al_2O_3 SiO_2, *J Am. Ceram. Soc.*, **58**, 507 (1988).
2   F.J. Klug, S. Prochazka, and R.H. Doremus, Alumina–Silica Phase Diagram in the Mullite Region, *J. Am. Ceram. Soc.*, **70**, 750–759 (1987).
3   W.E. Cameron, Mullite a Substituted Alumina, *Am. Miner.*, **62**, 747–755. (1977)

4   R.A. Angel, C.T. Prewitt, Crystal Structure of Mullite: a Re-examination of the Average Structure, *Am. Miner.*, **71**, 1476–1482 (1986).
5   R. Saliger, T. Heinrich, T. Gleissner, J. Fricke, Sintering Behavior of Alumina-modified Silica Aerogels, *J. Non-Crystal. Solids*, **186**, 113 (1995).
6   B. Himmel, Th. Gerber, H. Burger, G. Holzhuter, A. Olbertz, Structural Characterization of SiO_2-Al_2O_3 aerogels, *J. Non-Crystal. Solids*, **186**, 149 (1995).
7   P.R. Aravind, P. Mukundan, P. K. Pillai, K.G.K. Warrier, Mesoporous Silica-alumina Aerogels with Thermal Pore Stability Through Hybrid Sol-Gel Route Followed by Supercritical Drying, *Micropor. Mesopor. Mater.*, **96**, 14 (2006).
8   D.W. Hoffman, R. Roy, and S. Komarneni, Diphasic Xerogels, A New Class of Materials: Phases in the System Al_2O_3-SiO_2, *J. Am. Ceram. Soc.*, **67** [7] 468-71 (1984).
9   S. Komarneni, Y. Suwa, and R. Roy, Application of Compositionally Diphasic Xerogels for Enhanced Densification: The System Al_2O_3-SiO_2, *J. Am. Cerarn. Soc.,* **69** [7] C-155-C-156 (1986).
10 M. B. M. U. Ismail, Z. Nakai, K. Minegishi, and S. Somiya, Synthesis of Mullite Powder and Its Characteristics, *Int. J. High Technol. Ceram.*, **2**, 123 (1986).
11 Y.M.M. Al-Jarsha, K.D. Biddle, A.K. Das, T.J. Davies, H.G. Emblem, K. Jones, J.M. McCullough, M.A. Rahman, A.N.A. Deen, and R. Wakefield, Mullite Formation from Ethyl Silicate and Aluminum Chlorides, *J. Mater. Sci.*, **20**, 1773-81 (1985).
12 K. Okada and N. Otsuka, Characterization of the Spinel Phase from SiO_2-Al_2O_3 Xerogels and the Formation Process of Mullite, *J. Am. Ceram. Soc.,* **69** [9] 652-56 (1986).
13 B.J. Clapsaddle, D.W. Sprehn, A.W. Gash, J.H. Satcher, Jr., R.L. Simpson, *J. Non-Crystal. Solids*, **350**, 173 (2004)
14 T.F. Baumann, A.E. Gash, S.C. Chinn, A.M. Sawvel, R.W. Maxwell, and J.H. Satcher, Jr., *Chem Mater.*, **17**, 395 (2005)
15 D.X. Li and W.J. Thomson, *J. Am. Ceram. Soc.*, **73**, 964 (1990)
16 D.X. Li and W.J. Thomson, *J. Am. Ceram. Soc.*, **74**, 574 (1991)
17 Power Diffraction File 15-776, International Centre for Diffraction Data
18 T. Ban and K. Okada, *J. Am. Ceram. Soc.*, **75**, 227 (1992)
19 K. Wefers and C. Misra, Oxides and Hydroxides of Aluminum, Alcoa Laboratories, 1987 http://www.alcoa.com/global/en/innovation/papers_patents/details/1987_paper_oxides_and_hydroxides.asp
20 R.L. Oréfice and W.L. Vasconcelos, *J. Sol-Gel Sci. Technol.*, **9**, 239 (1997)
21 P.R. Aravind, P. Shajesh, S. Smitha, P. Mukundan and K.G. Warrier, Nonsupercritically dried Silica-Alumina Aerogels-Effect of Gelation pH, *J. Am. Ceram. Soc.*, **91**, 1326 (2008)

DEVELOPMENT OF NOVEL MICROPOROUS ZrO$_2$ MEMBRANES FOR H$_2$/CO$_2$ SEPARATION

Tim Van Gestel, Doris Sebold, Wilhelm A. Meulenberg, Martin Bram, Hans-Peter Buchkremer, Detlev Stöver
Forschungszentrum Jülich GmbH, Institute of Energy Research, IEF-1: Materials Synthesis and Processing, Leo-Brandt-Strasse, D-52425 Jülich, Germany

ABSTRACT

This paper reports a study on the preparation of novel microporous separation membranes. These separation membranes show a typical graded structure in analogy with current ceramic gas separation membranes (macroporous support material, mesoporous intermediate layers, microporous functional layer) and are made using inexpensive, simple, scalable nano-dispersion and sol coating methods. The essential new features of the membranes include the application of 8Y$_2$O$_3$-ZrO$_2$ mesoporous intermediate membrane layers which show an improved thermal stability and ZrO$_2$ and TiO$_2$ based functional membrane layers, which display a much higher chemical and thermal stability than current SiO$_2$ based sol-gel gas separation membranes.

INTRODUCTION

A microporous ceramic separation membrane can be considered as a graded multilayer porous ceramic material, in which the last membrane layer shows a pore size < 2 nm. According to IUPAC notation, porous materials are classified into three kinds: microporous materials have a pore diameter < 2 nm, mesoporous materials have pore diameters between 2 nm and 50 nm and macroporous materials have a pore diameter > 50 nm.

The number of potential applications of such a microporous material which looks very simple at first sight is immense. Microporous membranes can be used for separation of all types of liquid – solid, liquid – liquid or gas – gas mixtures through effective exclusion mechanisms such as molecular sieving. The application field of the material is mainly determined by the pore size of the functional microporous layer, which implies that this layer must contain a minimal amount of defects and for certain applications the functional layer should be completely defect-free (e.g. gas separation). Possible application fields of a microporous ceramic membrane in different membrane processes are given in Table 1.

Table 1. Application fields of microporous ceramic membranes in different membrane processes

Appication field	Membrane process	Pore size functional layer	Complexity of preparation process
Separation Liquid – Dissolved Macromolecules	Nanofiltration	1 – 2 nm	Few larger defects are tolerated
Separation Liquid – Liquid	Pervaporation	< 1 nm	Few smaller defects or larger pores are tolerated
Separation Gas – Gas	Gas separation	< 0.6 nm	Defect-free membrane required

Membranes for nanofiltration (NF), pervaporation (PV) and gas separation (GS) can be classified into two major groups according to their material properties: organic polymeric membranes and inorganic microporous ceramic membranes. Organic polymeric membranes are relatively easy to prepare and can be produced cheaply at large scale. However, their application is limited to moderate temperatures and to feed streams which are not too corrosive.

Ceramic membranes with a toplayer made of microporous SiO$_2$ have been frequently considered as an alternative for polymeric membranes for a number of applications, because this material can also be rather easily synthesized, using common sol-gel technology [1]. Further, the desired pore size for the above mentioned applications (0.5 – 2 nm, dependent on the membrane process) can be easily obtained by using a common Si(OC$_2$H$_5$)$_4$ precursor and very simple synthesis methods [2]. A significant drawback of current microporous SiO$_2$ based membranes is however the limited chemical stability of the applied membrane materials towards compounds which are often present in gas and liquid feed streams such as water (vapour), acid and basic compounds [3].

Membrane materials based on zirconia and titania are generally recognized for their excellent chemical stability and – with the addition of a doping compound – also a high thermal stability can be obtained for such materials. In our research, the aim is to develop new graded multilayer membrane systems based on ZrO$_2$ and/or TiO$_2$, which are applicable for (advanced) nanofiltration (NF), pervaporation (PV) and gas separation (GS) and which show a good stability for the respective separation processes. The ultimate objective of our research is to provide a membrane based on the same materials, which can operate reliably in gas separation devices of fossil power plants.

EXPERIMENTAL
1. Substrate

The first substrate type is a graded substrate consisting of a NiO/8Y$_2$O$_3$-ZrO$_2$ plate with a pore size in the micron range and a macroporous NiO/8Y$_2$O$_3$-ZrO$_2$ membrane layer with a thickness of ~ 10 μm. The substrate plates are prepared according to a large-scale procedure, which has been applied in our institute for the manufacturing of solid oxide fuel cells during the last decades.

The first step in the preparation procedure includes the formation of a plate with a size of 25 x 25 cm² and a thickness of ~ 1.5 mm by a warm-pressing procedure. In order to reduce the roughness and the pore size for the deposition of very thin mesoporous and microporous membrane layers, an intermediate macroporous NiO/8Y$_2$O$_3$-ZrO$_2$ layer is deposited by vacuum slip-casting, starting from well-known commercially available NiO and 8Y$_2$O$_3$-ZrO$_2$ powders (J.T. Baker, Tosoh corporation). Partial sintering of the macroporous layer is done in air at 1000°C for 2 h.

In Figure 1a, the pore size distribution of the first substrate is shown. From this graph, a pore size of ~ 0.7 μm (large peak – support plate) and ~ 200 – 300 nm (small peak – intermediate macroporous membrane layer) is evident. Figure 1b shows a detail surface micrograph of the slip-casted intermediate membrane layer. This micrograph confirms a typical macroporous structure with a particle size in the range 200 – 400 nm and an average pore size in the range 200 – 300 nm. Further, it is also confirmed that this layer shows a wide pore size distribution, which is actually not the most optimal siutation for the deposition of thin-film membranes.

A second model substrate with a low surface roughness and a uniform pore size distribution was made by vacuum-casting a 8Y$_2$O$_3$-ZrO$_2$ powder suspension into disks with a thickness of ~ 3 mm. Each disk was sintered at 1100°C for 2h and then subjected to surface grinding with a diamond grinding wheel and polished very carefully with diamond particles (6 μm, 3 μm). The final size and thickness of the substrates was ~ 4 cm and ~ 2 mm, respectively. Due to their well-defined pore structure (very uniform pore size of ~ 120 nm) and small surface roughness, these substrates appear very suitable as a model substrate during the development process of the membranes; for example, these substrates are used for studying new sol-gel coating processes, before transferring them to large-scale substrates.

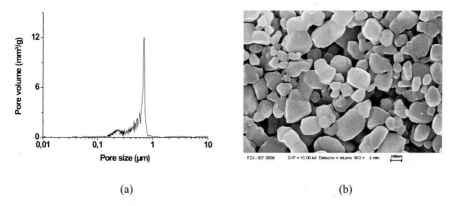

(a)   (b)

Fig. 1a. Pore size distribution of the first substrate type (gradual), consisting of a pressed support plate and a macroporous membrane layer. Fig. 1b. Surface micrograph of the macroporous layer.

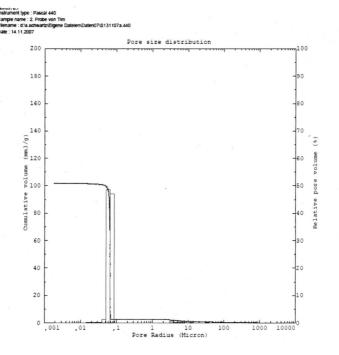

Fig. 1c. Pore size distribution of a model substrate with a very unifrom pore size

2. Preparation of mesoporous intermediate membrane layers

As shown in previous coating experiments [4], deposition of continuous membrane layers on said substrates – especially the assymmetric substrates – requires a coating liquid containing relatively large particles. In this paper, coating experiments with four different coating liquids - including two nano-dispersions and two sols - are described. This includes the deposition of two kinds of nano-dispersions and a sol with larger particles for the preparation of mesoporous intermediate layers and a special sol containing nano-particles for the preparation of the microporous separation layer.

A first nano-dispersion was made starting from a commercially available 8Y$_2$O$_3$-ZrO$_2$ nano-powder (Evonik Degussa) and an aqueous nitric acid solution. Characterization of the particle size was done by dynamic laser beam scattering (Horiba LB-550) in a similar way as in our previous work [1] and here an average particle size of ~ 85 nm was measured. A coating liquid was prepared from this dispersion by adding polyvinyl alcohol (PVA) as coating and drying controlling additive. Subsequently, a second nano-dispersion with a particle size of ~ 65 nm was prepared, using a 8Y$_2$O$_3$-ZrO$_2$ nano-powder supplied by Sigma-Aldrich.

The third coating liquid was a 'so-called' colloidal sol. The sol was prepared starting from a metal organic precursor (Zr(n-OC$_3$H$_7$)$_4$, Aldrich), by means of hydrolysis and condensation of this precursor in isopropanol-water-HNO$_3$ solutions at 98°C. Yttria-doping (8 mol% yttria) was done by adding the proper amount of Y(NO$_3$)$_3$.6H$_2$O to the zirconia sol. In this procedure, the size of the colloidal particles in the sol was controlled by the time of the aging process and the best results were obtained for particles with an average size of ~ 30 – 40 nm.

The fourth coating liquid in this study was a 'so-called' polymeric sol, which contains particles with a size in the nanometer range. This sol was produced by controlled hydrolysis of Zr(n-OC$_3$H$_7$)$_4$, in the presence of diethanol amine (DEA) as a precursor modifier/polymerization inhibitor and an yttrium precursor (Y(i-OC$_3$H$_7$)$_3$) as a doping compound. The essential feature of the preparation route is the addition of DEA, which leads to a reproducible formation of nano-particles with a size of 5 - 10 nm in the synthesis process and also acts as a coating and drying controlling additive. In Figure 2, an overview of the size distributions measured by dynamic laser scattering of the prepared sols and nano-dispersions is shown.

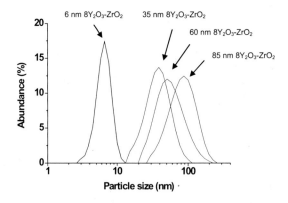

Fig. 2. Particle size distribution of nano-dispersions and sols described in this work

3. Membrane coating
Dip-coating experiments were performed using an automatic dip-coating device, equipped with a holder for 4 x 4 cm^2 square or disk-shaped substrates. The former square substrates were cut from 25 x 25 cm^2 plates obtained from our standard production process (see Fig. 1a and 1b); for the preparation of polished substrates, a disk shape was preferred (see Fig. 1c). In the dip-coating process, sol particles were deposited as a membrane film by contacting the upper-side of the substrate with the coating liquid, while a small under-pressure was applied at the back-side. The obtained supported gel-layers were fired in air at 450°C – 500°C for 2 h and then the coating – calcination cycle was repeated once, unless otherwise stated.

RESULTS AND DISCUSSION
1. Membrane preparation on graded large-scale supports
1.1 Membrane combination 1
Figures 3a and 3b show an overview and a detail micrograph of a membrane layer obtained by coating the first nano-dispersion with a particle size of ~ 85 nm (calcination temperature 500°C). From these micrographs, a typical graded membrane structure can be observed comprising subsequently the support plate, the macroporous NiO/8Y$_2$O$_3$-ZrO$_2$ layer and the calcined 8Y$_2$O$_3$-ZrO$_2$ layer. The coated 8Y$_2$O$_3$-ZrO$_2$ layer is in this back-scattering type of SEM micrograph visible as a brighter film, due to its much smaller pore size. Further, it appears that a separate and continuous layer was obtained, which uniformly covers the macroporous intermediate layer. By looking at the detail micrograph 3b, it appears clearly that infiltration of 8Y$_2$O$_3$-ZrO$_2$ particles into the macropores of the intermediate layer could be prevented, by applying a plastic compound (PVA) as an additive in the coating liquid. In this micrograph, a separation line between the successive membrane layers is also visible and it appears that the thickness of a single layer obtained by one coating – calcination step measures ~ 2 μm.

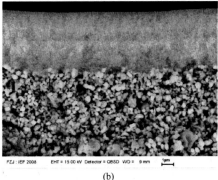

(a)   (b)

Fig. 3a. Micrograph of a calcined mesoporous 8Y$_2$O$_3$-ZrO$_2$ membrane layer, obtained by dip-coating nano-dispersion 1 on the standard substrate. Fig. 3b. Back-scattering detail micrograph
(Layer made by 2 x dip-coating and calcination at 500°C; (a) bar = 10 μm, (b) bar = 1 μm))

Attempts to deposit an ultra-thin microporous membrane layer on this mesoporous layer failed however, which was attributed to the pore size of the mesoporous layer which exceeded 10 nm, while the size of the nano-particles in the coating liquid (polymeric sol) measures ~ 5 nm. Therefore, an alternative mesoporous intermediate layer having a smaller pore size which is comparable to the particle size of the polymeric sol was developed.

1.2 Membrane combination 2

Figure 4 shows micrographs of a calcined 8Y$_2$O$_3$-ZrO$_2$ membrane layer which was prepared with a nano-dispersion containing smaller particles (size ~ 65 nm). In analogy with the previous coating experiment, a continuous and separate membrane layer was formed and infiltration of the smaller particles into the pores of the macroporous intermediate layer did also not occur. The main differences with the previous coating experiment included a decreased overall membrane thickness in the calcined state of ~ 2 μm and - as shown in surface micrographs 4c and 4d - a calcined layer with a significantly finer mesoporous structure was obtained.

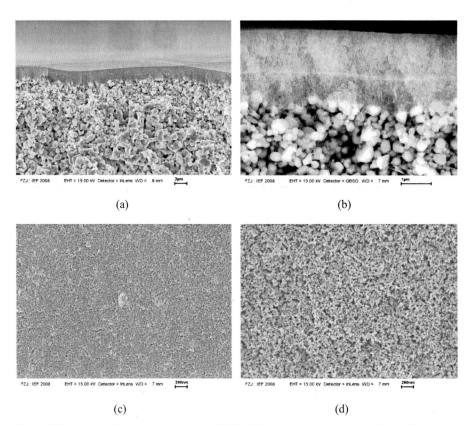

(a)

(b)

(c)

(d)

Fig. 4a. Micrograph of the cross-section of a 8Y$_2$O$_3$-ZrO$_2$ membrane, obtained by dip-coating nano-dispersion 2 on the standard substrate. Fig. 4b. Back-scattering detail micrograph. Fig. 4c and 4d. Comparison of the surface of 8Y$_2$O$_3$-ZrO$_2$ membranes, obtained by dip-coating nano-dispersion 2 (4c) and nano-dispersion 1 (4d). (Layers made by 2 x dip-coating and calcination at 500°C; (a) bar = 2 μm, (b) bar = 1 μm, (c,d) bar = 200 nm)

In further coating experiments with polymeric sols, it appeared however again that the substrate was not yet perfectly optimized for deposition of a continuous ultra-thin membrane layer starting from nano-particles. Therefore, an additional mesoporous membrane layer was coated in order to reduce the pore size and roughness of the substrate further.

1.3 Membrane combination 3

A significant optimization of the surface roughness and a reduction of the pore size below 5 nm can only be obtained by using a sol-gel coating procedure. In the first coating experiment, an additional layer was deposited on the previously coated nano-dispersion layer according to a so-called colloidal sol-gel coating procedure. As shown in the back-scattering micrograph given in Figure 5a, a rather thin membrane layer with an average thickness of ~ 0.3 – 0.4 µm was obtained by using a colloidal sol and the finer sol particles (average size ~ 35 nm) gave clearly a membrane layer with a smaller pore size (Figure 5b). The separation between the two successively coated sol-gel layers can also be easily detected and it can be estimated that a single sol-gel layer measures ~ 200 nm in thickness. Further, the difference in surface roughness before and after coating the sol-gel layers is clearly visible in Figures 5c and 5d.

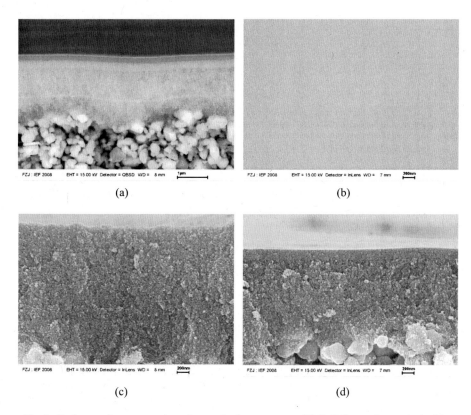

(a)

(b)

(c)

(d)

Fig. 5a. Back-scattering cross-section micrograph of a mesoporous 8Y$_2$O$_3$-ZrO$_2$ membrane, obtained by dip-coating subsequently nano-dispersion 2 and a colloidal sol. Fig. 5b. Surface micrograph of the same membrane. Fig. 5c and 5d. Comparison of the cross-section of 8Y$_2$O$_3$-ZrO$_2$ membranes, obtained by dip-coating nano-dispersion 2 (c) and by dip-coating subsequently nano-dispersion 2 and a colloidal sol (d). (Layers made by 2 x dip-coating and calcination at 500°C; (a) bar = 1 µm, (b,c,d) bar = 200 nm)

Pore analysis of the membrane material with N$_2$-adsorption/desorption measurements indicated a mesoporous structure (type IV isotherm) with a pore size maximum of ~ 3 - 4 nm (Figure 6), while the pore size of the first layer (nano-dispersion 2) measures ~ 7 nm. By comparing the surface of the first layer in Figure 4c and this of the second colloidal sol-gel derived layer in Figure 5b, a clear reduction in pore size and particle size after deposition of the sol-gel layer was also confirmed.

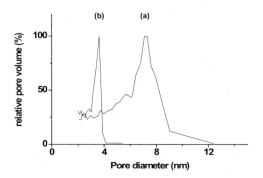

Fig. 6a and 6b. Pore size distribution of the mesoporous membrane materials after calcination at 500°C. ((a) Nano-dispersion material; (b) Colloidal sol-gel material)

In further coating experiments with polymeric sols, we found out that – after the previous coating step with a colloidal sol – the pore size was optimal for coating a layer with nano-particles. The unfavorable quality of the substrate (high surface roughness, number of large defects) prevented however the formation of a 100% continuous mesoporous layer with a pore size of ~ 3 – 4 nm. Therefore, in our current coating experiments, a graded combination of three mesoporous layers, with a first intermediate layer made from the 85 nm nano-dispersion, a second intermediate layer made from the 65 nm nano-dispersion and a third layer made from the 35 nm colloidal sol, is tested.

1.4 Membrane combination 4

Figures 7a, 7b and 7c show overview and detail SEM pictures of a graded membrane combination, consisting of three types of mesoporous layers. In these pictures, a gradual decrease in pore size and layer thickness is visible, with subsequently a mesoporous layer made from the 85 nm nano-dispersion, a mesoporous layer made from the 65 nm nano-dispersion and two mesoporous layers made from the 35 nm colloidal sol with an overall thickness of ~ 300 nm and a pore size of ~ 3 – 4 nm. To illustrate the effect of the sol-gel layer on the surface roughness, also a picture of a similar membrane system without an additional sol-gel layer is shown in Figure 7d.

In coating experiments with polymeric sols, it was however again not possible to form a continuous ultra-thin defect-free layer. Therefore, we adopt currently three approaches. First, it is attempted to obtain a completely defect-free mesoporous substrate with a pore size of 3 – 4 nm, by further optimizing the different intermediate layers. Second, a newly developed tape-casted support – which shows a considerably higher surface quality – is tested. Third, a model support with a well-defined surface quality and narrow pore size distribution is applied. This support enables an optimization of all coating procedures and the development of continuous layers on a well-defined surface; the developed layers can then afterwards be transferred to a large-scale substrate (e.g. pressed or tape-casted substrate).

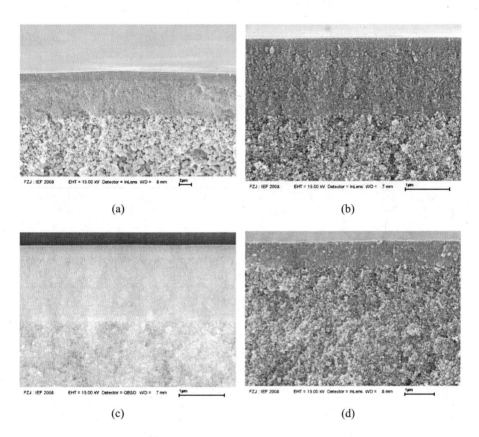

(a)

(b)

(c)

(d)

Fig. 7a and 7b. Micrographs of the cross-section of a 8Y$_2$O$_3$-ZrO$_2$ membrane, obtained by dip-coating subsequently nano-dispersion 1, nano-dispersion 2 and a colloidal sol on the standard substrate. Fig. 7c. Detail back-scattering micrograph. Fig. 7d. Micrograph of the cross-section of a 8Y$_2$O$_3$-ZrO$_2$ membrane, obtained by dip-coating subsequently nano-dispersion 1 and nano-dispersion 2 (Nano-dispersion layers made by 1 x dip-coating, colloidal sol layer 2 x dip-coating; calcination at 500°C; (a) bar = 2 μm, (b,c,d) bar = 1 μm)

2. Membrane preparation on lab-scale model substrates

Figure 8 shows micrographs of a graded 8Y$_2$O$_3$-ZrO$_2$ membrane combination, which was prepared with a nano-dispersion containing 65 nm particles and a colloidal sol containing 35 nm particles on a polished 8Y$_2$O$_3$-ZrO$_2$ support. In analogy with the previous coating experiments, continuous and separate membrane layer were formed and infiltration of the smaller particles into the pores of the support did also not occur. As shown in Figures 8a and 8b, the main difference with the previous coating experiments included the formation of very even membrane layers on the polished support. In the detail micrographs in 8c and 8d, the graded structure of the obtained membrane system can also be easily observed, with successively the macroporous support, a mesoporous layer with a pore size of ~ 7 nm and a finer mesoporous layer with a pore size of ~ 3 – 4 nm. Further, the back-scattering image in 8d illustrates that the thickness of a single layer measures ~ 1 μm for the layers made from a nano-dispersion and ~ 200 nm for the sol-gel layers.

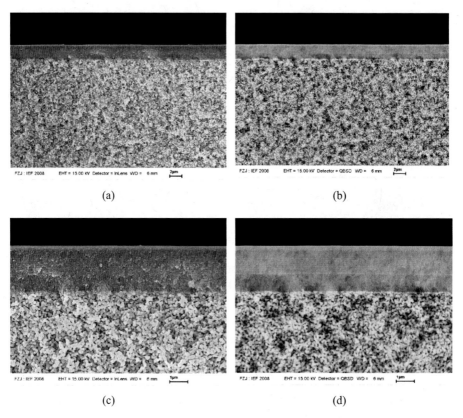

(a)   (b)

(c)   (d)

Fig. 8a and 8b. Micrograph and back-scattering micrographs of the cross-section of a 8Y$_2$O$_3$-ZrO$_2$ membrane, obtained by dip-coating nano-dispersion 2 and a colloidal sol on the model substrate. Fig. 8c. and 8d. Detail micrograph and detail back-scattering micrograph.
(Layers made by 2 x dip-coating and calcination at 500°C; (a,b,c,d) bar = 2 μm)

In further coating experiments with polymeric sols, it was observed that this kind of substrate – consisting of a polished plate with a uniform pore size distribution and the described mesoporous layers – is suitable for the deposition of ultra-thin nano-structured layers. Figure 9 shows detail micrographs of an Y$_2$O$_3$-doped and a TiO$_2$-doped ultra-thin ZrO$_2$ toplayer, deposited on the colloidal sol-gel derived Y$_2$O$_3$-doped ZrO$_2$ layer. The 8Y$_2$O$_3$-ZrO$_2$ toplayer shown in Figure 9a measures ~ 50 nm, but can not be so easily distinguished from the previous mesoporous layer in a picture with this small size, since the material composition is the same and the pore size is also similar. The TiO$_2$-doped ultra-thin ZrO$_2$ layer on the other hand is clearly visible – especially in the back-scattering image in Figure 9c – and from Figure 9b a thickness of ~ 100 nm was estimated.

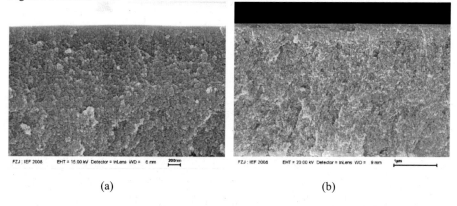

(a)   (b)

Fig. 9a. Micrograph of the cross-section of a multilayer-membrane, obtained by dip-coating a 8Y$_2$O$_3$-ZrO$_2$ polymeric sol (particle size ~ 5 nm) on the membrane system shown in Fig. 8. Fig. 9b. Micrograph of a multilayer-membrane, obtained by dip-coating a TiO$_2$-ZrO$_2$ (50-50 %) polymeric sol on the same membrane system.
(Polymeric sol layer made by 2 x dip-coating; calcination at 450°C; (a) bar = 200 nm, (b) bar = 1 μm)

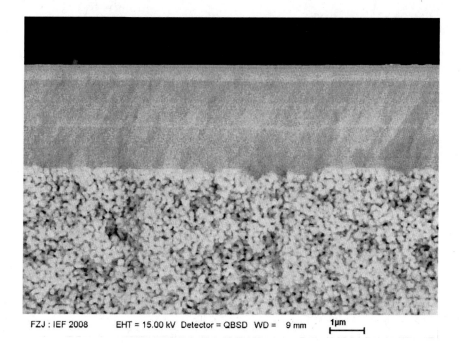

Fig. 9c. Back-scattering micrograph of the multilayer-membrane shown in Fig. 9b. (bar = 1 μm)

In order to visualise the Y$_2$O$_3$-doped ZrO$_2$ toplayer, additional TEM experiments were made, in which an alumina substrate with a similar pore size as the last mesoporous 8Y$_2$O$_3$-ZrO$_2$ membrane layer was coated with a 8Y$_2$O$_3$-ZrO$_2$ toplayer. In the TEM picture in Figure 10a, an ultra-thin toplayer with a thickness of ~ 50 nm can be clearly distinguished. Further, from the high resolution TEM picture in 10b, a crystalline structure is evident for the toplayer and a grain size of ~ 5 – 10 nm can be estimated. For the TiO$_2$-doped toplayer, a completely amorphous structure was observed (picture not shown).

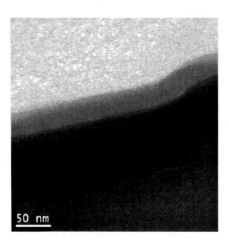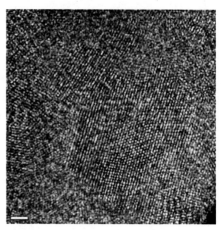

Fig. 10a. TEM micrograph of the cross-section of a multilayer-membrane, obtained by dip-coating a 8Y$_2$O$_3$-ZrO$_2$ polymeric sol (particle size ~ 5 nm) on an alumina substrate with a pore size of ~ 4 nm. Fig. 10b. High resolution TEM micrograph of the toplayer
(Polymeric sol layer made by 2 x dip-coating; calcination at 450°C; (a) bar = 50 nm, (b) bar = 2 nm)

These high resolution TEM pictures were initially taken in order to characterize/ visualize the pore structure of the toplayer, but none of the pictures showed any (micro)porosity. At the same time, gas permeation tests with He and N$_2$ showed no or a negligible gas flow. From this result, it was concluded that the combination of the previously described coating procedures was effective for the formation of a multilayer membrane with a continuous and defect-free ultra-thin nano-structured toplayer, but more research is required to optimize the porosity of the latter toplayer.

According to our current results, a lack of sufficient gas flow could be due to the formation of a dense crystallized ZrO$_2$ membrane layer, especially for the kind of membrane shown in Figure 10. Generally, the high gas permeability of current SiO$_2$-based membranes is attributed to the formation of an amorphous SiO$_2$-based network with micropores having a pore size of approximately 0.3 nm, which can act as a molecular sieve. Our first measurements on the TiO$_2$-doped ZrO$_2$ membrane toplayer shown in Figure 9 which shows an amorphous structure showed however also a very small gas flow, but these results must be further investigated and more work is currently conducted to examine the influence of the crystalline/amorphous character on the gas permeability of the described toplayers.

CONCLUSION

In this work, the development of microporous zirconia and titania based membranes is described. The first part of our research deals with the development of membrane layers on a cheap support material which can be produced at large-scale. Unfortunately, our standard support material, which is actually a support for the production of SOFC's contains still a number of large defects, which prevents the application of continuous ultra-thin sol-gel layers. In the second part, it is shown that model substrates with a polished surface and a very narrow pore size distribution enable the deposition of very homogeneous mesoporous 8Y$_2$O$_3$-ZrO$_2$ intermediate layers and subsequently continuous ultra-thin Y$_2$O$_3$-doped and TiO$_2$-doped ultra-thin ZrO$_2$ films could be deposited. Gas permeation tests with He and N$_2$ showed for our current membranes however no gas flow and apparently the porosity of the toplayers is not yet adapted for the desired application. Therefore, further research is currently devoted on all aspects which can determine the porosity of the toplayer (e.g. amorphous/crystalline character of the material, particle growth, remaining organic compounds after firing).

REFERENCES
[1] T.A. Peters, J. Fontalvo, M.A.G. Vorstman, N.E. Benes, R.A. van Dam, Z.A.E.P. Vroon, E.L.J. van Soest-Vercammen and J.T.F. Keurentjes, Hollow fibre microporous silica membranes for gas separation and pervaporation: Synthesis, performance and stability, J. Membr. Sci. 248 (2005) 73-80
[2] R. S. A. de Lange, J. H. A. Hekkink, K. Keizer, A. J. Burggraaf, Formation and characterization of supported microporous ceramic membranes prepared by sol-gel modification techniques, J. Membr. Sci. 99 (1995) 57-75
3 Andre Ayral, Anne Julbe, Vincent Rouessac, Stéphanie Roualdes, Jean Durand, Microporous silica membrane: Basic principles and recent advances, in: Inorganic Membranes Synthesis, Characterization and Applications, Edited by R. Mallada, M. Menéndez, European Membrane Society, 2008
[4] T. Van Gestel, D. Sebold, W.A. Meulenberg, H.-P. Buchkremer, Development of thin-film nanostructured electrolyte layers for application in anode-supported solid oxide fuel cells, Solid State Ionics 179 (2008) 428-437

Author Index